Quantum Measure Theory

Fundamental Theories of Physics

An International Book Series on The Fundamental Theories of Physics: Their Clarification, Development and Application

Volume 134

Quantum Measure Theory

by

Jan Hamhalter

Faculty of Electrical Engineering,
Czech Technical University, Prague, Czech Republic

KLUWER ACADEMIC PUBLISHERS

DORDRECHT / BOSTON / LONDON

A C.I.P. Catalogue record for this book is available from the Library of Congress.

ISBN 1-4020-1714-6

Published by Kluwer Academic Publishers,
P.O. Box 17, 3300 AA Dordrecht, The Netherlands.

Sold and distributed in North, Central and South America
by Kluwer Academic Publishers,
101 Philip Drive, Norwell, MA 02061, U.S.A.

In all other countries, sold and distributed
by Kluwer Academic Publishers,
P.O. Box 322, 3300 AH Dordrecht, The Netherlands.

Printed on acid-free paper

Printed in the Netherlands.

Table of Contents

PREFACE

This book has grown out of my research interests in the theory of operator algebras, orthomodular structures and mathematical foundations of quantum theory. It is based on a series of lectures on measure theory on nonboolean operator structures which I prepared for Ph.D. students in Workshops on Measure Theory and Real Analysis in Italy (Gorizia 1999, Grado 2001) and which I have delivered at the conferences of the International Quantum Structures Association in Berlin 1996, Cesena 2001, and Vienna 2002.

I have worked on these subjects in the framework of Prague's Seminar on Mathematical Formalism of Quantum Theory founded by P.Pták. Many results presented in the book were also obtained during my longer research stays abroad, in particular at the Department of Mathematics, Reading University, U.K. (1993) and at the Mathematical Institute of Erlangen University, Germany (1996-1997, 2000 and 2003).

Both the research activity and the work on the book was supported by a few international projects I participated in. I would like to acknowledge the support of European Community that supported my research stay in Reading University (Grant COST, Noncommutative Measure Theory and von Neumann Algebras, 1993). I am also very much grateful to the Alexander von Humboldt Foundation, Bonn, for long-term support of my research and for awarding me the grants that enabled me work at Erlangen University in 1996-1997, 2000 and 2003. Many ideas contained in the book came into existence during my von Humboldt grant stays. The research at my home institution, the Czech Technical University, was supported by the Grant Agency of the Czech Republic in the following large projects that helped our seminar enormously: GAČR 201/96/0117, Mathematical Formalism of Quantum Theories, 1996-1998, GAČR 201/00/0331, Operator Algebras, Ordered Structures with Orthocomplementations, 2000-2002,

GAČR 201/03/0455, Noncommutative Measure Theory, 2003-2005. Let me also acknowledge the support of the Czech Technical University, Grant No. MSM 210000010 "Applied Mathematics in Technical Sciences". Needless to say, thanks must also go to my university, Faculty of Electrical Engineering, Czech Technical University, for providing me good working conditions.

Many of the results presented in this book have grown out of highly successful and pleasant collaboration with L.Bunce (Department of Mathematics, University of Reading, U.K.). I have enormously benefited from his insight and wide knowledge. I would like to thank him for his enthusiasm which attracted me to the field of operator algebras. I am also grateful to D.Kölzow (Mathematical Institute of Erlangen University), who acted as scientific host during my von Humboldt stays in Erlangen, for his kind hospitality, helpful discussions and encouragement.

My research would not be possible without the help of my teacher and mentor P.Pták (Department of Mathematics, Faculty of Electrical Engineering, Czech Technical University) who acted as the supervisor of my Ph.D. Thesis and introduced me to the mathematical foundations of quantum theory. I have learned many from him and I value his constant interest in my work and encouragement to write this book.

Finally, I am obliged to my wife Zdeňka and son Aleš for understanding during my work on this monograph, for their infinite patience and help.

This book was partly written in Department of Mathematics, Faculty of Electrical Engineering, Czech Technical University in Prague and Mathematical Institute of University Erlangen-Nürnberg in 2003. I thank my colleagues from these institutions for helpful remarks concerning the material included. In particular, I would like to thank P.Pták for careful reading the entire text and for improving it in many respects.

Jan Hamhalter
Prague, August 2003

1. INTRODUCTION

The quantum measure theory is the theory of measures and states on projection lattices of von Neumann algebras. It has its origin in the effort to create mathematical formalism of quantum mechanics. The initial step can be found already in the first half of the last century when the theory went parallel with the infinite-dimensional analysis based on operators on Hilbert spaces.

In the standard measure theory, the basic role is played by a measure on the σ-field, $\mathcal{P}$, of subsets of some set. From an abstract point of view $\mathcal{P}$ forms a Boolean algebra with respect to usual set theoretic operations. In the quantum measure theory the basic entity is a measure on the projection lattice, $\mathcal{L}$, of a von Neumann algebra, or, equivalently, a measure on the structure of closed subspaces in a Hilbert space. The additivity is assumed with respect to mutually orthogonal subspaces. Obviously, the lattice $\mathcal{L}$ is not Boolean in general. The transition from the σ-field $\mathcal{P}$ to the projection lattice $\mathcal{L}$ consists in replacing the disjointedness of sets by a *geometric* concept of orthogonality of subspaces. It is a general goal of this book to show that this replacement and the study of "measure-theoretic" problems lead to a substantially new kind of measure theory.

The quantum measure theory has been steadily developed since its beginning. A main reason is its interdisciplinary character and so the pursuit attracted researches from many fields (notably from functional analysis, mathematical physics, theory of ordered structures and quantum probability theory). In particular, the last decades witnessed an enormous progress in this area (Gleason Theorem for von Neumann algebra and its applications, analysis of the continuity of states, etc.). Many of the results obtained have no analogy in the classical analysis and are mathematical reflections of the intrinsic character of "quantum realm". Our aim is to make them relatively easy accessible to an interested reader. Besides the exposition of the recent mathematical results of this part of functional analysis, we are concerned to (at least) indicate its relevance to quantum physics. The truth is that many mathematical problems studied in this book do find their origin in various parts of modern quantum theory (quantum field theory, quantum information theory, etc.) and, in particular, in the general program of axiomatization of quantum disciplines. As known, a basic model of quantum mechanics has the following components:

- the linear structure of observables given by the self-adjoint part of a von Neumann algebra,
- the convex structure of states given by the set of all states (positive normalized functionals) on a von Neumann algebra; the states are completely additive with respect to orthogonal projections,
- the group structure of transformations of the quantum system describing the time development and symmetries of the system given by the group of automorphisms of a von Neumann algebra.

There are many variants of this system of axioms. However, in all these approaches the most important role is played by the natural duality $\langle \cdot, \cdot \rangle$ between the structure of observables and the structure of states. This duality is given by the evaluation

$$\langle a, \varrho \rangle = \varrho(a) \,,$$

where a is an observable and ϱ is a state. The value $\varrho(a)$ represents the expectation value of the observable a when the system is prepared in the state ϱ. If a is a projection, then the corresponding observable can attain two possible numerical outcomes, 0 and 1, and it is thought of as yes-no experiment (or a proposition) on the quantum system. In this case $\varrho(a)$ amounts to a probability that the value 1 occurs. This can be interpreted as the fact that the given quantum proposition is true.

The system of axioms has proved to be very effective. Nevertheless, it has been primarily adopted for the pragmatic reason of mathematical convenience often without a direct physical justification. For example, the assumption that a state is completely additive (needed to represent a state by a density matrix) is a purely mathematical construction that has no physical justification because any physical observation may rely only on finitely many arithmetic operations. Therefore, as it was soon recognized by von Neumann, Segal, Birkhoff, Mackey, Varadarajan, Jauch, Piron, and others (see e.g. [34, 201, 248, 263, 224, 314]), an "appropriate" axiomatization of quantum theory should strive for recovering all aspects of the formalism outlined above from the concept of *finitely additive measure on projections*. Also, it should single out the projection lattice of a von Neumann algebra among all possible abstract ordered structures with orthocomplementation that may be relevant to the features of quantum system. This line of research was awarded with many deep results. For instance, one of the classical theorems in operator theory – the spectral resolution of a bounded self-adjoint operator – enables one to reconstruct an observable from a system of projections. To obtain a linear state from a measure on the projections in a similar way, which may be considered as "a dual spectral theorem", turned out to be much more difficult. An important breakthrough in this direction was the celebrated Gleason Theorem saying that any bounded σ-additive measure on the projection lattice of the algebra, $B(H)$, of all bounded operators on a separable Hilbert space, H, $\dim H \geq 3$, extends to a linear functional (state) on $B(H)$ [140]. After a considerable effort the Gleason Theorem was established in the early 90's for finitely additive vector measures on the projection lattices of von Neumann algebras. One part of this book is inspired by recent development concerning Gleason type theorems and their applications to various problems in quantum theory (hidden variables problem, Wigner type theorems on symmetries, velocity maps on von Neumann algebras, quantum history approach, quantum information theory, etc.).

Parts of this book deal with typically noncommutative phenomena that have no analogy in a classical measure or probability space. For example, it has turned out that the lattice homomorphisms on nonabelian von Neumann algebras are σ-additive, or that finitely additive measures on projection lattices whose kernels are lattice ideals enjoy many continuity properties. All these facts seem to be instances of the following general phenomenon that has both mathematical and physical interpretation. On mathematical level, there is a surprisingly strong interplay between the algebraic structure of the projection lattice and topological properties of measures and homomorphisms on them. It does not seem to occur so explicitly in the classical measure theory. In other words, unlike Kolmogorovian probability model, the continuity of quantum measures is already a consequence of the intrinsic character of the system of quantum propositions. From physical point of view, many technical assumptions (like σ-additivity, completeness of the underlying linear structure, etc.) in the probability structure of quantum physics can be derived only by dealing with discrete algebraic properties involving finitely many operations and having so a natural physical interpretation.

Most of the material included in our treatment will appear in a monograph for the first time. Since the recent development in quantum measure theory is not covered either by monographs on operator algebras or books on quantum logic or foundations of quantum theory, we would like to fill in this gap in the literature and draw attention to open problems in this area.

The problems discussed and their mathematical analysis should be accessible to both postgraduate students and research workers in operator algebras, mathematical physics, measure and probability theory and foundations of quantum theory. It may also be of some interest to researches in contemporary quantum disciplines such as quantum information theory and quantum field theory.

The essential prerequisite is some familiarity with the basic principles of the theory of Banach and Hilbert spaces, theory of operators on Hilbert spaces, measure theory and probability theory (first course level). Although all concepts and results from the theory of operator algebras (von Neumann algebras, C^*-algebras, Jordan algebras) are briefly surveyed, some knowledge of the theory of operator algebras and basic formalism of quantum mechanics may be helpful for better understanding. The first two chapters can be read without any special preparation, the subsequent chapters require more advanced technique of operator theory.

Let us now say a few words on the organization of the book. Our exposition is logically organized as follows. It starts with establishing the Gleason type theorems and deriving principles of measure and integration theory on projection lattices and goes on to presenting basic noncommutative phenomena like the Jauch-Piron states and independence of operator algebras. Nevertheless, the chapters are quite independent and may also be read independently according to the reader' s interest. Each chapter consists of the main part which is self-contained and which presents all proofs (except for unproved facts on operators algebras from the overview in Chapter 2 of course). At the end of each chapter one surveys historical development, states credits to authors, and remarks on related results. There usually is an extensive list of references. In this part one also formulates open problems believed to deserve a further study.

Let us now briefly comment on the content of the book. Chapter 1 is the introduction to the subject studied.

Chapter 2 is preparatory and surveys ideas of the theory of operator algebras needed later. Basic facts of the theory of Jordan algebras, ordered Banach spaces and orthomodular partially ordered sets are also provided. This chapter is a sort of non-technical introduction to the operator theory for the readers that are not so familiar with this part of functional analysis. In view of many outstanding monographs on the theory of von Neumann algebras and C^*-algebras, our book has no ambition to be a textbook of spectral functional analysis. For the proofs of folklore results and for further details we refer the reader to the standard monographs. Only those results that will be frequently used in the text and that are not so well known will be proved. We hope that this way of opening the subject will be instrumental in reading measure-theoretical chapters to follow.

In Chapter 3 we deal with Gleason Theorem that is proved here in the following form: Every bounded completely additive complex measure on the lattice, $P(H)$, of all projections acting on the Hilbert space H with $\dim H \geq 3$ extends to a normal functional on the von Neumann algebra, $B(H)$, of all bounded operators acting on the Hilbert space H. This theorem is a cornerstone of the measure theory built on the Hilbert space lattices. The proof presented in this book is based on the concept of the frame function and uses elementary but ingenious arguments from spherical geometry following the insight of Gleason, Piron, and Cook, Keane and Moran. Our organization of arguments enables us to eliminate advanced analysis in original Gleason's proof as well as deeper topological arguments in the hitherto known "elementary proofs". As we believe, it makes the proof readable for undergraduate students. We then prove the remarkable result by Dorofeev and Sherstnev which says that any completely additive measure on projections in an infinite-dimensional Hilbert space is automatically bounded. It allows one to relax the assumption on the boundedness of measure in the Gleason Theorem and to extend it, in the reformulation involving frame functions, to incomplete inner product spaces. Finally, we discuss the connection of ideas in the proof of Gleason Theorem to the problem of hidden variables and indeterminacy principle in the Hilbert space quantum mechanics.

One of the basic tools in quantum mechanics is the use of the infrastructure of Hilbert space. In Chapter 4 we study the role of completeness of Hilbert spaces in quantum measure theory. It was known for a long time that completeness can surprisingly be derived from the algebraic structure of the system of subspaces. The famous Amemiya-Araki Theorem says that an inner product space is complete if, and only if, the structure of its orthogonally closed subspaces is an orthomodular poset. We expose a new generation of completeness criteria of inner product spaces. They developed from the theorem by Hamhalter and Pták stating that an inner product space is complete if, and only if, its structure of orthogonally closed subspaces admits a nonzero completely additive measure. In other words, if we agree that an inner product space describes statistical correlations, we conclude that only Hilbert space can carry a reasonable measure and probability theory. This advocates completeness assumption in the axiomatics of quantum theory and gives it a natural physical foundation. Unlike original proof of this result that is based on Gleason Theorem we use much simpler arguments. As a by-product we show that Gleason Theorem can be extended to all inner product spaces of dimension at least three. Further, we provide many related recent results on subspace structures of inner product spaces (modernized Amemiya-Araki Theorem, algebraic and measure completeness criteria based on splitting subspaces, etc.) and relate them to the problem of characterizing Hilbert space lattices among ordered structures (Soler Theorem). We also provide a link to the problem of characterizing Hilbert spaces among topological linear spaces with distinguished structure of closed linear subspaces (Mackey-Kakutani Theorem).

In Chapter 5 we discuss the (highly nontrivial) extension of Gleason Theorem to finitely additive measures on projection lattices of von Neumann algebras. The main result, which is one of the most important principles in quantum measure theory and related part of mathematical physics, is the following Generalized Gleason Theorem: Let μ be a bounded finitely additive measure on the projection lattice of a von Neumann algebra, M, that does not contain Type I_2 direct summand. Let μ ranges in a normed space, X. Then μ extends to a bounded linear map of M into X. In this form the Generalized Gleason Theorem was proved by Bunce and Wright in the last decade as a culmination of more then thirty years research effort of many experts (Aarnes, Gunson, Christensen, Yeadon, Paszkiewicz, and others). The proof of this remarkable result is rather complex. We have tried to make it accessible to a broader scientific audience. We also comment on the problem of generalizing Gleason Theorem to C^*-algebras, formulated as a problem of linearity of a map additive with respect to commuting elements.

In Chapter 6 we address the question of to what extent the basic principles of standard measure theory hold if we replace the field of sets by the lattice of projections. In the first section we present a deep result, due to Dorofeev, that any real completely additive measure on the projection lattice of a von Neumann algebra that does not contain any nonzero finite-dimensional direct summand is automatically bounded. As a consequence, we obtain the Hahn-Jordan decomposition of completely additive measures on von Neumann projection lattices. We also prove the generalization of related Nikodym Boundedness Theorem on uniform boundedness of a system of measures on projection lattices. In the second section we deal with the decomposition of positive finitely additive measures into a σ-additive and a countably singular part. We formulate the Yosida-Hewitt type decompositions of measures, obtained by Bunce and Hamhalter, in the σ-additive measure theory on projection lattices. In the final section we investigate the convergence theorems for measures on projections inspired by the classical Vitali-Hahn-Saks Theorem.

In Chapter 7 we treat a few recent problems arising in quantum physics the solution of which requires, among others, the Generalized Gleason Theorem. The first part is devoted to multilinear form of Gleason Theorem obtained by Wright and Rudolf and its application to the description of decoherence functionals in newly developed quantum histories approach to quantum theory. Our discussion points out recent results in this field which allow one to obtain Isham-Linden-Schreckenberg representation of the decoherence functional as a very special case. In the second part we show, loosely speaking, that the derivations on von Neumann algebras are determined only by the their action on the projections. Results of Bhat, Bunce and Wright make it possible to recover the dynamics of the system, including Hamiltonian, entirely from the maps on the projections. Finally, as the last topic, we readdress the problem of the existence of 0-1 states and the states with small dispersions on the projections of von Neumann algebras and C^*-algebras. We provide a simple proof of the nonexistence of a finitely additive 0-1 state on the projection lattice of a von Neumann algebra without abelian part and Type I_2 part. Then we study dispersions of states and measures on projections. We show that in the case of C^*-algebras with sufficiently many projections the dispersions of states are uniformly bounded from below by a universal constant. It excludes the existence of approximate hidden variables on essentially nonboolean projection lattices. Further, we characterize the tracial states on von Neumann algebras as those which have the smallest possible dispersion on all 3×3 matrix substructures. We obtain in this way sharpening of existing no-go theorems on hidden variables for states on von Neumann algebras and quasi-linear maps on some important C^*-algebras.

In Chapter 8 we study the generalizations of Wigner Theorem to projection lattices and homomorphisms. Using the Generalized Gleason Theorem we show that every orthomorphism between von Neumann projection lattices (except for Type I_2 algebras), is a restriction of a $*$-Jordan homomorphism between corresponding algebras. This result, proved by Bunce and Wright, immediately implies Dye Theorem stating that any automorphism of the projection lattice of a von Neumann algebra without Type I_2 direct summand is a restriction of a $*$-Jordan isomorphism and thereby a restriction of a direct sum of a homomorphism and antihomomorphism. As another corollary we describe the structure of completely additive orthomorphisms between Hilbert space projection lattices which gives us the well-known Wigner Unitary-Antiunitary Theorem as a by-product. In the second part we state the theorem, proved by Bunce and Hamhalter, which says that any lattice homomorphism between projection lattices of von Neumann algebras (not containing either abelian or Type I_2 part) is σ-additive. This result makes it possible to explain the use of normal homomorphisms in quantum theory by referring only to the natural property of preserving suprema of finitely many projections. The results of this part have physical meaning – they describe the dynamics and symmetries of a quantum system purely in terms of elementary propositions on the system. In a sense they complete the axiomatization program within von Neumann algebras. In the rest, the automatic σ-additivity of orthogonally scattered vector measures is considered and new proofs of Wigner Theorem formulated for maps preserving transition probability are presented.

In Chapter 9 we study the restriction and extension properties of states on C^*-algebras. In its first part we derive the results on determinacy of the system of orthogonal pure states by "biorthogonal systems" of elements in the C^*-algebra. Moreover, we show that for finitely many orthogonal pure states on a C^*-algebra with separable G.N.S. representations there is a maximal abelian C^*-subalgebra such that all states are multiplicative (i.e. pure) on it. We construct an example showing that this result does not hold for infinitely many states. From the mathematical point of view we extend some principles of the topology of compact spaces, from the physical point of view we relate C^*-quantum mechanics to classical theory. In the second part we study the extensions of measures between von Neumann algebras and general ordered structures. Among others we show that any complex measure on the projection lattice of a von Neumann algebra without Type I_2 direct summand extends to a measure on an arbitrary larger orthomodular structure. A fairly general extension theorem by Pták is used in places. An involved extension theorem is obtained for positive Stone-algebra valued measures on the projection lattices. These results show the power of the Generalized Gleason Theorem and indicate the interplay between the validity of the Gleason Theorem and the possibility to obtain a linear extension in a natural order unit space linear structure associated with a given quantum logic. In the last section the restriction properties of extremal completely positive maps are studied.

Chapter 10 focuses on a new type of results that hold only in "proper" quantum measure spaces. The central topic is the one of Jauch-Piron state which is a state on a von Neumann algebra the kernel of which is closed under forming finite suprema of projections. We present the results mainly obtained by Bunce and Hamhalter showing that the discrete Jauch-Piron condition is strong enough to force a higher degree of continuity of states (nonsingularity, regularity, etc.). Among others it is shown that a pure state on essentially noncommutative von Neumann algebra and a factor state on a properly infinite algebra is σ-additive if, and only if, it is Jauch-Piron. Also the σ-additive states are characterized as the states for which the closure of transforms consists of Jauch-Piron states, and a geometrical analysis of the Jauch-Piron property is provided. The outcomes of the research in this field show that topological and algebraical aspects of the projection lattices are closely related. Roughly speaking, the more noncommutative the given algebra is, the more the topology of its dual space is determined by the lattice-theoretical character of its projection lattice. It also contributes to clarifying the system of axioms in the operator-algebraic approach to quantum theory.

The concluding Chapter 11 deals with recent results on the independence of von Neumann algebras and C^*-algebras. The independence is a well established concept in classical probability theory. However, it turned out that natural independence conditions imposed on operators that coincide for commutative algebras are far from being equivalent in noncommutative framework. We settle the position of major independence conditions arising in the formalism of local algebras in relativistic quantum field theory where they implement Einstein causality principle. Among others we present some new results on C^*-independence (simultaneous extension property in the category of C^*-algebras) and W^*-independence (simultaneous extension property in the category of von Neumann algebras), and their relevance to the structure theory of operator algebras and axiomatic quantum field theory.

2. OPERATOR ALGEBRAS

The aim of the present chapter is to survey fundamental results of the theory of operator algebras. They will play an important role in further exposition. Our overview is not completely systematic and most of the results will be stated without proof. However, we shall try to illustrate basic concepts and theorems by examples so that the reader not familiar with this part of functional analysis can understand basic ideas on an intuitive level. For a full treatment we refer the reader to the monographs [20, 35, 39, 91, 187, 206, 259, 290, 307, 308, 311].

The theory of operator algebras came into existence in connection with the theory of representations of groups and mathematical foundations of quantum mechanics. At the present time this discipline is a well developed and steadily growing mathematical theory that provides an infrastructure not only for quantum mechanics but also for noncommutative geometry, quantum information theory, quantum field theory, quantum probability, and other areas.

The operator algebras theory combines the techniques of its three main streams: theory of C^*-algebras, theory of von Neumann algebras and theory of Jordan structures. Each of these parts is of its own nature. It is appropriate to view the theory of C^*-algebras as noncommutative topology and the theory of von Neumann algebras as noncommutative measure theory. We shall follow this division in our overview. In the first part we shall survey basic properties of C^*-algebras from the abstract point of view. In the second part we state basic elements of the structure theory of von Neumann algebras. These results will be frequently used in the book. The third part deals with the Jordan algebras and related concepts of the theory of ordered Banach spaces and orthocomplemented ordered structures.

2.1. C^*-algebras

Abstract C^*-algebras. C^*-algebras can be thought of as a simultaneous generalization of commutative and infinite-dimensional function algebras and noncommutative and finite-dimensional matrix algebras. Both these structures form complex associative algebras with respect to arithmetic operations. In addition, the complex conjugate functions and adjoint matrices induce an involutive operation on the corresponding algebras. This generalizes to the concept of $*$-algebra. An associative complex algebra $(A, +, \cdot, *)$ is called a *$*$-algebra* if the operation $*$ obeys the following rules ($a, b \in A$ and $\lambda \in \mathbb{C}$) : (i) $a^{**} = a$ (ii) $(a+b)^* = a^* + b^*$ (iii) $(\lambda\, a)^* = \overline{\lambda}\, a^*$ (iv) $(a\, b)^* = b^*\, a^*$.

Let A be a complex $*$-algebra which is a Banach space with the norm $\|\cdot\|$. A is said to be an *involutive Banach algebra* if the following conditions are satisfied for all $a, b \in A$: (i) $\|a\, b\| \leq \|a\| \cdot \|b\|$ (ii) $\|a^*\| = \|a\|$. A *C^*-algebra* is an involutive Banach algebra satisfying (iii) $\|a^*\, a\| = \|a\|^2$ for all $a \in A$. Let us remark that the latter C^*-condition (iii) together with condition (i) already implies condition (ii). The *centre*, $Z(A)$, of a C^*-algebra A is the set of all elements commuting with every element in A.

2.1.1. Example. Let $M_n(\mathbb{C})$ be the complex algebra of $n \times n$ complex matrices endowed with the standard arithmetic operations and with the involution sending a matrix a into its adjoint matrix a^*. Then $M_n(\mathbb{C})$ with the matrix norm is a finite-dimensional C^*-algebra. It can be proved that all finite-dimensional C^*-algebras are direct sums of algebras of this type.

2.1.2. Example. Let X be a locally compact Hausdorff space. Let us say that a continuous function f on X vanishes at infinity, if for each positive number ε the set $\{x \in X \mid |f(x)| \geq \varepsilon\}$ is compact. Let us denote by $C_0(X)$ the $*$-algebra of all continuous complex functions on X vanishing at infinity, where the $*$-operation assigns to each function its complex conjugate. Let us endow $C_0(X)$ with the norm $\|f\| = \sup_{x\in X} |f(x)|$. Then $C_0(X)$ becomes an abelian (=commutative) C^*-algebra. In particular, if X is a compact space, then the algebra $C_0(X)$ coincides with the algebra $C(X)$ of all continuous complex functions on X.

2.1.3. Example. Let H be a Hilbert space and let $B(H)$ be the $*$-algebra of all bounded operators on H with usual arithmetic operations and the $*$ operation that assigns to each operator its adjoint operator. Then $B(H)$ endowed with the operator norm

$$\|a\| = \sup_{\xi \in H\,,\, \|\xi\|=1} \|a\,\xi\|\,,$$

is a C^*-algebra. Any normed closed $*$-subalgebra of $B(H)$ is also a C^*-algebra. As we shall see later, all C^*-algebras are (up to an isomorphism) of this form.

2.1.4. Example. Let $C(X, M_2(\mathbb{C})$ be the algebra of all continuous maps $f\colon\ X \to M_2(\mathbb{C})$, where X is a compact Hausdorff space with pointwise defined arithmetic operations and the $*$-operation given by $f^*(x) = f(x)^*$ $(x \in X)$. The norm $\|f\| = \sup_{x \in X} \|f(x)\|$ converts $C(X, M_2(\mathbb{C}))$ to a C^*-algebra.

2.1.5. Example. Let F be a subset of $B(H)$ that is closed under forming the adjoint operators. Let us denote by F' the commutant of F, i.e. the set $F' = \{a \in B(H) \mid a\,x = x\,a \text{ for all } x \in F\}$. Then F' is a C^*-subalgebra of $B(H)$. This is a basic example of von Neumann algebra. It is a special case of C^*-algebra that will be dealt with in the next section.

A very powerful tool in the C^*-theory is the function calculus that is a consequence of the following fundamental Gelfand Theorem. It describes abelian C^*-algebras. We shall need the following concepts. Let A be abelian C^*-algebra. A *character* ω on A is a nonzero linear map $\omega\colon\ A \to \mathbb{C}$ for which $\omega(a\,b) = \omega(a)\,\omega(b)$ and $\omega(a^*) = \overline{\omega(a)}$ for all $a, b \in A$. The *spectrum*, $\Omega(A)$, of an algebra A is the set of all characters on A endowed with the topology of pointwise convergence on elements of A. It can be proved that $\Omega(A)$ is a locally compact Hausdorff space. The *Gelfand transform* is the map $\tau\colon\ A \to C_0(\Omega(A))$ defined for all $a \in A$ by the formula

$$\tau(a)(\omega) = \omega(a) \qquad \text{for all } \omega \in \Omega(A)\,.$$

2.1.6. Theorem (Gelfand). *For each abelian C^*-algebra A the Gelfand transform is a $*$-preserving isometric isomorphism of A onto $C_0(\Omega(A))$.*

As a result, all abelian C^*-algebras are more the algebras of continuous functions on locally compact spaces.

If a C^*-algebra, A, contains a unit with respect to the multiplication, usually denoted by 1, then A is said to be *unital.* It is clear that $1^* = 1$ and $\|1\| = 1$. The algebra $C_0(X)$ is unital if, and only if, X is compact. For each C^*-algebra A there is a unital C^*-algebra $\tilde{A}$ containing A as a closed ideal and such that $A = \tilde{A}$ if A is unital and $\dim \tilde{A}/A = 1$ otherwise. We say that $\tilde{A}$ is the *unitalization* of A. If $A = C_0(X)$, the unitalization, $\tilde{A} \cong C(X \cup \{\infty\})$, corresponds to one-point compactification of X. In the rest of this paragraph A will always denote a C^*-algebra. Let $x \in A$. The *spectrum* of x (in symbols $\mathrm{Spec}(x)$) is defined by the equality

$$\mathrm{Spec}(x) = \{\lambda \in \mathbb{C} \mid (x - \lambda 1) \text{ is not invertible in } A\}$$

(here 1 denotes the unit in $\tilde{A}$). It can be proved that the spectrum is the same in all C^*-subalgebras of A containing x. We say that an element $x \in A$ is *normal* if $x^* x = x x^*$. In other words, a normal element is contained in some abelian C^*-subalgebra of A. The smallest C^*-subalgebra containing an element $x \in A$ will be denoted by $C^*(x)$. It is the norm closure of the set of all finite sums of the form

$$\sum_{i,j} a_{ij}\, x^i\, x^{*j}, \quad a_{ij} \in \mathbb{C}.$$

Since $C^*(x)$ is abelian if x is normal, we can apply the Gelfand Theorem and identify $C^*(x)$ with the continuous functions on its spectrum $X = \Omega(C^*(x))$. It turns out that X can be identified with $\mathrm{Spec}(x) \setminus \{0\}$. Indeed, let ω be a character on $C^*(x)$. Then ω is uniquely given by the value $\omega(x) \neq 0$. Any character ω extends canonically to the unitalization of $C^*(x)$ such that $\omega(1) = 1$. As $\omega(x - \omega(x)\,1) = 0$, $x - \omega(x)\,1$ cannot be invertible. Consequently, $\omega(x) \in \mathit{Spec}(x) \setminus \{0\}$. On the other hand, if $\lambda \in \mathit{Spec}(x) \setminus \{0\}$, then $(x - \lambda 1)$ is in a proper closed ideal, I, of $\tilde{A}$. The canonical map $x \in A \to x + I \in A/I$ gives us a character, ω, with $\omega(x) = \lambda$. In summary, the map $\omega \in X \to \omega(x) \in \mathrm{Spec}(X \setminus \{0\})$ establishes a one-to-one correspondence that can be shown to be a homeomorphism between X and $\mathrm{Spec}(x) \setminus \{0\}$. In this manner the algebra $C^*(x)$ can be identified with the algebra $C_0(Spec(x) \setminus \{0\})$. This result is crucial for establishing the function calculus for normal elements of C^*-algebras. For a normal element $x \in A$ let us denote by $f(x)$ the element in $C^*(x)$ that corresponds to the function $f \in C_0(Spec(x) \setminus \{0\})$. Then the assignment $x \to f(x)$ is the function calculus that has a natural meaning for the polynomials. If x is a normal operator on a Hilbert space H and $x\,\xi = \lambda\,\xi$ for a unit vector $\xi \in H$ and $\lambda \in \mathbb{C}$, then $f(x)\,\xi = f(\lambda)\,\xi$ for any $f \in C_0(\mathrm{Spec}(x) \setminus \{0\})$.

An element x of a C^*-algebra A is called *self-adjoint* if $x = x^*$. By A_{sa} we shall always denote the set of all self-adjoint elements in A. It is a real Banach space. A general element $x \in A$ can written as $x = y + i\,z$, where $y = \frac{x+x^*}{2} \in A_{sa}$ and $z = -\frac{i\,(x-x^*)}{2} \in A_{sa}$. The elements y and z are called the *real* and the *imaginary part* of x, respectively. As an example, the element $f \in C(X)$ is self-adjoint if, and only if, the function f is real valued. An operator x acting on a Hilbert space H is self-adjoint if, and only if, the spectrum of x consists of real numbers. An element x of a C^*-algebra A is called *positive* (or *nonnegative*) if $x = a^*\,a$ for some $a \in A$. Positive elements in $C(X)$ are precisely nonnegative functions. An operator x acting on a Hilbert space H is positive if, and only if, $(x\,\xi, \xi) \geq 0$ for all $\xi \in H$. In general, an element of a C^*-algebra is positive if, and only if, its spectrum consists of positive numbers. It is clear that a positive element is self-adjoint. We will denote by A_+ the set of all nonnegative elements of a C^*-algebra A. It can be proved that A_+ is norm closed and that it is a cone, i.e. the convex set satisfying (i) $x + y \in A_+$ whenever $x, y \in A_+$ (ii) $\lambda\,x \in A_+$ whenever $\lambda \geq 0$ and $x \in A_+$. The positive cone A_+ defines the order, $\leq$, on the space of self-adjoint elements A_{sa} by writing $x \leq y$ if, and only if, $y - x \in A_+$. The order $\leq$ defined above is a partial ordering that specializes to the usual pointwise order of functions in the case of function algebras. If $x \leq y$, then $a^*\,x\,a \leq a^*\,y\,a$ for all $a \in A$. Any self-adjoint element $x \in A$ can be uniquely written in the form

$$x = x^+ - x^-\,,$$

where x^+ and x^- are positive elements with $x^+\,x^- = 0$. To see this, we can use the function calculus and put $x^+ = f^+(x)$, $x^- = f^-(x)$, where $f^+(t) = \max(t, 0)$, $f^-(t) = \max(-t, 0)$, $t \in \mathbb{R}$. The elements x^+ and x^- are called the *positive* and the *negative part* of x, respectively. As a result, positive elements linearly span the algebra. For any positive element x the spectrum $\mathrm{Spec}(x)$ consists of nonnegative numbers and thus we can define the square root $\sqrt{x}$ of x. It can be proved that the function $t \to \sqrt{t} : \mathbb{R}^+ \to \mathbb{R}^+$ is *operator monotone* in the sense that $0 \le x \le y$ implies $0 \le \sqrt{x} \le \sqrt{y}$. Further important property of the positive elements is the fact that the set $A_+^1 = \{x \in A \mid 0 \le x \le 1\}$ forms an approximate unit for A. An *approximate unit* is an upwards directed system $(a_\alpha)_{\alpha \in I}$ of nonnegative elements of A such that $\lim_{\alpha \in I} \|x - x\, u_\alpha\| = \lim_{\alpha \in I} \|x - u_\alpha\, x\| = 0$, for each $x \in A$. Of course, for A_+^1 the set I coincides with A_+^1 ordered by $\le$.

An element u in the unital C^*-algebra A is said to be *unitary* if $u^*\, u = u\, u^* = 1$. The spectrum of any unitary element is contained in the unit circle in the complex plane. If $A = C(X)$, the unitary elements are just continuous mappings of X into the unit circle. If $A = M_n(\mathbb{C})$, then the set of unitary elements is nothing but the set of unitary (=orthonormal) matrices. The set of all unitary elements in A endowed with the multiplication forms the group, $U(A)$, called the *unitary group* of A. As for any self-adjoint element x with $\|x\| \le 1$ we find that the elements $u = x + i\,(1 - x^2)^{1/2}$ and $u^* = x - i\,(1 - x^2)^{1/2}$ are unitary with $x = \frac{1}{2}(u + u^*)$, we see that $U(A)$ linearly generates the whole of A. An important example of algebra generated by special unitary operators is so-called group C^*-algebra. Let Γ be a group and let λ_g, $\lambda \in \Gamma$, be a unitary operator acting on $l^2(\Gamma)$ by $\lambda_g\, f(x) = f(x\, g^{-1})$, $x \in \Gamma$. Then each λ_g is unitary and the C^*-subalgebra, $C_\lambda^*(\Gamma)$, of $B(l^2(\Gamma))$ generated by the set $\{\lambda_g \mid g \in \Gamma\}$ is called the *reduced group* C^*-algebra of Γ. An element s in the unital C^*-algebra is called *symmetry* if it is unitary and self-adjoint.

An important role in our analysis will be played by projections. A *projection* in a C^*-algebra A is a self-adjoint idempotent, i.e. an element $p \in A$ such that $p = p^2 = p^*$. If the algebra A acts on a Hilbert space H, then $p(H)$ is a closed subspace of H and p is an orthogonal projection of H onto $p(H)$. All projections arise in this way. In the algebra $C_0(X)$ the projections are exactly the characteristic functions of clopen subsets of X.

The following example describes all projections in the algebra $M_2(\mathbb{C})$.

2.1.7. Example. Any projection in the algebra $M_2(\mathbb{C})$ is of the form

$$\begin{pmatrix} a & \sqrt{a - a^2}\, e^{i\varphi} \\ \sqrt{a - a^2}\, e^{-i\varphi} & 1 - a \end{pmatrix},$$

where $a \in [0, 1]$ and $\varphi \in \mathbb{R}$. Indeed, let

$$p = \begin{pmatrix} a & b \\ c & d \end{pmatrix}$$

be a projection in $M_2(\mathbb{C})$. As $0 \leq p \leq 1$ we see that

$$0 \leq \begin{pmatrix} 1 & 0 \\ 0 & 1 \end{pmatrix} \begin{pmatrix} a & b \\ c & d \end{pmatrix} \begin{pmatrix} 1 & 0 \\ 0 & 0 \end{pmatrix} = \begin{pmatrix} a & 0 \\ 0 & 0 \end{pmatrix} \leq 1 .$$

Hence, $0 \leq a \leq 1$. The condition that p has one-dimensional range implies that the trace of p is 1, and thus $d = 1 - a$. The self-adjointeness of p gives us $b = \bar{c}$. The equality $p^2 = p$ implies $|b| = \sqrt{a - a^2}$, and hence $b = \sqrt{a - a^2}\, e^{i\varphi}$ for some real φ.

Throughout the book $P(A)$ will always denote the set of all projections in a C^*-algebra A. Let us note that if A is unital, then the map $p \to 2p - 1$ is a one-to-one correspondence between the projection structure and the set of all symmetries in A. We shall deal with the set of all projections in more detail later on.

We shall need the following facts on the geometry of the unit ball in C^*-algebras. The unit ball A_1 of a C^*-algebra A has an extreme point if, and only if, A is unital. In this case the set of all extreme points consists of all elements $v \in A$ (partial isometries) such that $v^* v = p \in P(A)$, $v v^* = q \in P(A)$ and $(1 - p) A (1 - q) = \{0\}$. Extreme points of the self-adjoin part $A_1 \cap A_{sa}$ of the unit ball are precisely the symmetries. Extreme points of the positive part $A_1 \cap A_+$ of the unit ball are precisely the projections.

Operations with C^*-algebras. Let $(A_\alpha)_{\alpha \in I}$ be a system of C^*-algebras. The direct sum $\sum_{\alpha \in I} \oplus A_\alpha$ is a C^*-algebra of all nets $(x_\alpha)_{\alpha \in I}$ with pointwise defined operations and the norm

$$\|x_\alpha\| = \sup_{\alpha \in I} \|x_\alpha\| < \infty .$$

Let A be a C^*-algebra. By $M_n(A)$ we shall denote the algebra of all $n \times n$ matrices with entries in A with the standard matrix operations. The algebra A can be embedded into $M_n(A)$ as the upper left corner. It can be shown that there exists exactly one norm on $M_n(A)$ that extends the C^*-norm on A and that makes $M(A)$ into a C^*-algebra.

Intrinsic structure of C^*-algebras. A C^*-subalgebra B of a C^*-algebra A is defined as the normed closed $*$-subalgebra of A. By the symbols $C^*(a_1, \ldots, a_n)$ we shall denote the C^*-subalgebra generated by the elements $a_1, \ldots, a_n \in A$. Such an algebra is always separable. A closed subset I of A is called *closed left ideal* if $a\,x \in I$ for each $a \in A$, and $x \in I$. Analogously, a closed subset I is called the *closed right ideal* if $x\,a \in I$ for each $a \in A$ and $x \in I$. A closed subset in A is called *closed (two-sided) ideal* if it is simultaneously left and right ideal. The matrix algebra has no nontrivial ideals. The C^*-algebra having no nontrivial ideals is called *simple*.

Let us consider an abelian algebra $A = C(X)$. Then there is a canonical correspondence between closed ideals in A and closed subsets of X as follows: Let I be a closed ideal in $C(X)$. Then there is a (unique) closed subset F of X such that $I = \{f \in C(X) \mid f \text{ vanishes on } F\}$. A vice versa assertion is evident.

A C^*-subalgebra B of A is called *hereditary* if its positive cone, B_+, is a *hereditary cone*, which means the following condition: If $0 \le x \le y$, where $x \in A$ and $y \in B$, then $x \in B$. A natural example of hereditary algebra is the algebra $p\,A\,p = \{p\,a\,p \mid a \in A\}$, where $p \in A$ is a projection. Any closed ideal is a hereditary subalgebra. In particular, any closed ideal is closed with respect to the $*$-operation, which is a nontrivial fact. Given a hereditary subalgebra B of A we shall write

$$L(B) = \{x \in A \mid x^*\,x \in B_+\}\,.$$

There is a one-to-one correspondence between hereditary C^*-subalgebras and closed left ideals in A given by the assignment

$$B \to L(B)\,, \qquad I \to I \cap I^*\,.$$

Let I be a closed ideal, then the quotient algebra A/I endowed with the following operations and norm:

$$(a+I)\,(b+I) = a\,b+I\,, \quad (a+I)^* = a^*+I\,, \quad \|a+I\| = \inf\{\|a+x\| \mid x \in I\}$$

becomes a C^*-algebra. For example, if F is a closed subset in a compact space X, then $I = \{f \in C(X) \mid f|F = 0\}$ is a closed ideal and $C(X)/I$ is isomorphic to the algebra $C(F)$.

Morphisms of C^*-algebras. A linear map $\pi\colon A \to B$ between C^*-algebras A and B is called a $*$-*homomorphism* if it preserves product and $*$-operation. It turns out that any $*$-homomorphism π is norm decreasing, i.e. $\|\pi(x)\| \leq \|x\|$ for all $x \in A$, and thereby it is continuous. Moreover, the image $\pi(A)$ is a C^*-subalgebra of B. The kernel $\operatorname{Ker}\pi = \{a \in A \mid \pi(a) = 0\}$ is a closed ideal in A, and $\pi(A)$ is isomorphic to $A/\operatorname{Ker}\pi$. An injective $*$-homomorphism of A onto B is called $*$-isomorphism. A $*$-isomorphism $\pi\colon A \to A$ is called a $*$-automorphism. Any $*$-isomorphism is an isometry. As a consequence, *there is only one C^*-norm on a C^*-algebra.* A natural example of a $*$-automorphism is a map $\pi : a \to u^* a u : A \to A$, where u is a unitary element of A. In this case we say that π is *implemented by* u. Note that not all $*$-automorphisms arise this way. A $*$-homomorphism of a C^*-algebra A into $B(H)$ is also called $*$-representation of A on Hilbert space H. If the kernel of a $*$-representation π is zero, the representation π is called *faithful.* As we shall see later any C^*-algebra can be faithfully represented on some Hilbert space. A linear map $\pi\colon A \to B$ between C^*-algebras A and B is called $*$-antihomomorphism if it preserves the $*$-operation and if $\pi(a\,b) = \pi(b)\,\pi(a)$ for all $a, b \in A$. An injective $*$-antihomomorphism of A onto B is called $*$-*antiisomorphism.* A $*$-Jordan homomorphism $\varrho\colon A \to B$ between C^*-algebras A and B is a linear map preserving the $*$-operation and fulfilling $\varrho(a^2) = \varrho(a)^2$ for all $a \in A_{sa}$. A $*$-Jordan isomorphism is an injective $*$-Jordan homomorphism of A onto B. It holds that any $*$-Jordan homomorphism $\varrho\colon A \to B$, where the C^*-algebra A has no one-dimensional quotient, is a direct sum of a $*$-homomorphism and a $*$-antihomomorphism in the sense that there is a $*$-homomorphism $\pi_1\colon A \to B$ and a $*$-antihomomorphism $\pi_2\colon A \to B$ such that $\varrho(A) = \pi_1(A) \oplus \pi_2(A)$ and $\varrho(x) = \pi_1(x) \oplus \pi_2(x)$ for all $x \in A$. Let us note that any $*$-homomorphism commutes with the function calculus in the sense that $\pi(f(x)) = f(\pi(x))$ for all normal $x \in A$ and any $*$-homomorphism $\pi\colon A \to B$.

States and positive functionals. In this paragraph we state basic facts on positive functionals on C^*-algebras. The concept of state is central in the present book. It generalizes the notion of probability measure and, on the other hand, enables one to realize a given C^*-algebra as algebra of operators on a Hilbert space. A linear form $f\colon A \to \mathbb{C}$ on a C^*-algebra A is called a *positive functional* if $f(a) \geq 0$ whenever $a \geq 0$. One of the remarkable properties of C^*-algebras is the result that *any positive functional is bounded.* Indeed, seeking a contradiction suppose that $f : A \to \mathbb{C}$ is positive and unbounded. Since A is a linear span of A_+, we can find a sequence $(a_n) \subset A_+$ such that $\|a_n\| \leq 1$ and $2^n \leq \varrho(a_n)$ for all n. Put $a = \sum_{n=0}^{\infty} \frac{1}{2^n} a_n$. Then $N \leq \sum_{n=0}^{N-1} \varrho(\frac{1}{2^n} a_n) \leq \varrho(a)$. Thus, $\varrho(a) \geq N$ for all N which is impossible.

There is another interplay between the C^*-norm and the positivity of functionals. Indeed, a bounded functional ϱ on a C^*-algebra A is positive if, and only if, $\|\varrho\| = \sup_\alpha \varrho(a_\alpha)$, where (a_α) is (any) approximate unit of A. In particular, if A is unital, then a bounded functional ϱ on A is positive if, and only if, $\|\varrho\| = \varrho(1)$. Making use of the Hanh-Banach Theorem we can therefore extend any positive functional from any C^*-subalgebra to a positive functional on the whole algebra without changing the norm. Moreover, this extension is unique if the subalgebra is hereditary. A positive functional ϱ on a C^*-algebra is called a *state* if $\|\varrho\| = 1$. The following example shows a few important types of states.

2.1.8. Example. (i) Let $A = C(X)$, where X is a compact Hausdorff space. Then the states on A are in one-to-one correspondence with Borel probability measures on X. If μ is such a measure, then the corresponding state, ϱ_μ, is given by the equality

$$\varrho_\mu(f) = \int_X f\, d\mu\,, \qquad f \in C(X)$$

(The Riesz Representation Theorem). In particular, any $x \in X$ gives rise to a state, ϱ_x, (corresponding to the Dirac measure δ_x) acting on $C(X)$ by the pointwise evaluation $\varrho_x(f) = f(x)$, $f \in C(X)$.
(ii) Let A be a C^*-algebra acting on a Hilbert space H. Let $\xi \in H$. A *vector functional,* ω_x, is defined by setting

$$\omega_\xi(a) = (a\,\xi, \xi) \qquad \text{for all } a \in A\,.$$

It is clear that ω_ξ is positive. Moreover, if ξ is a unit vector and A contains the identity operator, then ω_ξ is a state that is called the *vector state.*
(iii) It can be proved that any state on the matrix algebra $M_n(\mathbb{C})$ is a convex combination of the vector states $\omega_{e_1}, \omega_{e_2}, \ldots, \omega_{e_n}$, where $e_1, \ldots, e_n$ is an orthonormal basis of the underlying Hilbert space.

In C^*-algebraic quantum mechanics, the set of all observables of a physical system is given by the self-adjoint part A_{sa} of a C^*-algebra A. The spectrum of $a \in A_{sa}$ corresponds to all possible values of a given observable. Let ϱ be a state on A. The value, $\varrho(a)$, $a \in A_{sa}$, represents the expectation value of an observable a when the system is in the physical state represented by ϱ. This physical background explains the terminology. The next result shows that for any value λ that a given observable a can attain we can find a state of the whole system such that the expectation value of a in this particular state is precisely λ. Let x be a normal element in a C^*-algebra A. For any $\lambda \in Spec(x)$ there is a state ϱ of A such that $\varrho(x) = \lambda$. In particular, there is a state ϱ such that $|\varrho(x)| = \|x\|$.

Let $\varrho \in A^*$ be a bounded functional on A. Then the equation $\varrho^*(a) = \overline{\varrho(a^*)}$, $(a \in A)$ defines another functional called the *adjoint* of ϱ. If $\varrho = \varrho^*$, we call the functional ϱ *hermitean* (or *self-adjoint*). Each $\varrho \in A^*$ can be expressed, uniquely, in the form $\varrho = \varrho_1 + i\,\varrho_2$, where $\varrho_1 = \frac{1}{2}(\varrho + \varrho^*)$ and $\varrho_2 = \frac{1}{2}i\,(\varrho^* - \varrho)$ are hermitean. An analog of the Hahn-Jordan decomposition of measures holds also for any hermitean functional. Indeed, for any hermitean functional ϱ on a C^*-algebra A there are uniquely determined positive functionals ϱ^+ and ϱ^- on A such that $\varrho = \varrho^+ - \varrho^-$ and $\|\varrho\| = \|\varrho^+\| + \|\varrho^-\|$. As a consequence, any element in the dual is a combination of four positive functionals. An element $x \in A$ is said to be *strictly positive* if $\varphi(x) > 0$ for each nonzero positive functional φ. If A is unital, then the unit is a strictly positive element. In general, a C^*-algebra admits a strictly positive element if, and only if, it has a countably approximate unit. In particular, all separable algebras have a strictly positive element.

The Gelfand-Naimark-Segal construction. Let φ be a positive functional on the C^*-algebra A. The φ endows A with a semidefinite inner product

$$(x, y) = \varphi(y^*\,x) \qquad x, y \in A\,. \tag{2.1}$$

We say that φ is *faithful* if $\varphi(x^*\,x) = 0$ implies $x = 0$. In other words, φ is faithful if the product (2.1) is an inner product. By specializing the Cauchy-Schwarz inequality to the product (2.1), we obtain the following important inequalities for all $x, y \in A$:

$$|\varphi(x^*\,y)|^2 \;\leq\; \varphi(x^*\,x)\cdot\varphi(y^*\,y) \tag{2.2}$$

$$|\varphi(x^*\,y)|^2 \;\leq\; \|\varphi\|\cdot\|x^*\,x\|\cdot\varphi(y^*\,y)\,. \tag{2.3}$$

Whence,

$$\varphi(x)^2 \leq \|\varphi\| \cdot \varphi(x^2) \qquad \text{for all} \qquad x \in A_{sa}\,. \tag{2.4}$$

(The equation (2.4) follows from (2.3) by setting $x = 1$ and $y = x$.) Besides, the Cauchy-Schwarz equality stated above has the following consequence that will be frequently used in the sequel. Let $a \in A$. We say that a nonnegative functional φ on A is *definite at the element* $a \in A$ if $\varphi(a^* a) = |\varphi(a)|^2$. Then φ has the following multiplicativity property: $\varphi(a\,b) = \varphi(b\,a) = \varphi(a)\,\varphi(b)$ for all $b \in A$. We shall establish this inequality under the condition that the algebra A is unital. By virtue of (2.2),

$$|\varphi((a - \varphi(a)\,1)\,b)|^2 \leq \varphi[(a^* - \overline{f(a)}1)\,(a - \varphi(a)\,1)] \cdot \varphi(b^*\,b) = 0\,.$$

Then $\varphi(a\,b) = \varphi(a)\,\varphi(b)$ and the equality $\varphi(b\,a) = \varphi(a)\,\varphi(b)$ can be proved in the same way. Let us note that a state φ is definite at $a \in A$ for a self-adjoint a whenever $\varphi(a) = \|\varphi\| = 1$. Indeed, in this case (2.4) gives us $1 = \varphi(a)^2 \leq \varphi(a^2) \leq 1$, and in turn we have $\varphi(a)^2 = \varphi(a^2)$. A positive functional φ on A is called *multiplicative* if it is definite at all elements, i.e., if φ is a $*$-homomorphism of A into $\mathbb{C}$. If a multiplicative functional is nonzero, then it is a state and its kernel is a maximal proper closed ideal in A of codimension one. For this reason nonzero simple algebras do not admit nonzero complex-valued $*$-homomorphisms. After those preliminaries we pass to the Gelfand-Naimark-Segal construction that is fundamental to all operator theory. We outline its basic idea.

Let φ be a nonnegative functional on a C^*-algebra A. The *left kernel*

$$L_\varrho = \{a \in A \mid \varrho(a^*\,a) = 0\}$$

is a closed left ideal in A. By factorizing A endowed with the product (2.1) by the subspace L_ϱ, we obtain an inner product space A/L_ϱ. Let us denote its completion by H_ϱ. The subspace A/L_ϱ is dense in H_ϱ. The inner product of elements in A/L_ϱ is given by the equality

$$(a + L_\varrho, b + L_\varrho) = \varrho(b^*\,a)\,.$$

For each $x \in A$ let us define the map $\pi_\varrho \colon A/L_\varrho \to A/L_\varrho$ by the formula

$$\pi_\varrho(x)\,(a + L_\varrho) = x\,a + L_\varrho\,, \qquad a \in A\,.$$

Since L_ϱ is a left ideal, $\pi_\varrho(x)$ is a well-defined linear map. Furthermore,

$$\begin{aligned}\|\pi_\varrho(x)\,(a+L_\varrho)\|^2 &= \|x\,a+L_\varrho\|^2 = \varphi(a^*\,x^*\,x\,a)\\ &\le \|x^*\,x\|\cdot\varphi(a^*\,a) = \|x\|^2\cdot\,|a+L_\varrho\|^2\,.\end{aligned}$$

It means that $\pi_\varrho(x)$ is a bounded operator and can therefore be extended uniquely to bounded linear operator on H_ϱ. We shall denote this extension by $\pi_\varrho(x)$ again. It is a routine matter to verify that the mapping

$$\pi_\varrho : x \to \pi_\varrho(x) : A \to B(H_\varrho)$$

is a $*$-representation. Let (u_α) be an approximate unit of A. Then, for $\alpha \le \beta$,

$$\|(u_\alpha + L_\varrho) - (u_\beta + L_\varrho)\|^2 = \varrho((u_\alpha - u_\beta)^2) \le \varrho(u_\alpha - u_\beta)\,.$$

As $\varrho(u_\alpha) \to \|\varrho\|$, we see that the net $(u_\alpha + L_\alpha)_\alpha$ is convergent to an element $\xi_\varrho \in H_\varrho$. The $*$-representation of a C^*-algebra A on a Hilbert space H is called *cyclic* if there is a vector ξ, called the *cyclic vector* for the $*$-representation π, such that $[\pi(A)\,\xi] = \overline{\mathrm{sp}}\{\pi(a)\,\xi \mid a \in A\} = H_\varrho$. It is clear from the construction that ξ_ϱ is a cyclic vector for the $*$-representation π_ϱ. The representation π_ϱ constructed above is called the *Gelfand-Naimark-Segal representation* (in abbreviation G.N.S representation) corresponding to a positive functional ϱ on A. The following property of the G.N.S. representation is straightforward:

$$\varrho(x) = (\pi_\varrho(x)\,\xi_\varrho, \xi_\varrho) \qquad \text{for all } x \in A\,.$$

In other words, any positive functional can be obtained as a composition of a $*$-representation and a vector functional. We summarize this discussion in the following theorem:

2.1.9. Theorem. *For any positive functional ϱ on a C^*-algebra A there is a Hilbert space, H_ϱ, a $*$-homomorphism, $\pi_\varrho\colon A \to B(H_\varrho)$, and a cyclic vector, $\xi_\varrho \in H_\varrho$, such that*

$$\varrho(x) = (\pi_\varrho(x)\,\xi_\varrho, \xi_\varrho)\,. \tag{2.5}$$

Moreover, the triple $(\pi_\varrho, H_\varrho, \xi_\varrho)$ is unique up to a unitary transformation between the corresponding Hilbert spaces.

Before more thorough analysis of the G.N.S. representation, let us state a few illustrating examples. They can be verified easily by checking the cyclicity of the corresponding vectors and equality (2.5).

2.1.10. Example. (i) Let ξ be a unit vector in a Hilbert space H. Then the G.N.S. representation of the vector state ω_ξ on $B(H)$ is (up to a unitary equivalence) the identity map on $B(H)$.
(ii) Let $A = C(X)$, where X is a compact Hausdorff space. Let ϱ be a state on A given by the probability Borel measure μ on X by the equality

$$\varrho(f) = \int_X f(x)\, d\mu(x)\,.$$

For any $f \in C_0(X)$, let us define the multiplication operator M_f on $L^2(X,\mu)$ by

$$M_f(g) = f\,g\,, \qquad g \in L^2(X,\mu)\,.$$

Then the $*$-representation $\pi_\varrho\colon f \to M_f$ is (up to a unitary equivalence) the G.N.S. representation of ϱ with the cyclic vector equal to the unit function on X.
(iii) Let $A = L^\infty(X,\mathcal{P},\mu)$ be an abelian C^*-algebra corresponding to a finite measure space. Then the G.N.S. representation of the state given by the integration with respect to μ is (up to a unitary equivalence) a representation determined by assignment $f \to M_f$, where the operator M_f acts on $L^2(X,\mathcal{P},\mu)$ and it is defined in the same way as in the previous example. The unit function is again the G.N.S. cyclic vector.
(iv) Let $A = M_n(\mathbb{C})$ and τ be the normalized matrix trace on A. Let $e_1,\ldots,e_n$ be an orthonormal basis of the n-dimensional vector space H_n. Then the G.N.S. representation of τ is (up to a unitary equivalence) the direct sum of identity representations of A on the n-fold direct sum $H_n \oplus H_n \oplus \cdots \oplus H_n$ with the cyclic vector $\xi_\varrho = \frac{1}{\sqrt{n}} \sum_{i=1}^n \oplus e_i$.

Given a positive functional ϱ on a C^*-algebra A and given $a \in A$, the functional

$$\varrho_a(x) = \varrho(a^*\, x\, a)\,, \qquad x \in A,$$

is called the *transformed functional of* ϱ. In terms of the G.N.S. representation $(\pi_\varrho, H_\varrho, \xi_\varrho)$ of ϱ, the transformed functionals of ϱ are exactly the functionals $\omega_h \circ \pi_\varrho$, where $h = a + L_\varrho \in A/L_\varrho$. Indeed, $\omega_h \circ \pi_\varrho(x) =$

$(\pi_\varrho(x)\,(a+L_\varrho), a+L_\varrho) = \varrho_a(x)$. Let T_ϱ denote the norm closure of the set of all transformed functionals of ϱ. If $x_n \to x$ in a Hilbert space H, then $\|\omega_{x_n} - \omega_x\| \to 0$ as can be verified easily. It is clear from the G.N.S. construction and the preceeding remark that $\omega_h \circ \pi_\varrho \in T_\varrho$, where $h \in H_\varrho$. On the other hand, it can be shown that $T_\varrho = \{\omega_h \circ \pi_\varrho \mid h \in H_\varrho\}$. The following technical point will be needed later. Suppose that $\pi_\varrho(a_n)\,x_\varrho \to x_\varrho$ as $n \to \infty$ in H_ϱ. Then, for each a in A,

$$(\pi_\varrho(a)\,\pi_\varrho(a_n)\,x_\varrho, \pi_\varrho(a_n)\,x_\varrho) = (\pi_\varrho(a_n^*\,a\,a_n)\,x_\varrho, x_\varrho) = \varrho(a_n^*\,a\,a_n)$$

and, in turn, $\|\varrho_{a_n} - \varrho\| \to 0$ as $n \to \infty$.

Representations and pure states. The character of the G.N.S. representation reflects the geometric structure of the state space. Let π be a $*$-representation of a C^*-algebra A on a Hilbert space H. A closed subspace $F \subset H$ is called *invariant* for a representation π if $\pi(F) \subset F$. This is equivalent to saying that the orthogonal projection p of H onto F is in the commutant $\pi(A)'$. In this case π can be decomposed into the direct sum

$$\pi = p\,\pi \oplus (1-p)\,\pi\,.$$

The representation $p\pi$ is called a *subrepresentation of* π. Two $*$-representations π_1 and π_2 of a C^*-algebra A are called *equivalent* if there is a $*$-isomorphism $\tau\colon\ \pi_1(A)'' \to \pi_2(A)''$ between bicommutants such that $\pi_2 = \tau \circ \pi_1$. The representations π_1 and π_2 are called *unitarily equivalent* if the automorphism τ above is implemented by a unitary map. The representation π_1 is called *subequivalent* to a representation π_2 if it is equivalent to some subrepresentation of π_2. A $*$-representation is called *irreducible* if it has no nontrivial invariant subspace, i.e. if it cannot be written as nontrivial direct sum of $*$-representations. Two representations π_1 and π_2 of A are called *disjoint* if no subrepresentation of π_1 is equivalent to any subrepresentation of π_2. Two irreducible representations are either unitarily equivalent or disjoint.

A state ϱ on a C^*-algebra A is called *pure* if it cannot be written as a nontrivial convex combination of states, i.e. if the following implication holds: If $\varrho = \frac{1}{2}(\varrho_1 + \varrho_2)$ for states ϱ_1 and ϱ_2 on A, then $\varrho_1 = \varrho_2 = \varrho$. The positive part of the dual A^*, i.e., the set of all positive functionals on A, can be ordered in the following way

$$\varrho_1 \le \varrho_2 \iff \varrho_1(a) \le \varrho_2(a) \text{ for all } a \in A_+\,.$$

Using this order, we can say that ϱ is a pure state if, and only if, each positive functional τ with $\tau \leq \varrho$ is a scalar multiple of ϱ.

2.1.11. Example. (i) Any vector state ω_x on $B(H)$ is a pure state.
(ii) If $A = C_0(X)$, where X is a locally compact Hausdorff space, then for any pure state ϱ on $C_0(X)$ there is a uniquely determined point $x \in X$ such that

$$\varrho(f) = f(x) \text{ for all } f \in C_0(X).$$

In this way the set of all pure states on $C_0(X)$ endowed with the weak*-topology is homeomorphic to X.

If ϱ is a pure state on an abelian C^*-subalgebra B of a C^*-algebra A, then $\varrho(x\,y) = \varrho(x)\,\varrho(y)$ whenever $x \in B$ and $y \in A$ commutes with all elements of B. Any pure state remains pure when restricted to the centre $Z(A)$. Hence, $\varrho(x\,y) = \varrho(x)\,\varrho(y)$ for all $x \in Z(A)$ and $y \in A$.

The following important theorem relates pureness of states to irreducibility of the G.N.S. representations.

2.1.12. Theorem. *A state ϱ on a C^*-algebra A is pure if, and only if, the G.N.S. representation of ϱ is irreducible.*

The C^*-algebra A acting on a Hilbert space H is *irreducible* if its identity representation is irreducible. The following *Kadison Transitivity Theorem* says that topological irreducibility is equivalent to the algebraic irreducibility.

2.1.13. Theorem. *Let $\pi_1, \ldots, \pi_n$ be disjoint irreducible representations of a C^*-algebra A on the Hilbert spaces $H_1, \ldots, H_n$, correspondingly. Let $t_i \in B(H_i)$ and K_i be a finite-dimensional subspace of H_i $(1 \leq i \leq n)$. For each $\varepsilon > 0$ there is an element $a \in A$ such that*

$$\pi_i(a)|K_i = t_i|K_i \text{ and } \|a\| \leq \|t_i\| + \varepsilon \qquad \text{for each } i = 1, \ldots, n.$$

Moreover, if each t_i is self-adjoint, we can choose a to be self-adjoint, too.

Let ϱ be a pure state on a C^*-algebra A with the (irreducible) G.N.S. representation π_ϱ. Then any state $\omega_h \circ \pi_\varrho$ is a pure state unitarily equivalent to ϱ.

The G.N.S. construction allows one to represent faithfully a given C^*-algebra as a closed $*$-algebra of operators on some Hilbert space. If x is a nonzero positive element in a C^*-algebra A, then there is a pure state ϱ such that $\varrho(x) = \|x\|$. This immediately implies that $\pi_\varrho(x) \neq 0$ whenever x is nonzero. Therefore the *atomic representation*

$$\pi_{at} = \sum_{\varrho \in P(S)} \oplus \pi_\varrho ,$$

where $P(S)$ is the set of all pure states on A, is faithful. Also, the *universal representation*

$$\pi_u = \sum_{\varrho \in S(A)} \oplus \pi_\varrho ,$$

where $S(A)$ denotes the set of all states, is faithful. Both atomic and universal representation allow one to identify given C^*-algebra with the algebra of operators. Let us note that in the case of abelian algebra, A, all irreducible representations are complex-valued $*$-homomorphisms and hence the atomic representation is in fact the Gelfand transform. The algebra $\pi_u(A)''$ acting on the universal Hilbert space H_u, called the *enveloping von Neumann algebra of* A, can be identified with the double dual A^{**}. This aspect will be discussed later.

State spaces of C^*-algebras. The convex structure of the set of states in the dual of a C^*-algebra is closely related to the inner structure of the algebra. Let us review some of the most fundamental results. First, let us recall basic definitions of convexity theory. Let K be a subset of vector space, V. K is *convex* if $\alpha\, x + (1 - \alpha)\, y \in K$ whenever $x, y \in K$ and $\alpha \in [0, 1]$. A *face* F in K is a convex subset of K such that the following implication holds: If $\alpha\, x + (1 - \alpha)\, y \in F$ for $x, y \in K$ and $\alpha \in [0, 1]$, then $x, y \in F$. An element $x \in K$ is called an *extreme point* of the set K if the set $F = \{x\}$ is a face of K. A function $f\colon\ K \to \mathbb{C}$ is called *affine* if $f(\alpha\, x + (1 - \alpha)\, y) = \alpha\, f(x) + (1 - \alpha)\, f(y)$ for all $x, y \in K$ and $\alpha \in [0, 1]$. Let $A^b(K)$ be the real Banach space of all bounded affine functions on K endowed with the supremum norm. If $0 \in K$, then the symbol $A^b_0(K)$ will denote its subspace consisting of all real affine functions that vanish at zero.

If the underlying vector space V is equipped with a locally convex topology in which K is compact, then $A(K)$ and $A_0(K)$ will denote the space of all continuous functions on K and the space of all continuous functions on K vanishing at zero if K contains it, respectively.

Let A be a C^*-algebra. By the symbol $S(A)$ we shall denote the convex set of all states on A. $S(A)$ is called the *state space of* A. The extreme points of this set are pure states. If A is unital, then $S(A)$ is a compact set when endowed with the weak*-topology. The symbol $Q(A)$ will stand for the set of all positive functionals on A of norm not greater than one. This set is called the *quasi-state space.* The set $Q(A)$ is always convex and compact with respect to the weak*-topology. Let F be a face of $Q(A)$ or $S(A)$. Then its norm closure is again a face, which, surprisingly enough, is a nontrivial fact. Another important property of C^*-algebras is the one-to-one correspondence between closed left ideals and weak*-closed faces of $Q(A)$. Formally, the map

$$L \to L^{\perp} \cap Q(A) = \{\varphi \in Q(A) \mid \varphi(x) = 0 \text{ for all } x \in L\}$$

establishes order preserving bijective correspondence between the set of all closed left ideals in A and weak* closed faces of $Q(A)$ containing zero. According to the Krein-Milman Theorem, the set $Q(A)$ is well supplied by extreme points. All extreme points of this set, except for the zero functional, are pure states of A. For a pure state ϱ of A, the left ideal corresponding to the minimal face $\{\varrho\}$ is the left kernel of ϱ. Such ideals are characterized as being the maximal ones. Besides the G.N.S. construction, the following function representation is a noncommutative extension of the Gelfand transform. Any element $a \in A^{**}$ induces an affine function $\hat{a}$ on $Q(A)$ given by the equality $\hat{a}(\varrho) = \varrho(a)$ for all $\varrho \in Q(A)$. Then the map $a \to \hat{a}$ is an order preserving linear isometry of the self-adjoint part of A^{**} onto $A_0^b(Q(A))$ such that A_{sa} (canonically embedded into its second dual) is mapped onto $A_0(Q(A))$.

A subset S of the state space $S(A)$ is called *order determining* if, and only if, $\varrho(a) \geq 0$ for all $\varrho \in S$ implies that $a \geq 0$. If A is a unital C^*-algebra, then $S \subset S(A)$ is order determining if, and only if, the convex hull of S is weak*-dense in $S(A)$, or equivalently, if, and only if, the weak* closure of S contains all pure states.

Tensor products of C^*-algebras. One of the most important techniques in the operator algebra theory is that of forming tensor products of C^*-algebras. We shall start our review of this topic by discussing spatial tensor products. These types of products are based on the tensor products of operators. Let X and Y be linear spaces. Let us denote by $X \otimes_{alg} Y$ their algebraic tensor product. It is linearly spanned by the simple tensors $x \otimes y$, $x \in X, y \in Y$. Tensor products allow one to turn bilinear forms into linear maps. Indeed, for each bilinear form $b\colon X \times Y \to \mathbb{C}$ there is a unique linear form f on $X \otimes_{alg} Y$ such that

$$b(x, y) = f(x \otimes y) \quad \text{for all} \quad x \in X,\ y \in Y.$$

If B_1 and B_2 is a linear basis of X and Y, respectively, then the set $\{b_1 \otimes b_2 \mid b_1 \in B_1, b_2 \in B_2\}$ is a linear basis of $X \otimes_{alg} Y$.

Let H and K be Hilbert spaces. Then there is a unique inner product on $H \otimes_{alg} K$ such that

$$(x_1 \otimes y_1, x_2 \otimes y_2) = (x_1, x_2) \cdot (y_1, y_2),$$

where $x_1, x_2 \in X$ and $y_1, y_2 \in Y$. The tensor product, $H \otimes K$, of Hilbert spaces H and K is defined as the Hilbert space resulting by completion of $H \otimes_{alg} K$ with respect to this inner product. Let us note that for any two orthonormal bases $(e_i)_{i \in I}$ and $(f_j)_{j \in J}$ of H and K, respectively, the set $\{e_i \otimes f_j \mid (i, j) \in I \times J\}$ is an orthonormal basis of $H \otimes K$. Given two operators $a \in B(H)$ and $b \in B(K)$, one can show that there is a unique bounded operator on $H \otimes K$, denoted by $a \otimes b$, whose action on $H \otimes K$ is determined by the equality

$$(a \otimes b)(\xi \otimes \nu) = a\xi \otimes b\nu$$

for all $\xi \in H$, $\nu \in K$. It holds that $\|a \otimes b\| = \|a\| \cdot \|b\|$. Let us now suppose that A and B are two C^*-algebras and choose faithful representations π_1 and π_2 of A and B on the Hilbert spaces H and K, respectively. The C^*-subalgebra of $B(H \otimes K)$ generated by the set $\{a \otimes b \mid a \in \pi_1(A), b \in \pi_2(B)\}$ do not depend (as an abstract C^*-algebra) on the choice of the faithful representations π_1 and π_2 and it is called the *spatial tensor product* (or the *minimal tensor product*) of A and B.

2.1.14. Example. (i) Let $A = M_n(\mathbb{C})$, $B = M_m(\mathbb{C})$ be matrix algebras. Then $A \otimes B$ is the algebra $M_{m\,n}(\mathbb{C})$. For $a = (a_{ij}) \in A$ and $b = (b_{kl}) \in B$, where $b_{11} = 1$ and $b_{kl} = 0$ otherwise, the simple tensor $a \otimes b$ is the $n\,m \times n\,m$ matrix that have the $m \times m$ matrix a in the left upper corner and zero elements otherwise.
(ii) Let A be a C^*-algebra. The tensor product $l^\infty \otimes A$ is $*$-isomorphic to the direct sum $\sum_{n=1}^{\infty} \oplus A$, such that simple tensors $(\lambda_n) \otimes a$ correspond to $\sum_{n=1}^{\infty} \oplus \lambda_n\, a$. Tensor product construction thereby generalizes direct sums of algebras.
(iii) Let X be a compact Hausdorff space and let A be a C^*-algebra. The tensor product $C(X) \otimes A$ can be identified with the algebra $C(X, A)$ of all continuous A-valued functions on X in such a manner that the simple tensors, $f \otimes a$, $f \in C(X)$, $a \in A$, correspond to the functions $x \to f(x)\, a : X \to A$.
(iv) Let A be a C^*-algebra. The tensor product $A \otimes B(H_n)$, where $\dim H_n = n$, can be identified with the algebra $M_n(A)$ of all $n \times n$ matrices with entries in A and usual matrix operations. A more general than this is the tensor product $A \otimes B(H)$ which can be identified with the algebra of infinite matrices over A.
(v) (Special case (iii)) Let X and Y be compact Hausdorff spaces. Then $C(X) \otimes C(Y)$ is $*$-isomorphic to the algebra $C(X \times Y)$ such that the simple tensors $f \otimes g$ correspond to the functions $(x, y) \to f(x)\, g(y) : X \times Y \to \mathbb{C}$.

Let φ_1 and φ_2 be states on the algebras A and B, respectively. There is a state φ, called the *product state* of φ_1 and φ_2 (in symbols $\varphi_1 \otimes \varphi_2$) that is uniquely determined by the condition

$$(\varphi_1 \otimes \varphi_2)(a \otimes b) = \varphi_1(a)\, \varphi_2(b) \qquad a \in A\,, b \in B\,.$$

The product state $\varphi_1 \otimes \varphi_2$ is pure if, and only if, φ_1 and φ_2 are pure states. An analogous result holds for $*$-representations. Given two $*$-representations $\pi_1 \colon A \to B(H)$ and $\pi_2 \colon B \to B(K)$, there is a unique *product representation* $\pi_1 \otimes \pi_2 \colon A \otimes B \to B(H) \otimes B(K)$ such that

$$(\pi_1 \otimes \pi_2)(a \otimes b) = \pi_1(a) \otimes \pi_2(a)$$

for all $a \in A$ and $b \in B$. The product $\pi_1 \otimes \pi_2$ is irreducible if, and only if, π_1 and π_2 are irreducible.

The spatial tensor product is a special case of the tensor product in the category of abstract C^*-algebras. Constructions of more general tensor products start with algebraic tensor product $A \otimes_{alg} B$ of C^*-algebras A and B. It can be proved that there is a unique product and a unique $*$-operation on $A \otimes_{alg} B$ making it into a $*$-algebra such that for the simple tensors we have

$$(a_1 \otimes b_1)(a_2 \otimes b_2) = a_1\, a_2 \otimes b_1\, b_2$$

$$(a \otimes b)^* = a^* \otimes b^* .$$

A *C^*-norm* on $A \otimes_{alg} B$ is a norm β such that the completion of $A \otimes_{alg} B$ with respect to the norm β is a C^*-algebra. The algebras A and B can be identified with subalgebras $A \otimes \mathbb{C}1$ and $\mathbb{C}1 \otimes B$ in $A \otimes_{alg} B$, respectively. Obviously, the norm β then coincides with the norms on A and B. Moreover, a remarkable property of C^*-algebras is the fact that every C^*-norm on $A \otimes_{alg} B$ is a *cross norm*, i.e.

$$\beta(a \otimes b) = \|a\| \cdot \|b\| \qquad \text{for all } a \in A\,, b \in B\,.$$

The *tensor product*, $A \otimes_\beta B$ with respect to the C^*-norm β on $A \otimes_{alg} B$, is defined as the C^*-algebra that results by completing $(A \otimes_{alg} B, \beta)$. The C^*-norm β on $A \otimes_{alg} B$ is called *minimal* if it is given by the norm of the spatial tensor product of A and B. The *maximal* C^*-norm, $\|\cdot\|_{max}$, is defined by the equality

$$\|x\|_{max} = \sup_{\gamma \in \Gamma} \|x\|_\gamma\,,$$

where Γ is the set of all C^*-norms of $A \otimes_{alg} B$. (One can verify that the supremum above is finite.) One of the deepest results on tensor products says that for any C^*-norm β on $A \otimes_{alg} B$ the following inequality holds:

$$\|\cdot\|_{min} \le \beta \le \|\cdot\|_{max}\,.$$

If φ_1 and φ_2 are states on A and B, respectively, then there is a unique state, $\varphi = \varphi_1 \otimes_\beta \varphi_2$ on $A \otimes_\beta B$ such that

$$\varphi(a \otimes b) = \varphi_1(a)\,\varphi_2(b) \qquad a \in A, b \in B\,.$$

The state $\varphi_1 \otimes_\beta \varphi_2$ is called the *β-product state* of φ_1 and φ_2. Since the spatial norm is a minimal one, the identity map $i\colon\ A \otimes_{alg} B \to A \otimes_{alg} B$

extends uniquely to a $*$-homomorphism $\tau\colon A \otimes_\beta B \to A \otimes B$ for each C^*-norm β. In general, there are many C^*-norms on the algebraic tensor product. If there is only one, the algebra is called nuclear. More precisely, a C^*-algebra A is called *nuclear* if there is only one C^*-norm on each tensor product $A \otimes_{alg} B$ of A with an arbitrary C^*-algebra B. In this case the maximal norm coincides with the minimal one. It is known, e.g., that all abelian C^*-algebras and all C^*-algebras of compact operators are nuclear. The algebra $B(H)$, where $\dim H = \infty$, is not nuclear.

Projection structure of C^*-algebras. In this paragraph we summarize basic concepts and results concerning the set $P(A)$ of all projections in a C^*-algebra A. Let us note that it may happen that $P(A)$ consists of the zero projection only. For this, consider, e.g., the algebra $C_0(\mathbb{R})$. Projections e and f in a C^*-algebra A are called *orthogonal* if $e\,f = 0$. If the algebra is represented on a Hilbert space, then orthogonal projections project on orthogonal subspaces. If e and f are orthogonal projections, then the C^*-subalgebra generated by hereditary subalgebras $e\,A\,e$ and $f\,A\,f$ is $*$-isomorphic to their direct sum. First we explore the metric structure of $P(A)$. Let us observe that $\|e - f\| \le 1$ for all projections $e, f \in P(A)$. To see this, let us write $(e-f)^2 = e\,(1-f)\,e + (1-e)\,f\,(1-e)$. (The unit 1 is an element of the unitalization of A.) Hence, $(e-f)^2 \le e + (1-e) \le 1$ and thus $\|e-f\|^2 = \|(e-f)^2\| \le 1$. Another important inequality involving projections is the following one: Let A be a unital C^*-algebra. Then

$$\|e\,f\,(1-e)\| \le 1 \qquad \text{for all projections } e, f \in P(A)\,. \tag{2.6}$$

To see this let us write

$$(e\,f - f\,e)\,(f\,e - e\,f) = e\,f\,(1-e)\,f\,e + (1-e)\,f\,e\,f\,(1-e)\,.$$

As $e\,f\,(1-e)\,f\,e \in e\,A\,e$, $(1-e)\,f\,e\,f(1-e) \in (1-e)\,A\,(1-e)$, we obtain that

$$\|(e\,f - f\,e)\,(f\,e - e\,f)\| = \max(\|e\,f\,(1-e)\,f\,e\|, \|(1-e)\,f\,e\,f\,(1-e)\|)\,.$$

But,

$$\|e\,f\,(1-e)\,f\,e\| = \|(1-e)\,f\,e\|^2 = \|e\,f\,(1-e)\|^2 = \|(1-e)\,f\,e\,f\,(1-e)\|\,.$$

Hence, $\|e\,f\,(1-e)\| = \|e\,f - f\,e\|$. However, the commutant $e\,f - f\,e$ can be written as $e\,f - f\,e = \frac{1}{2}(s\,e - e\,s)$, where $s = 1 - 2\,f$ is a symmetry. Taking into account that $s\,e\,s \in P(A)$ we conclude by the above inequality that

$$\|e\,f - f\,e\| = \frac{1}{2}\,\|s\,e - e\,s\| = \frac{1}{2}\,\|e - s\,e\,s\| \leq \frac{1}{2}\,,$$

which is the desired result.

The set $P(A)$ with the order inherited from A becomes a partially ordered set. It holds that $e \leq f$ in $P(A)$ if, and only if, $e\,f = f\,e = e$. If A is unital, then the structure $P(A)$ is an orthomodular poset with the complement $p^{\perp} = 1 - p$. The structure $(P(A), 0, 1, \leq, \perp)$ is a Boolean algebra if, and only if, A is abelian. In this case, the supremum and infimum in $P(A)$ can be computed algebraically

$$e \vee f = e + f - e\,f\,, \qquad e \wedge f = e\,f\,.$$

We shall consider the following Murray-von Neumann equivalence on the set of all projections. An element $v \in A$ is called a *partial isometry* if $v^*\,v = p$, where $p \in P(A)$. In this case, $v\,v^* = q$, where q is another projection in $P(A)$. The projections p and q are said to be the *initial* and the *final projection* of v, respectively. The adopted terminology is explained by the following fact. If A is faithfully represented on a Hilbert space H, then $v|p(H)$ is a linear isometry mapping $p(H)$ onto $q(H)$ and $v|P(H)^{\perp}$ is zero. The adjoint map, v^*, maps isometrically $q(H)$ onto $p(H)$ and vanishes on $q(H)^{\perp}$. Two projections $p, q \in P(A)$ are called *(Murray-von Neumann) equivalent* (in symbols $p \sim q$) if

$$p = v^*\,v \quad \text{and} \quad q = v\,v^*$$

for some $v \in A$. The relation $\sim$ is an equivalence on the set $P(A)$. The projection p is said to be *subequivalent to a projection* q (in symbols $p \precsim q$) if $p \sim e$, where $e \leq q$. All distinct projections in an abelian algebra are not equivalent, the projections in the algebra $B(H)$ are equivalent if, and only if, their ranges have the same (orthonormal) dimension. If $\|p - q\| < 1$ for $p, q \in P(A)$, then $p \sim q$. The projections p and q in a unital C^*-algebra A are said to be *unitarily equivalent* (in symbols $p \sim_u q$) if there is a unitary map $u \in A$ such that $p = u^*\,q\,u$. It holds that $p \sim_u q$ if, and only if, $p \sim q$ and $1 - p \sim 1 - q$. Orthogonal equivalent projections are always unitarily equivalent. Further, if $p \sim q$, then

$$\begin{pmatrix} p & 0 \\ 0 & 0 \end{pmatrix} \sim_u \begin{pmatrix} q & 0 \\ 0 & 0 \end{pmatrix}$$

in $M_2(A)$.

Suppose now that $e_1, \ldots, e_n$ are orthogonal equivalent projections in a unital C^*-algebra A with $e_1+e_2+\cdots+e_n = 1$. Let v_{ij} be a partial isometry with the initial projection e_j and the final projection e_i. The system (v_{ij}) is called the system of $n \times n$ *matrix units.* The following rules are satisfied: $v_{ij}\, v_{kl} = 0$ if $j \neq k$ and $v_{ij}\, v_{jk} = v_{ik}$. Using these identities it can be shown that the map from the C^*-algebra generated by the system (v_{ij}) to the matrix algebra $M_n(\mathbb{C})$ that assigns to each v_{ij} the matrix unit with entry 1 at the position (i,j) and zero otherwise, extends to a $*$-isomorphism between $C^*(v_{ij})$ and $M_n(\mathbb{C})$. Moreover, it can be shown that in this case A is $*$-isomorphic to $e_1\, A\, e_1 \otimes M_n(\mathbb{C}) = M_n(e_1\, A\, e_1)$. Similar results hold for infinite sequence of orthogonal equivalent projections.

A projection p in a C^*-algebra A is said to be *infinite* if it is equivalent to a proper subprojection of itself, i.e. if there is a projection $q \in A$ such that $p \sim q < p$. If p is not infinite, then p is said to be *finite.* A unital C^*-algebra is said to be *finite* (resp. *infinite*) if its unit is a finite (resp. infinite) projection. A projection in $B(H)$ is finite if, and only if, it is finite-dimensional. Hence, $B(H)$ is a finite algebra if, and only if, $\dim H < \infty$. A C^*-algebra is finite if, and only if, all projections in A are finite. Further, a projection p in a C^*-algebra A is finite if, and only if, the hereditary subalgebra $p\, A\, p$ is finite. The finiteness of projection is related to the existence of a tracial state. A state τ on C^*-algebra is called *tracial* (or a *trace*) if $\tau(x^*\, x) = \tau(x\, x^*)$ for all $x \in A$. This is equivalent to the fact that $\tau(a\, b) = \tau(b\, a)$ for all $a, b \in A$. It is clear that the trace is constant on equivalent projections. There is only one tracial state on the matrix algebra $M_n(\mathbb{C})$ which is the standard normalized matrix trace. If τ is a trace of a simple algebra A, then A has to be finite. Indeed, τ is faithful in this case, and so $p \leq q$ and $p \sim q$ for projections p, q implies that $\tau(q-p) = 0$, inferring immediately $q-p = 0$. It turns out that if p and q are projections in a simple C^*-algebra and q is infinite, then $p \lesssim q$. Every infinite simple C^*-algebra contains an infinite sequence of projections which are mutually orthogonal and all equivalent.

As the last point of this section we briefly state important types of C^*-algebras that are rich in projections. A unital C^*-algebra is said to have *real rank zero* if the invertible self-adjoint elements are dense in the set of all self-adjoint elements. Equivalently, A has real rank zero if every self-adjoint element can be approximated by self-adjoint elements with finite spectrum. Consequently, the real rank zero algebra is a closed linear span of its projections. It has been proved that any simple unital C^*-algebra such that every nonzero projection in A is infinite is real rank zero. The following C^*-algebras are examples of real rank zero algebras.

Let T be a unit circle in the complex plane and θ an irrational number. We shall consider two unitary operators u and v on $L^2(T)$ acting by

$$\begin{aligned}(u\,f)(z) &= z\,f(z)\\(v\,f)(z) &= f(e^{-2i\pi\theta}\,z)\end{aligned}$$

for all $f \in L^2(T)$ and $z \in T$. Then a simple calculation yields

$$u\,v = e^{2\pi i\theta}\,v\,u\,.$$

The C^*-algebra, A_θ, generated by u and v is called the *irrational rotation algebra.* It can be proved that A_θ is $*$-isomorphic to any C^*-algebra generated by the unitaries u and v satisfying the equalities (2.7). The algebra A_θ is an example of a simple C^*-algebra with a unique faithful trace.

Let H be an infinite-dimensional Hilbert space. The *Cuntz algebra*, O_n, is defined as the C^*-algebra generated by the isometries $s_1, \ldots, s_n$ acting on H with mutually orthogonal ranges, i.e. satisfying

$$\sum_{i=1}^{n} s_i\,s_i^* = 1\,.$$

A C^*-algebra A is called *monotone complete* if each bounded increasing net in A_{sa} has a least upper bound in A_{sa}. A C^*-algebra A is said to be *AW^*-algebra* if each maximal abelian C^*-subalgebra of A is monotone complete. A unital C^*-algebra $C(X)$ is AW^*-algebra if, and only if, X is a extremely disconnected space, i.e. if the closure of each open set in X is open. For each nonempty subset S of a AW^*-algebra there is a projection $p \in A$ such that $S^0 = \{a \in A \mid x\,a = 0 \text{ for all } x \in S\} = p\,A$. The structure $P(A)$ of a AW^*-algebra is a complete lattice. An important examples of AW^*-algebras are von Neumann algebras that will be treated in the next paragraph.

2.2. von Neumann algebras

Compact and trace class operators. In this paragraph we deal with the structure of compact operators and trace class operators on Hilbert spaces that are important for identifying preduals of von Neumann algebras and will be used widely in the first two chapters. We continue to denote by $B(H)$ the C^*-algebra of all bounded operators acting on a Hilbert space H. Let $x \in B(H)$. The *range projection*, $r(x)$, of x is defined as the orthogonal projection of H onto $\overline{\mathrm{sp}}\{x\,\xi \mid \xi \in H\}$. If x is self-adjoint, then $1 - r(x)$ is the projection onto the kernel, $\mathrm{Ker}\, x$, of x. By $|x|$ we shall denote the positive operator $|x| = \sqrt{x^*\,x}$. For any $x \in B(H)$ there is a unique partial isometry $v \in B(H)$ such that

$$x = v\,|x| \tag{2.7}$$

and $\mathrm{Ker}\, v = \mathrm{Ker}\, x$. Moreover, $v^* x = |x|$. The decomposition (2.7) is called the *polar decomposition* of x. (The terminology is inspired by the factorization of a complex number as a product of a complex unit (phase) and nonnegative number (modulus).)

Let $F(H)$ denote the set of all operators in $B(H)$ that have finite-dimensional range. As an example, let us take vectors $\xi, \nu \in H$ and denote by $\xi \odot \nu$ the operator on H given by the formula

$$(\xi \odot \nu)(h) = (h, \nu)\,\xi \qquad \text{for all } h \in H\,.$$

Then $\xi \odot \nu$ is an operator with the one-dimensional range $\mathrm{sp}\{\nu\}$. In particular, $\xi \odot \xi$, where ξ is a unit vector, is an orthogonal projection onto $\mathrm{sp}\{\xi\}$. It can be shown that all one-dimensional operators arise in this way and that $F(H)$ is a linear hull of the set of all one-dimensional projections. The set $F(H)$ is a (nonclosed) ideal in $B(H)$. The following property of the set $F(H)$ is important: If I is an ideal (not necessarily closed) in $B(H)$, then I contains $F(H)$. The operator $x \in B(H)$ is called *compact* if it maps the unit ball of H onto a compact set. The results on the spectrum show that the spectrum $Spec(x)$ of a compact operator x is countable and that each nonzero point of it is an eigenvalue of x and an isolated point of $Spec(x)$. Moreover, each nonzero eigenvalue of x has finite-dimensional subspace of eigenvectors. If x is a compact operator, then x^* as well as $|x|$ is a compact operator. In particular, the real and the imaginary part of a compact operator is a compact operator. The operator $x \in B(H)_{sa}$ is compact if, and only if, $x = \sum_n \lambda_n p_n$, where (p_n) is a sequence of pairwise orthogonal one-dimensional projections and $(\lambda_n) \in c_0$ (i.e. $\lim_{n\to\infty} \lambda_n = 0$). (From this point of view the space of all compact operators can be thought of as the noncommutative generalization of c_0.) These facts imply that any compact operator can be approximated in the operator norm by a finite-dimensional operator. Let $K(H)$ denote the set of all compact operators on a Hilbert space H. Then $K(H)$ is a closed ideal in the C^*-algebra $B(H)$ and thereby it is a hereditary subalgebra. Moreover, $F(H)$ is dense in $K(H)$. By the property that any ideal contains $F(H)$ we infer that $K(H)$ is the smallest closed ideal in $B(H)$. Further, it is known that if H is separable, then $K(H)$ is the only nontrivial closed ideal in $B(H)$. It can be verified easily that the set $K(H)$ acts irreducibly on $B(H)$. The following deeper result implies that all irreducible C^*-subalgebras of $K(H)$ are equal to $K(H)$: Let A be a C^*-algebra acting irreducibly on a Hilbert space H and having nonzero intersection with $K(H)$. Then $K(H) \subset A$. One of the most important properties of the ideal $K(H)$ is the fact that the automorphism group of $K(H)$ is implemented by the unitary group of $B(H)$. More precisely, let H and H' be Hilbert spaces and suppose that the map $\varphi\colon K(H) \to K(H')$ is a $*$-isomorphism. Then there is a unitary map $u\colon H \to H'$ such that $\varphi(x) = uxu^*$ for all $x \in K(H)$. More generally, any $*$-Jordan isomorphism $\varphi\colon K(H) \to K(H)$ is implemented in the above way by a unitary or an antiunitary operator. As a consequence, the same result holds for $B(H)$. It can be proved that any C^*-subalgebra of $K(H)$, (i.e. any C^*-algebra of compact operators) is isomorphic to the algebra $\{(x_\alpha) \mid (x_\alpha) \in \sum_\alpha \oplus K(H_\alpha),\, \lim_\alpha \|x_\alpha\| = 0\}$. Further, each representation of a C^*-algebra A of compact operators acting on H is a sum of irreducible representations that are subequivalent to the identity representation $i\colon A \to B(H)$.

The following result, known as the *Weyl-von Neumann-Berg Theorem* says that any normal operator on $B(H)$ is a small compact perturbation of a diagonalizable normal operator. (By a diagonalizable operator we mean the operator whose (possibly infinite) matrix is diagonal with respect to some orthonormal basis.) For any normal operator $x \in B(H)$ and $\varepsilon > 0$ there is a diagonalizable normal operator d and a compact operator $k \in K(H)$ such that $x = d + k$, and $\|k\| \le \varepsilon$.

Now we pass to a special subclass of compact operators, called the trace class operators. Let $x \in B(H)$ and let us define the positive number (possibly ∞), $\|x\|_1$, by the formula

$$\|x\|_1 = \sum_{e \in E} (\,|x|\, e, e)\,, \tag{2.8}$$

where E is an orthonormal basis of H. It can be shown that the expression on the right-hand side of (2.8) is the same for all orthonormal bases of E. The operator $x \in B(H)$ is said to be the *trace class operator* if $\|x\|_1 < \infty$. Let us denote by $L^1(H)$ the set of all trace class operators on H. Then $(L^1(H), \|\cdot\|_1)$ is an involutive $*$-algebra with respect to the usual adjoint operation such that $L^1(H) \subset K(H)$ and such that $F(H)$ is dense in $L^1(H)$ with respect to the norm $\|\cdot\|_1$. The following inequalities for the norm $\|\cdot\|_1$ hold true for all $x, y \in B(H)$:

$$\begin{aligned} \|x\| &\leq \|x\|_1 = \|x^*\|_1 \\ \|x\,y\| &\leq \|y\| \cdot \|x\|_1 \,. \end{aligned}$$

It implies that $L^1(H)$ is a self-adjoint ideal in $B(H)$. A self-adjoint operator x is of trace class if, and only if, $x = \sum_n \lambda_n p_n$, where (p_n) is a sequence of pairwise orthogonal one-dimensional projections and $\sum_n |\lambda_n| < \infty$. This explains the notation $L^1(H)$. For each $x \in L^1(H)$ the series $\sum_{e \in E} (x\,e, e)$, where E is an orthonormal basis of H converges and its sum does not depend on the choice of E. The number

$$\operatorname{tr} x = \sum_{e \in E} (x\,e, e)\,.$$

is called the *trace* of x. It reduces to the standard matrix trace in case of H being finite-dimensional. If $x \in L^1(H)$, then $\operatorname{tr}(x\,y) = \operatorname{tr}(y\,x)$ for all $y \in B(H)$. The following inequality involving traces will often be applied in the sequel:

$$|\operatorname{tr}(x\,y)| \leq \|y\| \cdot \|x\|_1 = \|y\| \operatorname{tr}(|x|)\,, \tag{2.9}$$

for all $y \in B(H)$ and $x \in L^1(H)$. Besides, the following proposition will be instrumental.

2.2.1. Proposition. *Let $x \in B(H)$ be a self-adjoint operator. Suppose that $\sum_{e \in E} (x\,e, e)$ converges for each orthonormal basis E of H. Then $x \in L^1(H)$.*

Proof: Seeking a contradiction, suppose that x satisfies the assumption and it is not a trace class operator. Let us decompose $x = x^+ - x^-$, where x^+ and x^- is the positive and the negative part of x, respectively. It means that $x^+ x^- = 0$, and hence the range projections $r(x^+)$ and $r(x^-)$ are orthogonal. Since $L^1(H)$ is a space, either $x^+ \notin L^1(H)$ or $x^- \notin L^1(H)$. Without loss of generality we can assume that $x^+ \notin L^1(H)$. Let us put $p = 1 - r(x^+)$, which is a projection onto $\mathrm{Ker}\, x^+$. Then

$$x^+ = x^+ (1 - p) \notin L^1(H).$$

It implies that there is an orthonormal sequence $(e_n) \subset (1-p)(H)$ such that $\sum_n (x^+ e_n, e_n) = \infty$. However, $x^+ e_n = x\, e_n$, and hence the series $\sum_n (x\, e_n, e_n)$ does not converge – a contradiction. □

As it is well known, the dual of the Banach space c_0 is l^1 and the dual of l^1 is l^∞. The same is the relationship between $K(H)$ ("noncommutative" c_0), $L^1(H)$ ("noncommutative" l^1), and $B(H)$ ("noncommutative" l^∞). Indeed, the elements of $L^1(H)$ and $K(H)^*$ are in a one-to-one correspondence via the map assigning to each $x \in L^1(H)$ the linear functional $f_x \in K(H)^*$ given by the formula

$$f_x(y) = \mathrm{tr}(x\, y)\,, \qquad \text{for all } y \in K(H)\,.$$

The above stated correspondence establishes a linear isometry between the dual space $K(H)^*$ and $(L^1(H), \|\cdot\|_1)$. Moreover, for any self-adjoint (resp. positive) functional on $K(H)$ the representing trace class operator is self-adjoint (resp. positive). In this manner the states on $K(H)$ are in a one-to-one correspondence with positive trace class operators with trace one, that are called the *density matrices*. Consequently, any state φ on $K(H)$ corresponds to the density matrix ϱ such that

$$\varphi(x) = \mathrm{tr}(\varrho\, x) \qquad \text{for all } x \in K(H)\,.$$

By choosing a diagonal orthonormal basis for ϱ, we can write

$$\varphi = \sum_n \lambda_n\, \omega_{e_n}\,, \tag{2.10}$$

where (e_n) is an orthonormal sequence in H, $\lambda_n \geq 0$, and $\sum_n \lambda_n = 1$. If φ is a state of $K(H)$, then it extends uniquely to a state on $B(H)$ because

$K(H)$ is a hereditary subalgebra. In other words, the states of the form (2.10) are precisely the states on $B(H)$ concentrated at the ideal $K(H)$.

The same results hold for the duality between $L^1(H)$ and $B(H)$. The map

$$x \in L^1(H) \to f_x \in B(H)^* ,$$

where

$$f_x(y) = \mathrm{tr}(x\, y) \qquad y \in B(H) ,$$

is a linear $*$-preserving isometry of $(L^1(H), \|\cdot\|_1)$ onto the dual of $B(H)$.

Elements of the theory of von Neumann algebras. We first introduce some locally convex topologies on the algebra $B(H)$. The *weak operator topology* on $B(H)$ is given by the system of seminorms $a \in B(H) \to |(a\, x, y)|$, where $x, y \in H$. The convergence in this topology is the pointwise weak convergence. The *strong operator topology* is given by the system of seminorms $a \in B(H) \to \|a\, x\|$, where $x \in H$. The strong operator topology is the topology of pointwise norm convergence and it is finer then the weak operator topology. It can be proved that the closures of convex subsets of $B(H)$ in both topologies coincide. The *σ-weak topology* (or *ultraweak topology*) is given by the system of seminorms induced by the trace class operators: $a \in B(H) \to |\mathrm{tr}(t\, a)|$, where $t \in L^1(H)$. As seen in the previous paragraph, $L^1(H)$ can be identified with the predual of $B(H)$ and therefore the σ-weak topology is a weak*-topology on $B(H)$. The σ-weak topology is finer then the weak operator topology and both topologies coincide on the closed unit ball of $B(H)$. When H is infinite-dimensional, then none of the mappings

$$\begin{aligned} (s,t) \to s\,t &\ :\ B(H) \times B(H) \to B(H) \\ t \to t^* &\ :\ B(H) \to B(H) \end{aligned}$$

is strong operator continuous. However, the former map is strong operator continuous on bounded subsets of $B(H)$, while the latter is strong operator continuous on the set of normal operators. Moreover, each continuous complex function on $\mathbb{R}$ is strongly operator continuous on any bounded subset B of normal operators in the sense that if the net $(x_\alpha) \subset B$ converges to x in the strong operator topology, then $f(x_\alpha)$ converges to $f(x)$.

The *von Neumann algebra* is defined as the C^*-algebra that can be faithfully represented as a strongly operator closed $*$-subalgebra of $B(H)$.

(Sometimes the term *W^*-algebra* is also used.) The following conditions on a $*$-subalgebra M of $B(H)$ are equivalent: (i) M is weakly operator closed (ii) M is strongly operator closed (iii) M is ultraweakly closed. It can be verified easily that the commutant of any $*$-subalgebra of $B(H)$ is a von Neumann algebra. The following *von Neumann Bicommutant Theorem* says that all von Neumann algebras arise this way: A $*$-subalgebra M of $B(H)$ that contains the unit of $B(H)$ is a von Neumann algebra if, and only if, $M'' = M$. In other words, if M is a $*$-subalgebra of $B(H)$, then every element in the bicommutant M'' can be approximated in the strong operator topology by element in M. Another important approximation theorem is the *Kaplansky Density Theorem*: Let A be a C^*-subalgebra of operators that generate a von Neumann algebra M. Then the unit ball of A is strongly operator dense in the unit ball of M. (The same results hold for the sets of self-adjoint, positive, and unitary elements, correspondingly.) Let us now consider a compact Hausdorff space X with a finite strictly positive regular Borel measure μ on X. Then $C(X)$ can be faithfully represented on the Hilbert space $H = L^2(X, \mu)$ by the representation $\pi\colon f \in C(X) \to M_f \in B(H)$ specified above. Then the von Neumann algebra generated by $\pi(C(X))$ consists precisely of the multiplication operators M_f, where $f \in L^\infty(X, \mu)$. Therefore the transition from C^*-algebras to von Neumann algebras corresponds in this case to the transition from continuous functions to bounded measurable functions. Any unital abelian C^*-algebra can be represented as algebra of continuous functions on some compact space. It turns out that abelian von Neumann algebras correspond in this connection to compact spaces specified as follows: A compact Hausdorff space is called a *stonean space* if the closure of each open set in X is open. A positive functional ϱ on $C(X)$ is called *normal* if it preserves the increasing nets of real functions, i.e. if $\varrho(f_\alpha) \nearrow \varrho(f)$ whenever $f_\alpha \nearrow f$ in the self-adjoint part of $C(X)$. The compact space is called *hyperstonean* if it is stonean and if for any nonzero positive real function f in $C(X)$ there is a positive normal functional ϱ with $\varrho(f) \neq 0$. A C^*-algebra A is an abelian von Neumann algebra if, and only if, it is isomorphic to $C(X)$, where X is a hyperstonean space.

The fact that von Neumann algebra M is closed with respect to the strong operator topology ensures that M is monotone closed. In particular, the projection structure $P(M)$ is a complete lattice. Also, if $x = u\,|x|$ is a polar decomposition of $x \in M$, then $u \in M$. Further, if $x \in M$, then the range projection, $r(x)$, of x is again in M. It can be proved that if $0 \leq x1$, then $r(x)$ is a supremum of an increasing sequence $(x^{1/n})$. Suppose now that $x \in M_{sa}$. Then $W^*(x,1)$ is an abelian von Neumann subalgebra of M that can be represented by a function $f \in C(X)$, where X is a hypers nean space. Let e_λ ($\lambda \in \mathbb{R}$) be the projection in M corresponding to the characteristic function of the set $X \setminus \overline{f^{-1}(\lambda, \infty)}$. Then $e_\lambda\, x \leq \lambda\, e_\lambda$ and $(1-e_\lambda)\, x \geq \lambda\, (1-e_\lambda)$. The system $(e_\lambda)_{\lambda \in \mathbb{R}}$ is increasing and it is called the *spectral resolution* of x. It holds that

$$(x\,\xi, \xi) = \int_{\mathbb{R}} \lambda\, d\,\mu_\xi(\lambda)\,, \qquad \xi \in H\,,$$

where $\mu_\xi((-\infty, \lambda]) = \|e_\lambda\, \xi\|^2$. As a result, any element of a von Neumann algebra is a norm limit of finite linear combinations of projections. If $x \in M$ with $0 \leq x \leq 1$, then there is a sequence of projections, $(e_n) \subset W^*(x,1)$, such that $x = \sum_{n=1}^{\infty} \frac{1}{2^n}\, e_n$. As a consequence, if φ is a bounded functional on M, then $\varphi(x) \leq \sup\{\varphi(p) \mid p \in P(M)\}$ for all x with $0 \leq x \leq 1$. More generally, by expressing the norm-one element as a linear combination of four positive elements with the norm less than 1, we obtain

$$\|\varphi\| \leq 4 \sup\{\varphi(p) \mid p \in P(M)\} \qquad \text{for every positive functional } \varphi\,. \tag{2.11}$$

The important fact connecting the order of the self-adjoint part of a von Neumann algebra M and the strong operator topology on M is the following result: if $(u_\alpha) \subset M_{sa}$ is an increasing net with the supremum u, then the net (u_α) converges to u in the strong operator topology. In particular, if $(p_\alpha)_{\alpha \in F}$ is a system of mutually orthogonal projections, then $\bigvee_{\alpha \in I} p_\alpha = \sum_{\alpha \in I} p_\alpha$, where the sum converges in the strong operator topology. Let x be a self-adjoint element in a von Neumann algebra M and let O be an open subset of $\mathrm{Spec}(x)$. Since M is monotone complete, we can define

$$E(O) = \sup\{f(x) \mid f \in C(\mathrm{Spec}(x)), 0 \leq f \leq 1, f \text{ vanishes outside } O\}\,.$$

Extending E to all Borel subsets of $\mathrm{Spec}(x)$ by setting

$$E(S) = \inf\{E(O) \mid S \subset O\,,\, O \text{ is open }\}\,,$$

for each Borel subset $E \subset \mathrm{Spec}(x)$, we obtain so-called *spectral measure* of x on Borel subsets of $\mathrm{Spec}(x)$. Let us note that $E(\,(\lambda_1, \lambda_2] \cap \mathrm{Spec}(x)\,) =$

$e_{\lambda_2} - e_{\lambda_1}$ and $E([0,1] \cap \mathrm{Spec}(x)) = r(x)$ if $0 \le x \le 1$. We shall also use the notation $E(S) = 1_S(x)$.

Let M be a von Neumann algebra acting on a Hilbert space, H. Then its *centre*, $Z(M)$, is an abelian von Neumann subalgebra of M and $Z(M) = M \cap M'$. A von Neumann algebra M is called the *factor* if $Z(M)$ consists of scalar multiples of the unit of M only. For each projection $p \in P(M)$ we define the *central cover*, $c(e)$, of e as the smallest central projection majorizing e. If z is a central projection and $p \in P(M)$, then $c(zp) = z\,c(p)$. Moreover, $c(e) = \sup\{u^* e\, u \mid u \in U(M)\}$. The projections and central projections determine the structure of weakly closed ideals in a von Neumann algebra. (By saying that a set is weakly operator closed we mean its closedness in the weak operator topology.) More precisely, any weakly operator closed left ideal I in a von Neumann algebra M is of the form $I = M\,p$, where p is a uniquely determined projection.

Any weakly operator closed ideal in M is of the form $I = z\,M$, where z is a uniquely determined central projection in M. All weakly operator closed hereditary subalgebras of M are of the form $p\,M\,p = \{p\,x\,p \mid x \in M\}$, where p is a uniquely determined projection. If we consider the algebra $p\,M\,p$ as acting on the Hilbert space $p(H)$, then the following equalities hold: $(p\,M\,p)' = p\,M'\,p$ and $Z(p\,M\,p) = p\,Z(M)\,p$. This implies that any hereditary subalgebra of a factor is a factor, too.

Normal states and normal homomorphisms. Throughout this paragraph, M will be a von Neumann algebra acting on a Hilbert space H. A bounded functional φ on M is called *normal* if $\varphi(u_\alpha) \to \varphi(u)$ whenever $u_\alpha \nearrow u$ in M_{sa}. It turns out that the following conditions are equivalent: (i) φ is a normal functional (ii) φ is weakly operator continuous on the unit ball of M, (iii) φ is strongly operator continuous on the unit ball of M, (iv) φ is a completely additive measure on the projection lattice $P(M)$, i.e. $\varphi\left(\sum_\alpha p_\alpha\right) = \sum_\alpha \varphi(p_\alpha)$ for any system of pairwise orthogonal projections (p_α) in M (v) φ is continuous in the ultraweak topology.

It means that any normal functional on M is given by a trace class operator $t \in L^1(H)$ such that

$$\varphi(x) = \mathrm{tr}(t\,x) \qquad \text{for each } x \in M\,.$$

In particular, any normal state φ is a sum of vector functionals, $\varphi = \sum_n \omega_n$, where $\sum_n \|x_n\|^2 = 1$. In a special situation when M has a separating vector every normal state on M is a restriction of a vector state. (A vector ξ is *separating* for M if $x\,\xi \neq 0$ for all nonzero $x \in M$.)

Any von Neumann algebra has a predual consisting of normal functionals in the canonical duality. Moreover, von Neumann algebras can be characterized as C^*-algebras that are dual Banach spaces. Another important characterization of von Neumann algebras reads that von Neumann algebras are exactly monotone closed C^*-algebras with a separating set of normal functionals. Let φ be a state on M. The projection p is called the *support* of φ if $\varphi(1-p) = 0$ and $\varphi(q) > 0$ for any subprojection q of p. Any normal state has a support. In the sequel, we shall always denote by $S_n(M)$ the set of all normal states on M. It is a convex compact set that is weak*-dense in the state space $S(M)$ and that is norm closed. The following Jordan-Hahn decomposition and polar decomposition hold for the normal functionals. If ϱ is a normal self-adjoint functional, then there are normal positive functionals ϱ^+ and ϱ^- on M and a projection $p \in M$ such that $\varrho = \varrho^+ - \varrho^-$ and, for each $x \in M$,

$$\varrho^+(x) = \varrho(p\,x\,p)\,, \qquad \varrho^- = -\varrho((1-p)\,x\,(1-p))\,.$$

In other words, any completely additive measure on $P(M)$ can be decomposed into a difference of two completely additive positive measures on $P(M)$ with mutually orthogonal supports.

If ϱ is a normal functional on a von Neumann algebra M, then there is a positive normal functional φ and a unique partial isometry $u \in M$ such that

$$\varrho(x) = \varphi(u\,x) \qquad \text{for all } x \in M$$

and such that $u^*\,u$ is the support of φ.

Let $\pi\colon M \to N$ be a $*$-homomorphism between von Neumann algebras M and N. The map π is called *normal* if it preserves increasing nets of self-adjoint elements, i.e. if $u_\alpha \nearrow u$ in M_{sa} implies that $\pi(u_\alpha) \nearrow \pi(u)$ in N. Equivalently, π is normal if, and only if, $\omega_x \circ \pi$ is a normal functional on M for each vector functional ω_x on N. If π is a normal $*$-homomorhism, then the algebra $\pi(M)$ is a von Neumann subalgebra of N. If π is a faithful $*$-homomorphism of M onto N, then π is a normal $*$-isomorphism. Any state on a von Neumann algebra is normal if, and only if, its G.N.S. representation is normal. If $\pi\colon\ M \to N$ is a normal $*$-homomorphism, then $\operatorname{Ker}\pi$ is a weak operator closed ideal in M that corresponds to some central projection in M. It implies that any normal $*$-homomorphism on a factor is automatically faithful. If p is a projection in the commutant M' of a von Neumann algebra M such that $c(p) = 1$, then the map $M \to M p : x \to x p$ is a $*$-isomorphism of M onto $M p$.

Let us recall a few usual facts on normal homomorphisms on algebras $B(H)$. Any normal $*$-representation of $B(H)$ is equivalent to the sum of identity representations. Any normal $*$-automorphism of $B(H)$ is unitarily implemented. Finally, any $*$-representation of $B(H)$ is a direct sum of normal $*$-representation and a $*$-representation that vanishes on $K(H)$.

Tensor products of von Neumann algebras. Let $M_1, \dots, M_n$ be von Neumann algebras acting on the Hilbert spaces $H_1, \dots, H_n$, correspondingly. The *von Neumann tensor product* of algebras $M_1, \dots, M_n$, denoted by $M_1 \overline{\otimes} M_2 \overline{\otimes} \cdots \overline{\otimes} M_n$, is the von Neumann subalgebra of $B(H_1 \otimes H_2 \otimes \cdots \otimes H_n)$ generated by the spatial C^*-tensor product $M_1 \otimes M_2 \otimes \cdots \otimes M_n$. For example, if M is a von Neumann algebra acting on a Hilbert space H with $\dim H = \alpha$, then the von Neumann tensor product $M \overline{\otimes} B(K)$ can be identified with the algebra of all bounded operators on the direct sum $\sum_\alpha \oplus K$, the matrix entries of which are in M. Suppose that $\varphi_1, \dots, \varphi_n$ are normal states on von Neumann algebras $M_1, \dots, M_n$, correspondingly. Then there is a unique normal state φ on $M_1 \overline{\otimes} M_2 \overline{\otimes} \cdots \overline{\otimes} M_n$ such that

$$\varphi(x_1 \otimes x_2 \otimes \cdots \otimes x_n) = \varphi_1(x_1)\, \varphi_2(x_2) \cdots \varphi_n(x_n)$$

for all $x_1 \in M_1, x_2 \in M_2, \dots x_n \in M_n$. The state φ specified above is called the *normal product state* and it is denoted by the symbol $\varphi_1 \overline{\otimes} \varphi_2 \overline{\otimes} \cdots \overline{\otimes} \varphi_n$. Given normal $*$-representations $\pi_i : M_i \to B(H_i)$, where M_i are von Neumann algebras, there is a unique normal $*$-representation π from the von Neumann tensor product $M_1 \overline{\otimes} M_2 \overline{\otimes} \cdots \overline{\otimes} M_n$ into the von Neumann tensor product $B(H_1) \overline{\otimes} B(H_2) \overline{\otimes} \cdots \overline{\otimes} B(H_n)$ such that

$$\pi(x_1 \otimes x_2 \otimes \cdots \otimes x_n) = \pi_1(x_1) \otimes \pi_2(x_2) \otimes \cdots \otimes \pi_n(x_n)$$

for all $x_1 \in M_1, x_2 \in M_2, \dots, x_n \in M_n$. The representation π described above is called the *normal product representation*.

The following result is a deep fact describing the commutant of the tensor product: If M and N are von Neumann algebras acting on Hilbert spaces H and K, respectively, then $(M\overline{\otimes}N)' = M'\overline{\otimes}N'$. In particular, $Z(M\overline{\otimes}N) = Z(M)\overline{\otimes}Z(N)$. Hence, product of factors is a factor, too.

Projection structures of von Neumann algebras. In this paragraph we summarize basic facts from the comparison theory of projections in von Neumann algebras. Throughout this part, M will be a von Neumann algebra acting on a Hilbert space H with the commutant M'. The projection lattice $P(M)$ is a complete orthomodular lattice. We shall denote by $e \vee f$ and $e \wedge f$ the supremum and the infimum of the projections e and f in $P(M)$, respectively. For larger systems of projections we shall use the symbols $\bigvee_{\alpha\in I} e_\alpha$ and $\bigwedge_{\alpha\in I} e_\alpha$ to denote the supremum and and the infimum of the system $(p_\alpha)_{\alpha\in I}$, respectively. The projections are in a one-to-one correspondence with closed subspaces of H onto which they project. The supremum of projections in a von Neumann algebra corresponds to the closed linear span of their ranges and the infimum correspond to the set intersection of the ranges. Moreover, the lattice operations in $P(M)$ are connected with the ranges of operators as follows:

$$e \vee f = r(e+f)\,, \qquad r(e\,f\,e) = r(e\,f) = e - e \wedge (1-f)\,,$$

for all $e, f \in M$. The projection $e \in M$ is said to be *cyclic* for M if its range fulfils $[M'\xi] = \overline{\mathrm{sp}}\{x\,\xi \mid x \in M'\}$, where $\xi \in H$. In this case, ξ is said to be a *generating vector* (or *cyclic*) for e. A vector ξ is said to be *generating* (or *cyclic*) for M if $[M'\,\xi] = H$. A vector ξ is cyclic for M if, and only if, it is separating for its commutant M'. By the standard Zorn's lemma argument, every projection is a sum of cyclic projections. Each subprojection of a cyclic projection is a cyclic projection, too. A projection $e \in M$ is called *σ-finite* (relative to M) if each orthogonal family of nonzero subprojections of e is countable. M is said to be a *σ-finite algebra* if the unit of M is a σ-finite projection. Every algebra acting on the separable Hilbert space is σ-finite, the converse is not true. (Let us note that von Neumann algebras acting on separable Hilbert spaces can be characterized as those which have separable preduals.) A projection $e \in M$ is σ-finite if, and only if, it is a support of some normal state on M. Any countable union of cyclic projections is a σ-finite projection.

Besides the order inherited from the C^*-algebra, the structure of projections is endowed with equivalence an relation, $\sim$. The subequivalence relation introduces another order on projections reflecting their dimensions. The following Cantor-Bernstein type result holds: $e \sim f$ if, and only if, $e \precsim f$ and $f \precsim e$. For any $x \in M$, the range projections $r(x)$ and $r(x^*)$ are equivalent. If $e, f \in M$, then

$$e \vee f - f \sim e - e \wedge f\,.$$

This relation is referred to as the *Kaplansky formula* or the *parallelogram law.* Two projections e, f in M have nonzero equivalent subprojections if, and only if, their central covers are not orthogonal, i.e. exactly when $c(e)\,c(f) \neq 0$. If $(e_\alpha)_{\alpha\in I}$ and $(f_\alpha)_{\alpha\in I}$ are two families of pairwise orthogonal projections such that $e_\alpha \sim f_\alpha$ for each α, then $\sum_{\alpha\in I} e_\alpha \sim \sum_{\alpha\in I} f_\alpha$. The following result is called the *Comparability Theorem*: For any pair $e, f \in P(M)$, there exists a central projection $z \in M$ such that

$$z\,e \precsim z\,f \qquad \text{and} \qquad (1-z)\,e \succsim (1-z)\,f\,.$$

In particular, if M is a factor, then for $e, f \in P(M)$, either $e \precsim f$ or $f \precsim e$. In other words, the order on the projection lattice of a von Neumann factor induced by the subequivalence relation is total. Let us note that the equivalence relation preserves cyclicity and σ-finiteness of projections.

We shall now deal with finite and infinite projections in von Neumann algebras. The following deep facts hold true: The set of all finite projections in M is an ideal in the lattice $P(M)$. This ideal is a modular lattice and it is invariant with respect to the equivalence relation. A projection $e \in M$ is finite if, and only if, the lattice $P(e\,M\,e)$ is a modular lattice. If $e \in M$ is a finite projection and $c(e)$ is σ-finite relative to the centre $Z(M)$, then e is a sum of countably many cyclic projections. In particular, a finite von Neumann algebra is σ-finite if, and only if, its centre is σ-finite. A projection $e \in P(M)$ is called *properly infinite* if $z\,e$ is either zero or infinite for any central projection $z \in M$. In other words, the hereditary subalgebra $e\,M\,e$ contains no nonzero finite central projection provided that e is properly infinite. The algebra M is called *properly infinite* if the unit is a properly infinite projection. The algebra $B(H)$ is properly infinite if, and only if, $\dim H = \infty$. Any infinite projection in a factor is properly infinite. For any infinite projection $e \in M$ there is a unique central projection z with $z \leq c(e)$ such that $z\,e$ is properly infinite and $(1-z)\,e$ is finite. If

e is a properly infinite projection in M and f is a σ-finite projection such that $c(f) \leq c(e)$, then $f \precsim e$. In particular, all infinite projections in a σ-finite factor are equivalent. If e is a properly infinite projection, then for each n there are pairwise orthogonal projections $e_1, \ldots, e_n$ such that each e_i is equivalent to 1 and such that $\sum_{i=1}^{n} e_i = 1$. In the same way the unit in a properly infinite algebra can be written as a sum of countably many pairwise orthogonal projections that are all equivalent to 1. A projection e in M is called *abelian* if the hereditary subalgebra $e\,M\,e$ is abelian. A projection is abelian if, and only if, it is minimal with respect to the set of all projections with the same central cover. Each abelian projection is finite. Let us note that the projection e in $B(H)$ is abelian if, and only if, e is minimal, i.e. a one-dimensional projection. Abelian projections are preserved by equivalence relations. If $c(e) \leq c(f)$, where e is abelian, then $e \precsim f$. In particular, abelian projections that have the same central cover are equivalent. If (e_α) is a family of abelian projections such that $e_\alpha \leq z_\alpha \in Z(M)$ for each α, where (z_α) are pairwise orthogonal, then the projection $\sum_\alpha e_\alpha$ is abelian.

Classification of von Neumann algebras. The nature of the projection lattice enables us to decompose any von Neumann algebra into direct summands of certain types that are of different character and express "degree" of noncommutativity of the algebras. A von Neumann algebra is said to be *Type I* algebra if there is an abelian projection $e \in M$ such that $c(e) = 1$. This is equivalent to saying that each nonzero central projection in M majorizes a nonzero abelian projection. Let n be cardinal number. If the unit in a von Neumann algebra M can be written as a sum of n equivalent abelian projections, then M is said to be of *Type* I_n. (It can be shown in this case that all systems of orthogonal equivalent abelian projections with sum 1 have the same cardinality.) Let us note that Type I_1 algebras are just abelian von Neumann algebras. Each algebra of Type I_n is called *homogeneous Type I algebra.* For every von Neumann algebra M of Type I and each cardinal less then $\operatorname{card} M$ there is a unique central projection z_α such that $z_\alpha M$ is zero or Type I_α and such that $\sum_\alpha z_\alpha = 1$. Hence, Type I algebras are built of homogeneous ones. By the symbol I_∞ we shall denote the fact that given algebra is properly infinite and of Type I. The most important result on Type I algebras reads that Type I_α algebras are exactly the $*$-isomorphic copies of the tensor products $A \overline{\otimes} B(H_\alpha)$, where A is an abelian von Neumann algebra and H_α is a Hilbert space of dimension α. Consequently, the Type I factors are nothing but algebras of all bounded operators on Hilbert spaces. Type I algebra is finite if, and only if, it is a direct sum of (countably many) Type I_n, $n < \infty$, algebras. The finite Type I homogeneous algebras can be identified with the algebras $M_n(C(X))$, where X is a hyperstonean space. Another way of looking at this algebra is to represent it as algebra $C(X, M_n(\mathbb{C}))$ of all continuous functions on a hyperstonean space X. A function $f \in C(X, M_n(\mathbb{C}))$ is called *locally constant* if $f(X)$ is finite, i.e. if there exists a partition $O_1, \ldots, O_k$ of X consisting of finitely many disjoint clopen sets such that $f|O_i$ is constant for each i. Locally constant functions form a dense $*$-subalgebra of $C(X, M_n(\mathbb{C}))$. Indeed, for a general $f \in C(X, M_n(\mathbb{C}))$ and $\varepsilon > 0$ one can find, by compactness of X, a finite open covering $U_1, \ldots, U_k$ of X such that, for each i, $\|f(x) - f(y)\| < \varepsilon$ for all $x, y \in U_i$. Set $O_1 = \overline{U_1}$, $O_i = \overline{U_i} \setminus \bigcup_{j=1}^{i-1} \overline{U_j}$. Then $O_1, \ldots, O_k$ is a disjoint open covering of X. Let us choose arbitrary points $x_1 \in O_1, x_2 \in O_2, \ldots, x_k \in O_k$ and set $g(x) = f(x_i)$ if $x \in O_i$. Obviously, g is a locally constant function such that $\|f - g\| \leq \varepsilon$. A similar result holds for projections. Let us note that $f \in C(X, M_n(\mathbb{C}))$ is a projection if, and only if, $f(x)$ is a projection in $M_n(\mathbb{C})$ for each $x \in X$. By the previous reasoning, any projection in $C(X, M_n(\mathbb{C}))$ can be approximated by a locally constant projection.

Type I_2 algebras are useful for describing subalgebras generated by two projections and hence for analysing the geometric position of two closed subspaces of a Hilbert space. Let e, f be projections in a von Neumann algebra M. Suppose that

$$e \wedge f = (1 - e) \wedge f = e \wedge (1 - f) = (1 - e) \wedge (1 - f) = 0 . \tag{2.12}$$

It turns out that the projection e is abelian in the von Neumann algebra $N = W^*(e, f)$ generated by e and f, and hence that $e\,M\,e$ can be identified with $C(X)$, where X is a hyperstonean space. It can be shown that N is $*$-isomorphic to the algebra $M_2(C(X))$ in such a way that, after identification,

$$e = \begin{pmatrix} 1 & 0 \\ 0 & 0 \end{pmatrix}, \quad f = \begin{pmatrix} a(x) & \sqrt{a(x) - a^2(x)} \\ \sqrt{a(x) - a^2(x)} & 1 - a(x) \end{pmatrix},$$

where $a \in C(X)$ is such that $0 \leq a(x) \leq 1$ for all $x \in X$. If the projections e and f do not satisfy (2.12), then it can be shown that by discarding "common parts" we can obtain position described in (2.12). In summary, for projections e, f in a von Neumann algebra M there is a unique central projection $z \in N = W^*(e, f)$ such that $z\,N$ is a finite-dimensional abelian von Neumann algebra and $(e \vee f - z)\,N$ is a Type I_2 algebra such that $(1 - z)\,e$ and $(1 - z)\,f$ satisfy (2.12).

As seen before any projection e in a properly infinite algebra can be halved in the sense that $e = f + (e - f)$, where $f \sim e - f$. This is not possible for Type I finite algebras as the simple example of $M_3(\mathbb{C})$ algebra shows. However, it follows from the structure theory of Type I algebras that projections can be halved up to an abelian remainder. Formally, let $e \in P(M)$ where M is a Type I algebra. Then there are orthogonal projections e_1, e_2, e_3 such that $e = e_1 + e_2 + e_3$, $e_1 \sim e_2$ and e_3 is abelian.

The algebras of Type I are close to abelian von Neumann algebras, one reason being that they contain many abelian hereditary subalgebras. We shall now briefly discuss other types of von Neumann algebras that have no nonzero abelian hereditary subalgebra. A von Neumann algebra M is said to be of *Type II* if it has no nonzero abelian projection but has a finite projection e such that $c(e) = 1$. Equivalently, the algebra is of Type II if it has no nonzero abelian projection and if every nonzero central projection majorizes a nonzero finite projection. The finite algebras of Type II are called Type II_1, the infinite ones are called Type II_∞. A von Neumann algebra is called *semifinite* if it is a Type I or Type II or a direct sum of algebras of these types. If M is a properly infinite and semifinite von Neumann algebra, then there exists an orthogonal family (z_α) of central projections in M indexed by infinite cardinals less then $\operatorname{card} M$ with $\sum_\alpha z_\alpha = 1$, and a family (N_α) of finite von Neumann algebras such that $z_\alpha M$ is $*$-isomorphic to $N_\alpha \overline{\otimes} B(H_\alpha)$, where $\dim H_\alpha = \alpha$ and z_α may be zero. The family (z_α) is uniquely determined. In addition, if M is σ-finite, then for any finite projection f with $c(f) = 1$, M is isomorphic to $f\,M\,f \overline{\otimes} B(H)$, where H is separable. While the examples of Type I algebras are obvious, it is not so easy to construct examples of Type II algebras. One of the canonical examples of finite Type II factor is the *group von Neumann algebra* $W^*_\lambda(\Gamma)$, where Γ is a discrete group that has all nontrivial conjugacy classes infinite. (The group von Neumann algebra is defined as the von Neumann algebra acting on $l^2(\Gamma)$ that is generated by the reduced group C^*-algebra $C^*_\lambda(\Gamma)$.) An example of Type II_∞ factor is the tensor product $W^*_\lambda(\Gamma) \overline{\otimes} B(H)$, where H is infinite-dimensional.

A fundamental property of finite algebras is the existence of the centre-valued trace. Let M be a von Neumann algebra with the centre $Z(M)$. The centre-valued trace $T\colon M \to Z(M)$ is a linear mapping such that (i) $T(x\,y) = T(y\,x)$ for all $x, y \in M$ (ii) T is identity of $Z(M)$ (iii) $T(x) > 0$ whenever $x \in M_+$ is positive. If such a mapping exists, then M has to be finite. Conversely, a deep theorem on finite algebras says that any finite von Neumann algebra admits exactly one centre-valued trace, T. Moreover, T has the following additional properties: (iv) $T(z\,x) = z\,T(x)$ for all $z \in Z(M)$ and $x \in M$ (v) $T(e) \leq T(f)$ for projections e, f if, and only if, $e \precsim f$ (vi) $\|T\| \leq 1$ (vii) T is ultraweakly continuous. The following nontrivial result characterizes types of von Neumann algebras in terms of the traces. Let $T\colon M \to Z(M)$ be a centre-valued trace on M. Then the following holds: (i) If M is of Type I_n, then $T(P(M))$ consists of all elements of the form

$$\frac{1}{n}\,z_1 + \frac{2}{n}\,z_2 + \cdots + \frac{n-1}{n}\,z_{n-1} + z_n\,,$$

where $z_1, \ldots, z_n$ are pairwise orthogonal central projections.

(ii) If M is of Type II_1, then $T(P(M))$ consists of all positive elements in the unit ball of $Z(M)$.

If φ is state on $Z(M)$, then $\varphi \circ T$ is a unique tracial state on M extending φ. All tracial states on finite von Neumann algebras arise this way. Let us illustrate these results by the examples of finite factors. If M is a factor, then $Z(M)$ identifies with $\mathbb{C}$ and the centre-valued trace identifies with a faithful tracial state. Such a state is unique and automatically normal. Type I_n factors are just matrix algebras $M_n(\mathbb{C})$. In this case the centre-valued trace is nothing but the normalized matrix trace that attain discrete values $\frac{k}{n}$, $0 \leq k \leq n$, on projections. On the other hand, the faithful tracial state on Type II_1 factor attains all values in the interval $[0, 1]$ when restricted to projections. In other words, equivalent classes of projections (i.e. dimensions) are labeled by uncountably many elements from [0,1]. For this reason, Type II_1 factors may be thought of as "continuous matrix algebras".

Finally, a von Neumann algebra is said to be of Type III (or *purely infinite*) if it contains no nonzero finite projection. Type III algebras are highly noncommutative because their projection lattice is far from being distributive. Indeed, there is no nonzero projection in Type III algebra such that the lattice of its subprojections is modular. Examples of Type

III algebras are most complicated and result from cross products and special G.N.S. representations of inductive limits of matrix algebras.

Every von Neumann algebra is uniquely decomposable into the direct sum of algebras of Type I, II_1, II_∞ and III. In particular any von Neumann factor is one of the Type I, II_1, II_∞, and III. If von Neumann algebra has zero Type I finite part, then for every projection e in M and any integer n there are orthogonal equivalent projections $e_1, \ldots, e_n$ such that $e = e_1 + e_2 + \cdots e_n$. In particular, for any projection e in an arbitrary von Neumann algebra there are projections e_1, e_2 and e_3 such that $e = e_1 + e_2 + e_3$, $e_1 \sim e_2$, and e_3 is abelian.

A von Neumann algebra is called *locally σ-finite* if it is a direct sum of σ-finite algebras. A projection e in a von Neumann algebra M is called locally σ-finite the hereditary subalgebra eMe is locally σ-finite. Every finite von Neumann algebra is locally σ-finite. A properly infinite von Neumann algebra is said to be of *uniform Type* λ if it contains a family $(e_i)_{i\in I}$ of orthogonal equivalent locally σ-finite projections with sum 1 and $\operatorname{card}(I) = \lambda$. The von Neumann algebras of uniform Type $\aleph_0$ are precisely the locally σ-finite properly infinite algebras. Algebras of Type I_∞ or II_∞ that are of uniform Type λ are precisely algebras $N\overline{\otimes} B(H_\lambda)$, where N is finite and $\dim H_\lambda = \lambda$. Any von Neumann algebra M can be decomposed into $M = M_0 \oplus \left(\sum_{\lambda\in\Gamma} \oplus M_\lambda\right)$, where M_0 is the largest locally σ-finite direct summand and each M_λ is of uniform Type λ if not nonzero. If M is semifinite and locally σ-finite, then $M = N_1 \oplus (N_2\overline{\otimes} B(H))$, where N_1 and N_2 are finite von Neumann algebras.

Let us now summarize basic facts on types of commutants and tensor products. Let M be a von Neumann algebra acting on a Hilbert space H. If M is of Type I (or Type II, or Type III), the same is true of its commutant M'. If M and N are von Neumann algebras of Type I_n and I_m, respectively, then the tensor product $M\overline{\otimes} N$ is of type I_{mn}. A tensor product $M\overline{\otimes} N$, where M and N are finite and at least one of them is of Type II_1 is of Type II_1. The tensor products of semifinite algebras are semifinite. In addition, if at least one of the semifinite algebras is of Type II, then their tensor product is of Type II. Finally, the tensor product of von Neumann algebras is of Type III if one of them is of Type III.

In the later investigation we shall need results on diagonalizing maximal abelian subalgebras and normal elements in von Neumann algebras that were proved by Kadison and Fillmore [208, 135]. Let us note that the character of the "diagonalizing basis" depends on the type of a given algebra. It is well known that normal matrix x in a matrix algebra $M_n(\mathbb{C})$ is diagonalizable in the sense that there is an orthonormal basis in the underlying Hilbert space such that x has diagonal matrix with respect to this basis. More generally, it is a consequence of a deep *Kadison's Diagonalization Theorem* that any maximal abelian von Neumann subalgebra A in a von Neumann algebra M of Type I_n, $n < \infty$, contains n equivalent abelian projections $e_1, \ldots, e_n$ such that $\sum_{i=1}^{n} e_i = 1$. In other words, M can be realized as the algebra $M_n(e_1 \, M \, e_1)$ in such a way that all elements in A have a diagonal matrix. For algebras not containing any finite Type I homogeneous direct summand the following holds: Let A be a maximal abelian von Neumann subalgebra of a von Neumann algebra M such that M has zero Type I finite part. Then, for each integer n, there are n equivalent projections $e_1, \ldots, e_n$ with $\sum_{i=1}^{n} e_i = 1$ and such that each e_i $(i = 1, \ldots, n)$ commutes with A. Equivalently, after identifying M with $M_n(e_1 \, M \, e_1)$, A becomes an algebra of diagonal matrices in M. As a result, any normal element in a von Neumann algebra can be embedded into maximal abelian von Neumann subalgebra and therefore it can be diagonalized in the above way. In particular, if M does not contain any Type I_n direct summand, where n is an odd integer, then for each normal element $x \in M$ there is a projection e commuting with x such that $e \sim 1 - e$.

In the category of C^*-algebras we have seen that any $*$-homomorphism is automatically continuos. As opposed to this, a $*$-homomorphism between von Neumann algebras need not be normal. It turns out that the automatic continuity of $*$-homomorphisms in the category of von Neumann algebras requires some type and size considerations. It is well known [134, 306, 307] that a $*$-homomorphism $\pi\colon\ M \to N$ between von Neumann algebras M and N is weak*-continuous provided that one of the following conditions holds: (i) M is of Type II_1 and σ-finite and N acts on a separable Hilbert space (ii) M is properly infinite and M and N are σ-finite. Let us note that in neither case it is not enough to assume only that M is σ-finite. For example, it is known that when M is of Type II_1 and has infinite-dimensional centre, then M admits a $*$-representation onto a factor that is not normal.

The enveloping von Neumann algebra. As seen before, the von Neumann algebra generated by the image of a C^*-algebra A in its universal representation can be identified with the double dual A^{**} of A into which A is embedded as a weak*-dense subalgebra. Let us recall that the double dual of A is called the *enveloping von Neumann algebra* of A. The normal state space, $S_n(A^{**})$, of the von Neumann algebra A^{**} is identified with the state space, $S(A)$, of the C^*-algebra A. In other words, for each bounded functional φ on A there is a unique normal functional $\tilde{\varphi}$ on A^{**} extending φ. For every $*$-representation $\pi\colon A \to B(H)$ there is a unique normal $*$-representation $\tilde{\pi}\colon A^{**} \to B(H)$ extending π. The kernel $\operatorname{Ker}\tilde{\pi}$ is a weak*-closed ideal in A^{**} and hence there is a unique central projection $c(\pi) \in A^{**}$, such that $\operatorname{Ker}\tilde{\pi} = (1 - c(\pi))\, A^{**}$. It is clear that the von Neumann subalgebra of $B(H)$ generated by $\pi(A)$ is $*$-isomorphic to $c(\pi)\, A^{**}$. Two representations π_1 and π_2 are equivalent if, and only if, $c(\pi_1) = c(\pi_2)$ and disjoint if, and only if, $c(\pi_1)\, c(\pi_2) = 0$. Let ϱ be a state on A with the G.N.S representation π_ϱ. We put $c(\varrho) = c(\pi_\varrho)$. If ϱ is a pure state, then $c(\varrho)\, A^{**}$ is $*$-isomorphic to $B(H_\varrho)$.

A projection $p \in A^{**}$ is called *open* if there is an increasing net (a_α) of positive elements in A such that $a_\alpha \nearrow p$ in A^{**}. The open projections correspond to hereditary subalgebras and therefore to closed left ideals in A. Indeed, the following conditions are equivalent: (i) p is an open projection (ii) p belongs to the strong operator closure (in A^{**}) of the hereditary subalgebra $p\,A^{**}\,p \cap A$ of A (iii) p belongs to the strong operator closure of the closed left ideal $A^{**}\,p \cap A$ in A. All open projections are in fact projections representing in the above way the weak*-closures of left ideals in A. If $A = C(X)$, then each ideal I in A corresponds to a unique open subset O of X such that I is the ideal of all continuous functions vanishing outside O. The characteristic function of O is a supremum of all continuous functions with values in [0,1] that are in I. This fact explains the terminology. It is known that the product of two commuting open projections is open, too. The projection p in A^{**} is called *closed* if $1 - p$ is open. If ϱ is a pure state on A, then the support projection, $s(\varrho)$, of ϱ is known to be a minimal closed projection in A^{**}.

2.3. Jordan algebras and ordered structures

In the present book aspects of quantum measure theory are discussed mainly in the context of C^*-algebras and von Neumann algebras. However, most of the topics have further development in the theory of Jordan algebras. For this reason, let us briefly outline basic concepts of the theory of Jordan Banach algebras and related theory of ordered structures.

Jordan algebras. A *Jordan algebra* is a real algebra $(A, \circ)$ such that the product $\circ$ has the following properties: (i) $a \circ b = b \circ a$, (ii) $a \circ (b \circ a^2) = (a \circ b) \circ a^2$, for all $a, b \in A$. A *JB algebra* is a Jordan algebra $(A, \circ, \|\cdot\|)$ where the complete norm $\|\cdot\|$ satisfies the following conditions: (i) $\|a \circ b\| \leq \|a\| \cdot \|b\|$, (ii) $\|a^2\| = \|a\|^2$, (iii) $\|a^2\| \leq \|a^2 + b^2\|$, for all $a, b \in A$. The important examples of JB algebras are self-adjoint parts of C^*-algebras endowed with the anticommutant product $a \circ b = \frac{1}{2}(a\,b + b\,a)$. More generally, a JB subalgebra is called the *JC algebra* if it is a JB subalgebra of a self-adjoint part of an abelian C^*-algebra endowed with the anticommutant product. The JB algebras are nonassociative generalizations of C^*-algebras. A JB algebra is associative if, and only if, it is isomorphic to the self-adjoint part of an abelian C^*-algebra. A *JBW algebra* is a JB algebra that is a dual Banach space. A self-adjoint part of a C^*-algebra A is a JBW algebra if, and only if, A is a von Neumann algebra. Let us note that all JBW algebras can be represented as algebras of operators on Hilbert spaces. For example, this is the case of the algebra of all 3×3 matrices over Cayley numbers. JBW algebras have rich projection structure and are classified into types similarly as von Neumann algebras. However, there are new aspects in the structure theory of JBW algebras. For example, there exists an infinite-dimensional Type I_2 factor called the spin factor that is a reflexive Banach space. It is defined as follows: Let H be a real Hilbert space of dimension at least 2. The *spin factor* is the JB algebra $A = H \oplus \mathbb{R}1$ with the product

$$(a + \lambda\,1) \circ (b + \mu\,1) = \mu\,a + \lambda\,b + ((a, b) + \lambda\,\mu)\,1\,,$$

and the norm $\|a + \lambda\,1\| = \|a\| + \|\lambda\|$ $(a, b \in A,\ \mu, \lambda \in \mathbb{R})$.

Ordered Banach spaces. The ordered vector spaces generalize the JB algebras and are important for conceptual foundations of quantum mechanics. A *partially ordered vector space* is a real vector space A with a proper convex cone A_+. The partial order $\leq$ on A is induced by A_+ in the standard way: $a \leq b$ if, and only if, $b - a \in A_+$. An ordered vector space is said to be *boundedly complete* if any bounded increasing net of elements has supremum. An element $e \in A$ is said to be an *order unit* for A if for all $a \in A$ there is a $\lambda > 0$ such that $-\lambda e \leq a \leq \lambda e$. A is said to be an *order unit space* if $n\,a \leq e$ for each $n \in \mathbb{N}$ implies $a \leq 0$. An order unit space has the norm $\|a\| = \inf\{\lambda > 0 \mid -\lambda e \leq a \leq \lambda e\}$. If $(A, \|\cdot\|)$ is complete, the space A is called *complete order unit space.* Any unital JB algebra with a unit and any self-adjoint part of a unital C^*-algebra form a complete order unit spaces. Let A be a complete order unit space which is a Jordan algebra for which the distinguished order unit acts as an identity element, and suppose that $-1 \leq a \leq 1$ implies $0 \leq a^2 \leq 1$ for all $a \in A$. Then A is JB algebra in the order norm. Let A and B be order unit vector spaces. A linear map $F\colon\ A \to B$ is called positive if $F(a) \in B_+$ whenever $a \in A_+$. A bounded linear map $F\colon\ A \to B$ preserving unit is positive if, and only if, it is bounded and norm-one. By a *state* on an order unit space we mean a positive norm one functional.

Orthocomplemented partially ordered sets. By a *lattice* $(L, \leq)$ we mean a partially ordered set (poset in the abbreviation) such that there is an infimum, $e \wedge f$, and supremum, $e \vee f$, for any pair of elements $e, f \in L$. A lattice L is called *complete* if any subset of L has infimum and supremum. A subset K of L is called *sublattice* if $e \vee f$ and $e \wedge f$ belong to K whenever e and f do. A subset K is called *ideal* if it is a sublattice and $e \wedge f \in K$ whenever $e \in K$ and $f \in L$. A lattice L is said to be *modular* if $e \leq g$ implies $(e \vee f) \wedge g = e \vee (f \wedge g)$ for all $f \in L$. Suppose that the lattice L has a least element, 0, and a greatest element, 1. If $e \in L$ we say that an element $e' \in L$ is complement for e if $e \wedge e' = 0$ and $e \vee e' = 1$. A lattice is called *complemented* if each element has a complement. A map $\varphi\colon\ L \to K$ between the lattices L and K is called the *lattice homomorphism* if it preserves the lattice operations: $\varphi(e \vee f) = \varphi(e) \vee \varphi(f)$ and $\varphi(e \wedge f) = \varphi(e) \wedge \varphi(f)$ for all $e, f \in L$.

Let $(P, \leq, 0, 1)$ be a poset with a least element, 0, and a greatest element, 1. A nonzero element $a \in P$ is called an *atom* if $x \leq a$ implies $x = 0$ or $x = a$ for all $x \in P$. P is called *atomic* if every nonzero element in P dominates an atom. A poset P is called *irreducible* if it cannot be written as a nontrivial direct product of other posets. The poset $(P, \leq, 0, 1, \perp)$ is called *orthocomplemented* if $\perp$ is a binary operation on P satisfying the following conditions $(a, b \in P)$: (i) $a^{\perp\perp} = a$, (ii) $a \leq b$ implies $b^\perp \leq a^\perp$, (iii) $a \vee a^\perp = 1$. Two elements in $e, f \in P$ are called *orthogonal* (in symbols, $e \perp f$) if $e \leq f^\perp$. An orthocomplemented poset P is called *orthoposet* if any pair of orthogonal elements has a supremum. An orthoposet P is called *orthomodular poset* (or *quantum logic*) if $e \leq f$ implies

$$f = e \vee (f \wedge e^\perp).$$

An important example of orthomodular lattices are projection lattices of JBW algebras and von Neumann algebras.

A *(finitely additive) measure* μ on an orthoposet P is a map $\mu\colon P \to \mathbb{C}$ such that

$$\mu(e \vee f) = \mu(e) + \mu(f) \text{ whenever } e \perp f.$$

(The real measure and the positive measure on P are defined in the same way.) A measure μ on P is called a *state* if μ has values in the unit interval [0,1] and $\mu(1) = 1$. Moreover, a measure $\mu\colon P \to \mathbb{C}$ is called *σ-additive* if

$$\mu\left(\bigvee_{n=1}^{\infty} e_n\right) = \sum_{n=1}^{\infty} \mu(e_n)$$

for any sequence (e_n) of pairwise orthogonal elements in P. Finally, a measure $\mu\colon P \to \mathbb{C}$ is called *completely additive* if

$$\mu\left(\bigvee_{\alpha \in I} e_\alpha\right) = \sum_{\alpha \in I} \mu(e_\alpha)$$

for any system $(e_\alpha)_{\alpha \in I}$ of pairwise orthogonal elements in P.

3. GLEASON THEOREM

In the Hilbert space model of quantum mechanics, the states of a physical system are typically identified with probability measures on the lattice $P(H)$ of all orthogonal projections in a Hilbert space, H, which embodies the "logic" of the quantum system. A natural example of a probability measure on projections is the restriction of a normed positive linear functional on the von Neumann algebra $B(H)$ of all bounded operators acting on H. Whether or not all probability measures are of this very form was one of the basic questions of mathematical foundations of quantum mechanics. This question has led to many deep mathematical results known today as Gleason type theorems. The aim of this chapter is to provide a proof of Gleason Theorem on linear extension of bounded completely additive measure on a Hilbert space projection lattice and its ramifications in the light of recent results of quantum measure theory. The contents is organized as follows. In the first section we show how one can reduce the problem to the real three-dimensional Hilbert space. In the second step we prove Gleason Theorem for the real three-dimensional Hilbert space by using the concept of the frame function. Finally, in the last section we show that the boundedness of measure can be relaxed in Gleason Theorem whenever the corresponding Hilbert space has infinite dimension, which is a rather deep result by Dorofeev and Sherstnev. We conclude with historical remarks and some links with recent investigations.

3.1. Reduction to three-dimensional space

3.1.1. Theorem (Gleason). *Let H be a Hilbert space with* $\dim H \geq 3$. *Then any bounded completely additive measure μ on the projection lattice $P(H)$ extends uniquely to a normal functional on the algebra $B(H)$ of all bounded operators acting on H.*

As we saw in Chapter 1 (Section 1.2, Normal states and normal homomorphisms), any normal functional on a von Neumann algebra can be represented by a trace class operator. In this regard Gleason Theorem gives a lucid characterization of any bounded completely additive measure μ on $P(H)$ in the form

$$\mu(p) = \mathrm{tr}(Tp)\,, \qquad p \in P(H)\,, \tag{3.1}$$

where T is a trace class operator on H. This version of Gleason Theorem is well known to physicists. The operator T appearing in the formula above is called the *density matrix* of given physical state. Equivalently, the equation (3.1) can be rewritten in the form

$$\mu(p) = \sum_{n=1}^{\infty} \alpha_n\, \omega_{x_n}(p)\,, \qquad \alpha_n \in \mathbb{C},\ p \in P(H)\,, \tag{3.2}$$

where ω_{x_n} are vector states, $\omega_{x_n}(p) = (p\,x_n, x_n) = \|p\,x_n\|^2$, $(p \in P(H))$, corresponding to the orthonormal sequence (x_n) of eigenvectors of T. In particular, any completely additive probability measure on the Hilbert space structure is a σ-convex combination of pure vector states, which is the geometrical content of Gleason Theorem. In this connection the density matrix describes how to decompose given state into a mixture of pure states.

We will now discuss the hypotheses formulated in Gleason Theorem. First, let us observe that the assumption on dimension of H is indispensable. Consider the two-dimensional structure $P(H_2)$ of all projections acting on the two-dimensional Hilbert space H_2. Any probability measure μ on $P(H_2)$ is a function $\mu\colon P(H_2) \to [0,1]$ such that $\mu(0) = 0$, $\mu(1) = 1$ and

$$\mu(p) + \mu(q) = 1\,, \tag{3.3}$$

whenever p and q are pairwise orthogonal one-dimensional projections in $P(H_2)$. Hence, the value of μ can be defined on each pair of pairwise orthogonal atomic projections separately. Let e and f be nonorthogonal one dimensional projections in $P(H_2)$. Then the system $\{e, 1-e, f\}$ is a linear basis of the real space X of all self-adjoint elements in $B(H_2)$. Since any linear form on X is uniquely determined by its value at this basis, we can easily choose μ satisfying (3.3) but not being of a form of a restriction of any linear functional on X.

Another assumption is the complete additivity of μ. This assumption is necessary for getting *normal* linear extension. However, a corresponding version of Gleason Theorem also holds for finitely additive bounded measures. The transition to finitely additive measures is highly nontrivial and will be treated in Chapter 6.

Finally, let us study the role of the boundedness assumption. It is well known that there are additive functions on $\mathbb{R}$ that are unbounded on the unit interval. For this it suffices to view the real line as a vector space over rationals and to take an unbounded linear form on this space. Pick up now one additive function φ on $\mathbb{R}$ that is unbounded on the unit interval and consider a vector state ω_x on $P(H)$ where $\dim H < \infty$. Then the composition $\varphi \circ \omega_x$ is a (completely additive) unbounded measure on $P(H)$. Therefore a completely additive measure may be unbounded and hence not extendable to a bounded linear functional. Nevertheless, we will see that complete additivity and boundedness may differ only in a finite-dimensional case. More precisely, we shall prove later that any completely additive measure on the projection lattice $P(H)$ of an infinite-dimensional Hilbert space H has to be bounded.

The following concept of frame function will often be useful in the sequel.

3.1.2. Definition. Let S_1 be the unit sphere of a (real or complex) Hilbert space H. A function $f: S_1 \to \mathbb{R}$ is called the *frame function on H* if there is a number, $w(f)$, called the *weight* of f, such that

$$\sum_{\alpha \in I} f(x_\alpha) = w,$$

whenever (x_α) is an orthonormal basis of H. A function $f: S_1 \to \mathbb{R}$ is called the *weak frame function* if its restriction to any finite-dimensional subspace of H is a frame function.

It follows immediately from the definition that the property of being frame function is hereditary to closed subspaces. Indeed, let f be a frame function on a Hilbert space H and let G be its closed subspace. Choose two orthonormal bases (u_α) and (v_α) of G and extend them by adding a fixed orthonormal basis of the orthogonal complement $G^\perp$ of G to orthonormal bases of the whole of H. The frame condition now implies $\sum_\alpha f(u_\alpha) =$

$\sum_\alpha f(v_\alpha)$. In particular, it means that

$$f(x) = f(\alpha x)$$

for any unit vector x and $\alpha \in \mathbb{C}$ with $|\alpha| = 1$.

If μ is a completely additive measure on the lattice $P(H)$ of all orthogonal projections acting on H, then the function $f\colon S_1 \to \mathbb{R}$, defined on the unit sphere S_1 of H by setting

$$f(x) = \mu(p_x)\,, \qquad x \in S_1\,, \tag{3.4}$$

where p_x denotes the orthogonal projection of H onto the one-dimensional subspace generated by x, is a frame function. Vice versa, given a frame function $f\colon S_1 \to \mathbb{R}$ on the Hilbert space H, we obtain a completely additive measure on $P(H)$ by assigning

$$\mu(p) = \sum_\alpha f(e_\alpha), \quad p \in P(H),$$

where (e_α) is (any) orthonormal basis of the subspace $p(H)$. This establishes a one-to-one correspondence between frame functions and completely additive measures. Let us also remark that if μ is a finitely additive measure on $P(H)$, then the function f defined by (3.4) is a weak frame function.

3.1.3. Definition. A weak frame function f on the Hilbert space H is called *regular* if there is a bounded self-adjoint operator T acting on H such that

$$f(x) = (Tx, x) \text{ for any unit vector } x \in H\,.$$

Note that if f is a regular frame function, then the operator T in the Definition 3.1.3 has to be a trace class operator with $\operatorname{tr} T = w(f)$ (Proposition 2.2.1). In that case the corresponding completely additive measure extends obviously to a normal functional. Making use of the notion of frame function we can now restate Gleason Theorem in the following form. Before formulating it let us recall a few concepts of linear algebra. By a *conjugate-bilinear form* on a Hilbert space H we mean a complex-valued function B on $H \times H$ that is linear in the first variable and conjugate-linear in the second variable. B is called *bounded* if there is a constant $C > 0$ such that $|B(x,y)| \le C$ for all unit vectors $x, y \in H$. A conjugate-bilinear form B is called *hermitean* if $B(x,y) = \overline{B(y,x)}$ for all $x, y \in H$. The *quadratic form on* H is a function $F\colon H \to \mathbb{C}$ such that $F(x) = B(x,x)$ $(x \in H)$ where B is a conjugate-bilinear form on H.

3.1.4. Theorem. *Let H be a Hilbert space with* $\dim H \geq 3$. *Then any bounded weak frame function on H is regular.*

This rephrasing of Gleason Theorem has the advantage of describing the behaviour of finitely additive measures on atomic parts of the projection lattices, which is the essential content of Gleason Theorem. The following proposition gives equivalent conditions to regularity of frame functions and allows us to reduce the problem to the three-dimensional space.

3.1.5. Proposition. *Let f be a weak frame function on a Hilbert space H,* $\dim H \geq 2$. *Then the following conditions are equivalent:*

(i) *f is regular.*

(ii) *There is a bounded hermitean conjugate-bilinear form, B, on H such that*

$$f(x) = B(x,x) \text{ for all unit vectors } x \in H.$$

(iii) *There is a bounded quadratic form, F, on H such that*

$$f(x) = F(x) \text{ for all unit vectors } x \in H.$$

(iv) *f is bounded and regular when restricted to any two-dimensional subspace of H.*

Proof: The equivalence (i)⇔(ii) is the well known fact on representing bounded hermitean conjugate-bilinear form by a self-adjoint bounded operator (see e.g. [206]). The equivalence of (ii) and (iii) is an elementary correspondence between quadratic forms and hermitean bilinear forms. So the main part of the theorem is the equivalence (iv)⇔(i).

Suppose that $f\colon S_1 \to \mathbb{R}$ is a regular frame function on every two-dimensional subspace of H. Set

$$F(x) = \|x\|^2 f\left(\frac{x}{\|x\|}\right), \qquad x \neq 0,$$

and

$$F(0) = 0.$$

For each two-dimensional subspace $X \subset H$, there is a hermitean conjugate-bilinear form, B_X, such that

$$F(x) = B_X(x,x) \qquad \text{for all } x \in X.$$

Define now a function B on H by setting

$$B(x, y) = B_X(x, y)\,, \tag{3.5}$$

whenever X is a two-dimensional subspace containing x and y. (Note that this definition depends only on x and y.) Because B is a hermitean bilinear form on all two-dimensional subspaces we have

(i) $B(\alpha x, y) = \alpha B(x, y)$,

(ii) $B(x, y) = \overline{B(y, x)}$,

(iii) $4 \operatorname{Re} B(x, y) = F(x + y) - F(x - y)$,

(iv) $2F(x) + 2F(y) = F(x + y) + F(x - y)$.

Hence, by (iii) we have

$$8 \operatorname{Re} B(x, z) + 8 \operatorname{Re} B(y, z) = 2F(x+z) - 2F(x-z) + 2F(y+z) - 2F(y-z)\,.$$

By (iv) and (iii) applied successively we obtain

$$\begin{aligned} &8 \operatorname{Re} B(x, z) + 8 \operatorname{Re} B(y, z) \\ &= F(x + y + 2z) + F(x - y) - F(x + y - 2z) - F(x - y) \\ &= 4 \operatorname{Re} B(x + y, 2z) = 8 \operatorname{Re} B(x + y, z). \end{aligned}$$

Replacing x and y by ix and iy, we also find that

$$\operatorname{Im} B(x, z) + \operatorname{Im} B(y, z) = \operatorname{Im} B(x + y, z)\,.$$

It gives us

$$B(x + y, z) = B(x, z) + B(y, z)\,,$$

for all $x, y, z \in H$. Therefore B is a bilinear form. Since the quadratic form canonically induced by B is F, B has to be bounded. □

We shall mainly make use of the fact that a frame function on a Hilbert space of dimension at least three is regular if, and only if, its restrictions to all three-dimensional subspaces are regular. The proof of this fact is straightforward. The reduction of the problem to finite-dimensional subspaces allows to extend Theorem 3.1.4 to general inner product spaces. Let S be an inner product space. A function f defined on its unit ball is called the *weak frame function* on S, if it is a frame function when restricted to any finite-dimensional subspace of S. The following ramification of Gleason Theorem will often be used in the sequel.

3.1.6. Theorem. *Let f be a bounded weak frame function on an inner product space S, $\dim S \geq 3$. Then there is a bounded self-adjoint operator T acting on the completion H of S such that*

$$f(x) = (T\,x, x) \qquad \textit{for all unit vectors } x \in S\,.$$

Proof: By the very same argument as in Proposition 3.1.5 we can find a bounded conjugate-bilinear form $B\colon\ S \times S \to \mathbb{C}$ such that

$$f(x) = B(x, x) \qquad \text{for all unit vectors } x \in S\,.$$

By continuity, the hermitean bilinear form B can be extended to a bounded hermitean conjugate-bilinear form, denoted again by B, that is defined on $H \times H$. Let T be a bounded self-adjoint operator acting on H representing B, i.e.

$$B(x, y) = (T\,x, y) \qquad \text{for all } x, y \in S\,.$$

Then T is the desired operator. □

The following lemma says that the proof of Gleason Theorem reduces to real Hilbert space. This seemingly innocent simplification is very powerful in that it allows us to use the methods of spherical geometry. We say that a real subspace X of a complex Hilbert space H is *completely real* if the inner product (x, y) is real for all $x, y \in X$. Typically, a closed completely real subspace will be a real closed linear span of its orthonormal basis.

3.1.7. Lemma. *Let f be a bounded frame function on a two-dimensional Hilbert space H_2. Then f is regular as soon as f is regular when restricted to any completely real subspace of H_2.*

Proof: Suppose that f is regular on all completely real subspaces of H_2. Let us show that f attains its supremum M on the unit sphere S_1 of H_2. To this end, take a sequence $(x_n) \subset H$ with $f(x_n) \to M$. By the compactness of S_1 we can assume that $x_n \to y \in S_1$. There is a sequence of complex units (α_n) such that $(\alpha_n\, x_n, y)$ is real for all n. Using the compactness argument again, we can assume that $\alpha_n \to \alpha$ ($|\alpha| = 1$). Then $(\alpha_n\, x_n, y) \to \alpha \in \mathbb{R}$ and since the vectors $\alpha\, y$, y, $\alpha_n\, x_n$, are contained in some completely real subspace, we see from the continuity of f on this subspace that

$$f(y) = f(\alpha\, y) = \lim_{n\to\infty} f(\alpha_n\, x_n) = \lim_{n\to\infty} f(x_n) = M\,.$$

As f has maximum at y it follows from the elementary theory of quadratic forms that

$$f(z) = M\,|(z,y)|^2 + (w(f) - M)\,(1 - |(z,y)|^2)\,,$$

for all $z \in S_1$ lying in a completely real subspace containing also y. Take now an arbitrary $x \in S_1$. By choosing a suitable complex unit α we may arrange for $(\alpha\, x, y)$ being real. Then

$$\begin{aligned} f(x) &= f(\alpha\, x) = M\,|(\alpha\, x, y)|^2 + (w(f) - M)\,(1 - |(\alpha\, x, y)|^2) \\ &= M\,|(x,y)|^2 + (w(f) - M)\,(1 - |(x,y)|^2)\,. \end{aligned}$$

Hence, f is a restriction of a quadratic form. □

The following proposition summarizes our discussion.

3.1.8. Proposition. *If any bounded frame function on a real Hilbert space of dimension 3 is regular, then any bounded weak frame function on a Hilbert space of dimension at least 3 is regular.*

Next section will be devoted to proving the regularity of frame functions on three-dimensional real Hilbert spaces, which is the main part of the proof of Gleason Theorem. Theorems 3.1.1 and 3.1.4 will then follow from this result.

3.2. Regularity of frame functions on $\mathbb{R}^3$

In this part we establish the regularity of frame functions on the real three-dimensional space. The arguments are based on spherical geometry and for this reason we introduce basic geometric terminology and collect basic properties of linear and affine objects on the sphere. In the sequel, S will always denote the unit sphere in the three-dimensional real Hilbert space $\mathbb{R}^3$. If $s, t \in S$, then $\theta(s,t)$ will stand for the angle between vectors s and t. For a fixed vector $p \in S$ we will use the following geographic terminology and notation. The set

$$N_p = \{s \in S \mid \theta(p,s) \leq \pi/2\}$$

is called the *northern hemisphere* with respect to p, and the set

$$E_p = \{s \in S \mid s \perp p\}$$

is called the *equator* with respect to p. The *latitude* of $s \in N_p$ is defined as $\pi/2 - \theta(s,p)$. The *latitude function* is a function l_p on N_p given by

$$l_p(s) = \cos^2\theta(p,s) = (p,s)^2 .$$

If the north pole p is fixed, we usually drop the subscript p in the previous notation.

A *frame* is an ordered pair (r,s,t) of pairwise orthogonal vectors from N_p. By expressing the north pole p in the orthonormal basis (r,s,t) we obtain

$$1 = l(r) + l(s) + l(t) . \tag{3.6}$$

Conversely, given numbers $a,b,c \in [0,1]$ such that

$$a + b + c = 1 ,$$

we always find a frame (r,s,t) with

$$a = l(r) , \quad b = l(s) , \quad \text{and } c = l(t) . \tag{3.7}$$

The *great circle* on S is the intersection of S with a plane containing the centre of S. Let $s \in N_p \setminus \{p\}$ with $l(s) > 0$. There is exactly one great circle, denoted by $C(s)$, having s as its northernmost point. Indeed, $C(s)$ is characterized by the property that the plane containing $C(s)$ intersects the plane containing the parallel $T_s = \{t \in S \mid l(t) = l(s)\}$, in the line tangent to $C(s)$ at s. In other words, $C(s)$ intersects the equator in points orthogonal to s. The great half-circle

$$D(s) = C(s) \cap N_p$$

will be called the *descent* through s.

Let us recall that a frame function is a function $f\colon S \to \mathbb{R}$ such that

$$f(r) + f(s) + f(t)$$

has a constant value, $w(f)$, for each frame (r,s,t). The following properties of the frame function follow immediately from the definition.

3.2.1. Lemma. *Let $f: S \to \mathbb{R}$ be a frame function. Then,*

(i) $f(s) = f(-s)$ *for all* $s \in S$.

(ii) $f(s) + f(t) = f(s') + f(t')$ *whenever* s, t, s', t' *lie on the same great circle and* $s \perp t$, $s' \perp t'$.

The property (ii) in the previous lemma has the following consequence that will be used frequently in the sequel.

3.2.2. Lemma. *Let $f: S \to \mathbb{R}$ be a bounded frame function with $\sup_{s \in S} f(s) = M$ and $\inf_{s \in S} f(s) = m$. If, for an $\varepsilon > 0$, $f(s) > M - \varepsilon$, then there is $t \in S$, $t \perp s$, such that $f(t) < m + \varepsilon$.*

Proof: If $f(s) > M - \varepsilon$ we can find $\delta > 0$ such that $f(s) > M - \varepsilon + \delta$. Choose $u \in S$ with $f(u) < m + \delta$. Then take vectors $t, v \in S$ with $t \perp s$, $v \perp u$, and moreover, with the property that t and v lie in the great circle passing through s and u. Then

$$f(t) + f(s) = f(v) + f(u)$$

and so

$$\begin{aligned} f(t) &= f(v) + f(u) - f(s) < M + m + \delta - (M - \varepsilon + \delta) \\ &= m + \varepsilon . \end{aligned}$$

□

Our next goal is to prove the continuity of frame functions. The following lemma says that a frame function is continuous if it has an arbitrarily small oscillation on some open set. Let $U \subset S$. We say that a function $f: S \to \mathbb{R}$ has the *oscillation* less then $\delta > 0$ on U if

$$\sup_{s \in U} f(s) - \inf_{s \in U} f(s) < \delta .$$

3.2.3. Lemma. *Let $f: S \to \mathbb{R}$ be a bounded frame function. If there is a neighbourhood $U(p)$ of a point $p \in S$ such that the oscillation of f on $U(p)$ is less then $\delta > 0$, then any point $q \in S$ admits a neighbourhood $U(q)$ such that f has oscillation less then $4\,\delta$ on $U(q)$.*

Proof: Let us view p as the north pole and assume that $U(p)$ contains all points with the latitude strictly greater then $\pi/2 - \alpha$, where $0 < \alpha < \pi/2$. Let us take a point y on the equator E_p and a point r in the south hemisphere having the same altitude as y and the latitude $-\alpha/2$. It is clear that there exists a neighbourhood $U(y)$ of y with the following property: If $x \in U(y)$, then there are points $r', x' \in U(p)$ lying on the great circle, passing through r and x, and satisfying $r' \perp r$ and $x' \perp x$ (see Figure 3.2.1)

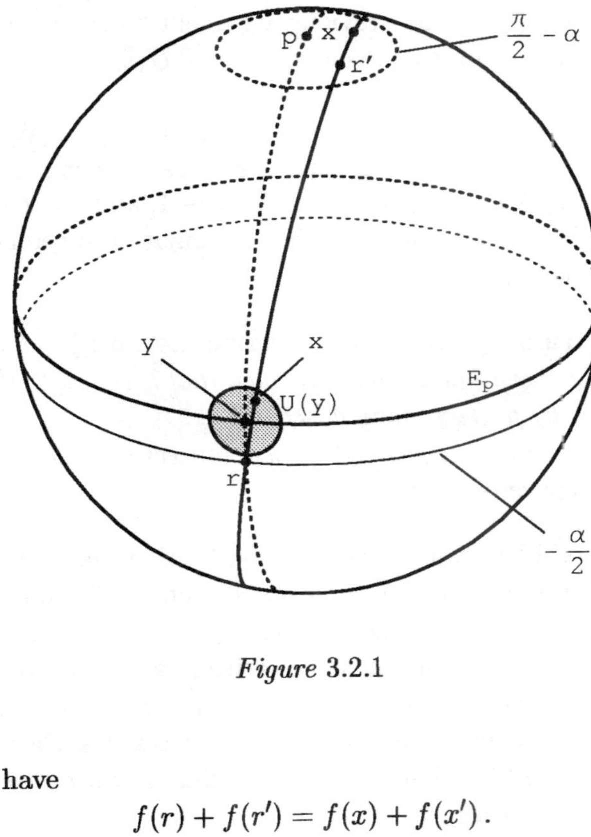

Figure 3.2.1

Then we have

$$f(r) + f(r') = f(x) + f(x') .$$

Consider now two points $x_1, x_2 \in U(y)$. By the above reasoning we can find vectors $x_1', x_2', r_1', r_2' \in U(p)$ with

$$\begin{aligned} f(r) + f(r_1') &= f(x_1) + f(x_1') \\ f(r) + f(r_2') &= f(x_2) + f(x_2') . \end{aligned}$$

It implies that

$$f(x_1) - f(x_2) = f(r'_1) - f(r'_2) - (f(x'_1) - f(x'_2))\,.$$

Hence,

$$|f(x_1) - f(x_2)| < |f(r'_1) - f(r'_2)| + |f(x'_1) - f(x'_2)| < 2\delta\,.$$

In other words, the oscillation of f on $U(y)$ is less then 2δ. It is easily seen that any point $q \in S$ can be connected by two arcs of great circles of arc-length $\pi/2$. Applying the previous argument twice we infer that f has oscillation less then 4δ on some neighbourhood $U(q)$ of q. □

Suppose that $s \in N_p$ lies south of r (i.e. $l(s) < l(r)$). We say that s is *reachable from r by n descents* if there exists a sequence $x_1, \ldots, x_{n-1}$ such that $x_1 \in D(r)$, $x_i \in D(x_{i-1})$ for $i = 2, \ldots, n-1$, and $s \in D(x_{n-1})$.

In the next Proposition we explicitly formulate the geometric idea behind Gleason Theorem.

3.2.4. Proposition. (Piron's Geometric Lemma). *Let r and s be points on the northern hemisphere N_p such that s is south of r. Then s is reachable from r by finitely many descents.*

Moreover, if s lies between $D(r)$ and the equator E_p, then s can be reached by two descents from r.

Proof: The problem can be transformed to plane geometry by taking a map T projecting each point of N_p from the centre of S onto the plane P tangent to S and passing through the pole p.

It is clear that points of the same latitude are projected onto circles centred at p, while the descent through $r \in N_p$ is mapped into the straight line through $T(r)$ tangent to the circle centred in p and such that it contains $T(r)$ (see Figure 3.2.2). First, let us show that s can be reached by two descents from r whenever

$$(T(r) - p) \cdot (T(s) - T(r)) > 0\,. \tag{3.8}$$

Indeed, walking along the line $T(D(r))$ sufficiently far away (see Figure 3.2.3) we can obtain $y \in N_p$ such that $T(y) \in T(D(r))$ and

$$(T(y) - p) \cdot (T(s) - T(y)) < 0\,.$$

By the continuity of the inner product (formally, by Rolle theorem) there must be a point $T(x) \in T(D(r))$ with

$$(T(x) - p) \cdot (T(s) - T(x)) = 0,$$

meaning that s lies on the descent through x. Therefore, we can move from

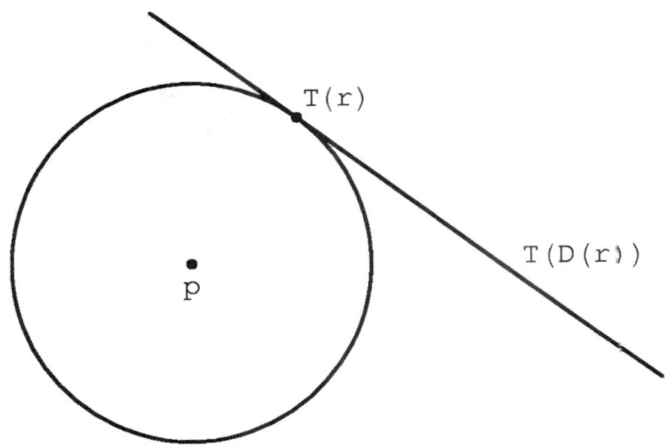

Figure 3.2.2

r to s by two descents. The set of all points in N_p lying (strictly) between $D(r)$ and the equator projects onto the half-line given by the border line $T(D(r))$ and such that it does not contain $p\,(= T(p))$. Since any point $T(s)$ from this half-plane satisfies (3.8), we can get by two descents from r to s.

This establishes the second part of Proposition 3.2.4. As a special case of the previous arguments we obtain that one can connect by two descents the points whose images under T lie in the same half-line in P containing p. For proving the general case we can construct a sequence of points $x_0 = r, x_1, \ldots, x_n = s \in S$ such that $x_i \in D(x_{i-1})$ and the angle between $T(x_i) - p$ and $T(x_{i+1}) - p$ (measured in the plane P) is π/n (see Figure 3.2.4). For the distances ϱ_i of $T(x_i)$ from p (measured again in P) we have

$$\frac{\varrho_{i+1}}{\varrho_i} = \frac{1}{\cos(\frac{\pi}{n})}.$$

Hence,

$$\varrho_n = \frac{\varrho_0}{\cos^n(\frac{\pi}{n})}.$$

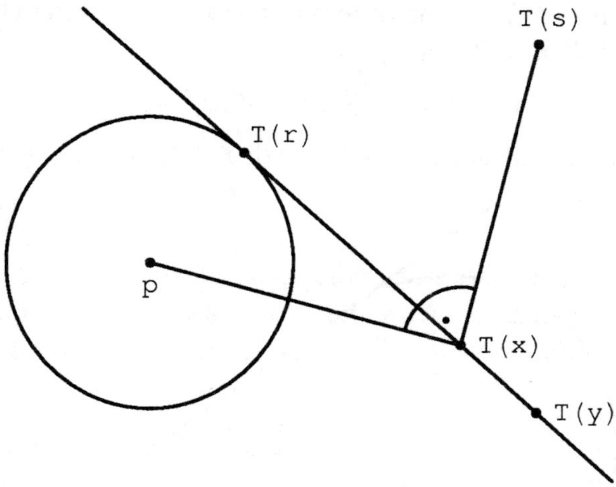

Figure 3.2.3

Since $\lim_{n\to\infty}(\cos(\pi/n))^{-n} = 1$, we can arrange for ϱ_n being arbitrarily close to ϱ_0. In other words, we can get from r by finitely many descents to a point whose projection onto P is arbitrarily close to the point in the plane P symmetric to $T(x)$ with respect to the point p. Considering now arbitrary $s \in S$ with $l(s) < l(r)$ we can get by finitely many descents from r to a point whose projection lies in the half-line containing $T(s)$ and p. Then we manage to reach s by additional two descents by the first part of the proof. □

The proceeding lemma is known as the Piron's Geometric Lemma and its second part is often referred to as the Weak Piron's Geometrical Lemma. Using these results we shall establish important properties of frame functions constant on some great circle.

3.2.5. Lemma. *Let $f\colon S \to \mathbb{R}$ be a bounded frame function constant on the equator E_p. If, for some $\varepsilon > 0$,*

$$f(p) > \sup_{s\in S} f(s) - \varepsilon\,,$$

then

$$f(r) \geq f(s) - \varepsilon$$

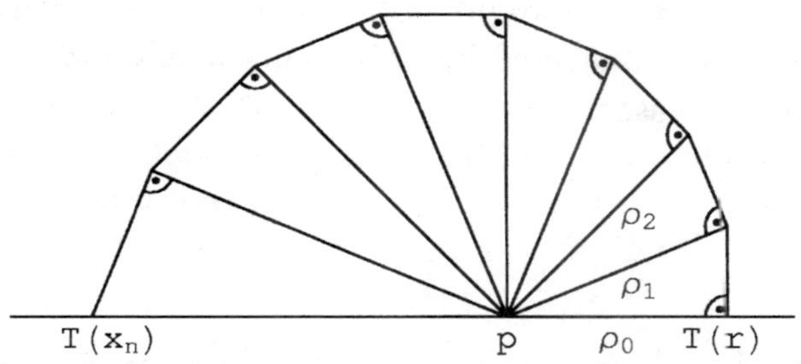

Figure 3.2.4

whenever $s \in D(r)$.

In particular, if $f(p) = \sup_{s \in S} f(s)$, *then*

$$f(r) \geq f(s)$$

whenever $l(r) > l(s) \geq 0$.

Proof: Write $M = \sup_{s \in S} f(s)$ and $m = \inf_{s \in S} f(s)$. By Lemma 3.2.2, there is a point $x \in E_p$ such that $f(x) < m + \varepsilon$. Hence,

$$f(e) \leq m + \varepsilon$$

for all $e \in E_p$. Suppose that $s \in D(r)$. Take a vector $r' \in C(r)$ orthogonal to r. Then $r' \in E_p$ and therefore

$$f(r') \leq m + \varepsilon\,.$$

By selecting a vector $s' \in C(r)$ with $s' \perp s$ we have

$$f(r) + f(r') = f(s) + f(s')$$

and therefore

$$f(r) - f(s) = f(s') - f(r') \geq m - (m + \varepsilon) = -\varepsilon\,.$$

Since ε can be taken arbitrarily small if $f(p) = M$ we obtain

$$f(r) \geq f(s)\,,$$

whenever $s \in D(r)$. As any two points in the northern hemisphere with different latitude can be connected by finitely many descents (Lemma 3.2.4), we infer that in this case

$$f(r) \geq f(s),$$

whenever $l(r) > l(s)$. □

3.2.6. Proposition. *Any bounded frame function on S is continuous.*

Proof: Let f be a bounded frame function. By adding a suitable constant function we can suppose that $\inf_{s \in S} f(s) = 0$. Fix $\varepsilon > 0$ and take a point $p \in S$ such that $f(p) < \varepsilon$. Let us regard p as the north pole of S. Consider an auxiliary frame function

$$g(x) = f(x) + f(\hat{p}\, x), \qquad x \in S,$$

where $\hat{p}\, x$ denotes the right-hand $\pi/2$-rotation of x about the line connecting p and the origin. It is clear that g is a frame function constant on the equator E_p by Lemma 3.2.1 (ii). Moreover,

$$g(p) = 2\, f(p) < 2\, \varepsilon .$$

Since g is nonnegative, it implies that

$$g(p) < \inf_{s \in S} g(s) + 2\, \varepsilon . \tag{3.9}$$

Suppose first that there is a point $s \in N_p \setminus \{p\}$ with

$$g(s) < \inf_{x \in S} g(x) + \varepsilon . \tag{3.10}$$

Applying Lemma 3.2.5 to the function $-g$ we get from (3.10) that

$$g(r) \leq g(s) + 2\, \varepsilon , \tag{3.11}$$

whenever $s \in D(r)$. Let G be the set of all points x in the northern hemisphere such that s is reachable by two descents from x. The set G has a nonempty interior. Indeed, using the notation of Proposition 3.2.4, all points x for which the central projection $T(x)$ lies in the disc containing p and $T(s)$ as antipodal points on its diameter are in G. (This is just the set of points fulfilling (3.8) for $r = x$.) Therefore we can find a nonempty open subset H of G. By (3.11)

$$g(x) < g(s) + 4\varepsilon$$

for all $x \in H$. So,

$$0 \leq \inf_{y \in S} g(y) \leq g(x) < \inf_{y \in S} g(y) + 5\varepsilon$$

for all $x \in H$. Hence, g has the oscillation less then 5ε on H. By Lemma 3.2.3 there is a neighbourhood O of p such that g has the oscillation less then 20ε on O. For all $x \in O$ we then have

$$0 \leq g(x) \leq g(p) + 20\varepsilon < 2\varepsilon + 20\varepsilon = 22\varepsilon .$$

Hence,

$$0 \leq f(x) \leq g(x) \leq 22\varepsilon$$

and so f has the oscillation less then 22ε on O. Hence, for any prescribed $\delta > 0$ there is an open set A in S such that the oscillation of f on A is less then δ. In view of Lemma 3.2.3, this is sufficient for proving the continuity of f.

Finally, suppose that there is no point $s \in N_p \setminus \{p\}$ satisfying (3.10). It gives us that

$$g(p) = \inf_{s \in S} g(s) .$$

By virtue of the second part of Lemma 3.2.5 we see that g is nonincreasing with respect to the latitude. Hence, by boundedness of g there must be points $a, b \in N_p \setminus \{p\}$ with $l(a) > l(b)$ and with

$$g(b) - g(a) < \varepsilon .$$

Since for any point $c \in S$ with $l(a) > l(c) > l(b)$ we have $g(c) \in [g(a), g(b)]$, we infer that g has the oscillation less then ε on the open set $\{t \in S \mid l(a) > l(t) > l(b)\}$. By Lemma 3.2.3 there is a neighbourhood O of p such that g has oscillation less then 4ε on O. Since f is a nonnegative function dominated by g, f has the oscillation less then 4ε on O. It gives us that f is continuous. □

3.2.7. Remark. Let f be a frame function on a Hilbert space H. Let us observe that if f is continuous then it has to be automatically uniformly continuous. Indeed, suppose that f is continuous at one point, say p, in the unit sphere S_1 of H. Then, given $\varepsilon > 0$, there is a neighbourhood $U(p)$ such that $|f(x) - f(y)| < \varepsilon$ whenever $x, y \in U(p)$. Observe that in Lemma 3.2.3 the size of the neighbourhood $U(q)$ depends only on the size of the neighbourhood $U(p)$ and not on f (it is also universal for all three-dimensional subspaces in H). Let us denote this size (i.e. the corresponding radius) by $\delta > 0$. Take now two points u, v in S_1 with $\|u - v\| < \delta$. Upon applying Lemma 3.2.3 to the three-dimensional subspace containing p, u, v, we see that $|f(u) - f(v)| < 4\varepsilon$. Thus, f is really uniformly continuous.

We say that a bounded frame function f is *simple* if it attains its supremum at a point, p, and it is simultaneously constant on the equator E_p. As we shall see later such functions correspond to extreme points in the convex set of all frame functions. An important step toward proving Gleason Theorem, achieved in the next proposition, is to prove the regularity of simple frame functions.

3.2.8. Proposition. *Let f be a simple frame function having maximum at point $p \in S$ and constant on the equator E_p. Then*

$$f(s) = m + (M - m)\cos^2\theta(s, p) \qquad \textit{for all } s \in S\,,$$

where $M = \sup_{s\in S} f(s)$ and $m = \inf_{s\in S} f(s)$.

Proof: If $m = M$ we are done. Let $M \neq m$. Then we may suppose that $m = 0$ and $M = 1$ by replacing f with $\frac{f-m}{M-m}$ if necessary. By Lemma 3.2.5,

$$f(r) \geq f(s)$$

whenever $l(r) > l(s) \geq 0$. We will show that f is constant on each parallel $L_h = \{s \in S \mid l(s) = h\}$ $(1 \geq h \geq 0)$. Suppose the contrary. If

$$\underline{f}(h_0) = \inf\{f(s) \mid s \in L_{h_0}\} < \overline{f}(h_0) = \sup\{f(s) \mid s \in L_{h_0}\}$$

for some $h_0 \in [0,1]$, then by continuity of f,

$$\underline{f}(h) < \overline{f}(h)$$

for all $h \in I$ from some open interval $I \subset [0, 1]$. On the other hand, from the monotonicity of f with respect to the latitude, we have

$$\sum_{h \in I} (\overline{f}(h) - \underline{f}(h)) \leq 1 ,$$

which is a contradiction with the fact that I is uncountable. Hence, there is an increasing function $g\colon [0, 1] \to [0, 1]$ such that

$$f(s) = g(l(s)) \qquad \text{for all } s \in N_p .$$

The function g has the property

$$g(a) + g(b) + g(c) = 1 \tag{3.12}$$

whenever $a, b, c \in [0, 1]$ with $a + b + c = 1$. Indeed, for such a, b, c there is a frame (r, s, t) consisting of elements of N_p such that

$$l(r) = a , \quad l(s) = b , \quad l(t) = c ,$$

and so $f(l(r)) + f(l(s)) + f(l(t)) = 1$ or, equivalently,

$$g(a) + g(b) + g(c) = 1 .$$

For any $a \in [0, 1]$ we have by (3.12)

$$g(a) + g(1 - a) = 1 .$$

Consequently, for $0 \leq a, b \leq a + b \leq 1$ one derives

$$g(a) + g(b) = 1 - g(1 - (a + b)) = g(a + b) .$$

This implies that $g(a) = a$ for each rational $a \in [0, 1]$ and so, by continuity of g, g is the identity on the unit interval. Finally,

$$f(s) = g(l(s)) = l(s) = \cos^2 \theta(p, s) = (p, s)^2$$

for all $s \in S$. □

We are in the position to prove the main result.

3.2.9. Theorem. *Any bounded frame function on S is regular.*

Proof: Let f be a bounded frame function on S. By continuity of f and Lemma 3.2.2 there are orthogonal vectors $p, r \in S$ such that $f(p) = \sup_{s\in S} f(s) = M$ and $f(r) = \inf_{s\in S} f(s) = m$. Choose $q \in S$ orthogonal to both p and r, and set $\alpha = f(q)$. We shall suppose that $m < \alpha < M$, for otherwise f or $-f$ is a simple frame function and we can complete the proof by employing Proposition 3.2.8 for f or $-f$. Denote by $\hat{p}, \hat{q}, \hat{r}$ the $\pi/2$ right-hand rotations about the lines containing the origin and points p, q, and r, correspondingly. The function

$$f(s) + f(\hat{p}\, s)$$

takes the constant value $m + \alpha$ on the equator E_p. Also, it attains its supremum $2\,M$ at p. We can therefore apply Proposition 3.2.8 to obtain

$$\begin{aligned} f(s) + f(\hat{p}\, s) &= m + \alpha + (2\,M - m - \alpha)\cos^2\theta(s,p) \\ &= 2\,M\cos^2\theta(s,p) + (m+\alpha)\,(1 - \cos^2\theta(s,p))\,. \end{aligned}$$

Using the notation

$$g(s) = M\cos^2\theta(s,p) + m\cos^2\theta(s,r) + \alpha\cos^2\theta(s,q)\,,$$

we obtain

$$f(s) + f(\hat{p}\, s) = g(s) + g(\hat{p}\, s) \qquad \text{for all } s \in S\,. \tag{3.13}$$

Applying the same arguments to $-f$ we also have

$$f(s) + f(\hat{r}\, s) = g(s) + g(\hat{r}\, s) \qquad \text{for all } s \in S\,. \tag{3.14}$$

Now, let (x, y, z) denote the coordinates with respect to the frame (p, q, r). Our next goal is to show that the frame functions f and g coincide on the great circles $x = y$, $x = z$, and $y = z$. Observe that $\hat{r}(x, y, z) = (-y, x, z)$, $\hat{p}(x, y, z) = (x, -z, y)$. By straightforward calculations we see that

$$\hat{p}\,\hat{p}\,\hat{r}(x, x, z) = (-x, -x, -z) \tag{3.15}$$

$$\hat{p}\,\hat{r}\,\hat{r}(x, z, z) = (-x, -z, -z) \tag{3.16}$$

$$\hat{r}\,\hat{p}\,\hat{p}\,\hat{p}\,\hat{r}(x, y, x) = (-x, -y, -x) \tag{3.17}$$

and

$$f(s)+f(\hat{r}\,s) = g(s)+g(\hat{r}\,s) \tag{3.18}$$
$$f(\hat{r}\,s)+f(\hat{p}\,\hat{r}\,s) = g(\hat{r}\,s)+g(\hat{p}\,\hat{r}\,s) \tag{3.19}$$
$$f(\hat{p}\,\hat{r}\,s)+f(\hat{p}\,\hat{p}\,\hat{r}\,s) = g(\hat{p}\,\hat{r}\,s)+g(\hat{p}\,\hat{p}\,\hat{r}\,s)\,. \tag{3.20}$$

Using (3.15) and the rules $f(s) = f(-s)$, $g(s) = g(-s)$, we obtain for $s = (x,x,z)$, by subtracting (3.19) from the sum of (3.18) and (3.20), that $f(s) = g(s)$.

Similarly, applying (3.16) and (3.17) we can prove that $f(s) = g(s)$ whenever $y = z$ or $x = z$.

The previous arguments imply immediately that $f = g$ also on the great circles $x = -y$, $x = -z$, and $y = -z$. For this, take $s = (x,-x,z)$. Then $\hat{r}\,s = (x,x,z)$ and $f(\hat{r}\,s) = g(\hat{r}\,s)$. Now, from (3.14) we infer that $f(s) = g(s)$. The other two cases can be proved in the same way.

Let us now define the frame function

$$h = g - f\,.$$

Then $h(p) = h(q) = h(r) = 0$ and so the weight of h is zero. We also know that h vanishes on six great circles: $x = \pm y$, $x = \pm z$, $y = \pm z$. The proof will be completed when we show that h is identically zero. For a contradiction, assume that the opposite is true. Like as in the beginning of the proof, we can find a frame (p',q',r') such that

$$M' = \sup_{s\in S} h(s) = h(p')$$
$$m' = \inf_{s\in S} h(s) = h(r')$$
$$\alpha' = h(q').$$

First we show that $M' = -m'$. Assume that $m' > -M'$, and hence $\alpha' < 0$ because $M' + m' + \alpha' = 0$. By Lemma 3.2.1 (ii) α' is the maximal value of h on the great circle orthogonal to p'. However, the great circle $x = y$ must intersect the former great circle at at least two points at which h is zero – a contradiction. Applying these arguments to $-h$, we derive a contradiction from the assumption $m' < -M'$.

Thus, we have $-m' = M'$, implying immediately that $\alpha' = 0$. We shall denote by (x', y', z') the coordinates with respect to the frame (p', q', r'). Proceeding as in the initial part of the proof, we can verify that

$$h(x', x', z') = M'(x'^2 - z'^2).$$

It means that h attains the zero value on the great circle $x' = y'$ precisely at the following four points:

$$(x', x', x'), (x', x', -x'), (-x', -x', x'), (-x', -x', -x').$$

The great circles $x = y$, $x = z$ and $y = z$ intersect at points (x, x, x), $(-x, -x, -x)$. The great circle $x' = y'$ must pass through these points, for otherwise there would be six points on the great circle $x' = y'$ at which h takes zero. The great circles $x = -y$ and $x = -z$ intersect at points $(x, -x, -x)$ and $(-x, x, x)$. These points must belong to the great circle $x' = y'$, too, because otherwise we would obtain six zero points of h on the great circle $x' = y'$. In summary, the great circle $x' = y'$ contains four points $\pm(x, x, x), \pm(x, -x, -x)$ and so it has to coincide with the great circle $x = y$. But h is identically zero on the circle $x = y$, which is a contradiction. Completing the proof, $h = 0$ and so f is a regular frame function. □

3.3. Boundedness of frame functions

In the first part of this chapter we commented on the fact that a frame function on a finite-dimensional space can be unbounded and so it may not be representable by a linear operator in the sense of Gleason Theorem. What is more, there are many unbounded real-valued measures on the projection structure of a finite-dimensional Hilbert space. This phenomenon illustrates one of the differences between classical and quantum measure theory. In classical case, any real-valued completely additive measure is a difference of two positive measures and thereby bounded. This is far from being true in measure theory built on general ordered structures. However, the surprising result of Dorofeev and Sherstnev [100] tells us that if the Hilbert space is infinite-dimensional, then any frame function on it is bounded. The present section will be devoted to the exposition of this remarkable result and its consequences. Similar questions concerning projection lattices in von Neumann algebras will be dealt with in Chapter 6.

First we extend the definition of the frame function to a slightly more general context of inner product spaces.

3.3.1. Definition. Let S be an inner product space. The function $f\colon S_1 \to \mathbb{R}$ defined on the unit sphere S_1 of S is said to be a *frame-type function* on S if the following conditions are satisfied:

(i) the restriction of f to any finite-dimensional subspace of S is a frame function,

(ii) the sum $\sum_i f(x_i)$ exists for all orthonormal sets (x_i) in S.

Moreover, f is called the *frame function on* S if there is a constant, $w(f)$, called the *weight* of f, such that $\sum_\alpha f(x_\alpha) = w(f)$ for all maximal orthonormal systems (x_α) in S.

It is easy to see that in case of finite-dimensional space S the frame functions and frame-type functions coincide because the condition (ii) of the previous definition is automatically satisfied. If f is a bounded frame-type function on a Hilbert space, H, then it can be by Theorem 3.1.4 represented by a bounded self-adjoint linear operator T on H. Since $\sum_i (Tx_i, x_i)$ converges for all orthonormal bases (x_i) of H, the operator T has to be of trace class (see Proposition 2.2.1). Henceforth, f is also a frame function. It will follow from our analysis that the validity of this statement does not require the completeness of the underlying linear structure. Studying the properties of frame-type functions, we start with a series of auxiliary lemmas. We will use the notation $M^\perp = \{y \in H \mid (x,y) = 0 \text{ for all } x \in M\}$ for the set M in an inner product space S.

3.3.2. Lemma. *Suppose that an infinite-dimensional inner product space V admits an unbounded frame-type function. Then there is an infinite-dimensional inner product space S possessing a frame-type function $f\colon S_1 \to \mathbb{R}$ satisfying the following properties:*

(i) *If $x \in S_1$ with $|f(x)| > 1$, then the restriction of f to the hyperplane $\{x\}^\perp$ is bounded.*

(ii) *There is a unit vector $a \in S$ such that $|f(a)| \le 1$ and*

$$\sup\{|f(b)| \mid b \in S_1, (b,a) = 0\} \le 1\,.$$

Proof: (i) Suppose, for a contradiction, that (i) is not satisfied for any frame-type function on any infinite-dimensional space. It implies that for a given frame-type function f on V there is a unit vector x_1 such that $|f(x_1)| > 1$ and f is unbounded on $\{x_1\}^\perp \cap S_1$. Since the function $f|\{x_1\}^\perp$ is a frame-type function on an infinite-dimensional space again, there exist a unit vector $x_2 \in \{x_1\}^\perp$ such that $|f(x_2)| > 1$ and f is an unbounded frame-type function when restricted to $\{x_1, x_2\}^\perp$. Continuing this process, we can produce an orthonormal sequence (x_n) in V with $\sum_n |f(x_n)| = \infty$. This contradicts the hypothesis.

(ii) We can take any frame-type function on S satisfying (i) and fix a unit vector $a \in S$ with $|f(a)| > 1$. Then, by scaling the function f appropriately, we obtain the condition (ii). □

3.3.3. Lemma. *Let f be an unbounded frame-type function on an inner product space S satisfying conditions (i) and (ii) of Lemma* 3.3.2. *Then there are orthogonal unit vectors e_1, e_2, e_3 such that $|f(e_i)| > 1$ for all $i = 1, 2, 3$.*

Proof: By unboundedness of f we can find a unit vector e_1 in S with $|f(e_1)| > 6$. Fix a vector a from the condition (ii) of Lemma 3.3.2. Select now unit vectors g and b in the linear span $\mathrm{sp}\{e_1, a\}$ such that $(g, e_1) = 0$ and $(b, a) = 0$. Then $f(e_1) + f(g) = f(a) + f(b)$. Since by the condition (ii) of Lemma 3.3.2, $|f(a)|, |f(b)| \leq 1$ we obtain,

$$\begin{aligned} |f(g)| &= |f(e_1) - f(a) - f(b)| \\ &\geq |f(e_1)| - |f(a)| - |f(b)| > 4\,. \end{aligned}$$

Find now a unit vector h in $\{e_1, a\}^\perp$. Since $(h, a) = 0$, we conclude that $|f(h)| \leq 1$. Moreover, taking into account that f is a bounded frame function on any three-dimensional subspace of $\{e_1\}^\perp$ containing g and h, we infer that f is continuous on the unit sphere of $\mathrm{sp}\{g, h\}$ (Proposition 3.2.6). It implies that there is a unit vector $e_2 \in \mathrm{sp}\{g, h\}$ such that $|f(e_2)| = 2$. Finally, take a unit vector e_3 lying in $\mathrm{sp}\{g, h\}$ and orthogonal to e_2. By the frame property of f we have

$$f(e_2) + f(e_3) = f(g) + f(h)$$

and so

$$\begin{aligned} |f(e_3)| &= |f(g) + f(h) - f(e_2)| \\ &\geq |f(g)| - |f(h)| - |f(e_2)| > 1\,. \end{aligned}$$

This inequality completes the proof. □

3.3.4. Lemma. *Let $\{e_1, e_2, e_3\}$ be an orthonormal set in a real four dimensional Hilbert space H_4. Suppose that e is a unit vector in H_4 such that $(e, e_1) \neq 0$ and $(e, e_2) \neq 0$. Then we can decompose e into a sum*

$$e = x + y\,, \tag{3.21}$$

where x and y are nonzero orthogonal vectors satisfying

$$(x, e_1) = (y, e_2) = (y - \|y\|^2 e, e_3) = 0\,. \tag{3.22}$$

Proof: We will introduce the coordinates in H_4 with respect to a fixed basis $\{e_1, e_2, e_3, e_4\}$ that extends the orthonormal set $\{e_1, e_2, e_3\}$. Write $e = (\alpha, \beta, \gamma, \delta)$. By the hypotheses, $\alpha \neq 0$ and $\beta \neq 0$. We will find x and y in the form

$$\begin{aligned} x &= (0, \beta, \gamma - u, \delta - v) \\ y &= (\alpha, 0, u, v)\,, \end{aligned}$$

where u and v are suitable parameters.

It is apparent that both x and y is nonzero,

$$e = x + y, \qquad (x, e_1) = 0, \qquad (y, e_2) = 0\,.$$

By writing the remaining conditions $(x, y) = (y - \|y\|^2\, e, e_3) = 0$ in terms of coordinates, we obtain the following system of nonlinear equations.

$$u\,(\gamma - u) + v\,(\delta - v) = 0 \tag{3.23}$$

$$u - \gamma\,(\alpha^2 + u^2 + v^2) = 0\,. \tag{3.24}$$

It follows from (3.23) and (3.24) that

$$u = \frac{\gamma\,(\alpha^2 + \delta\, v)}{1 - \gamma^2}\,. \tag{3.25}$$

(Notice that $\gamma < 1$ since $\alpha \neq 0$ and $\|e\| = 1$.) Substituting (3.25) into (3.23) we obtain the following quadratic equation for the unknown v:

$$A v^2 + B v + C = 0, \tag{3.26}$$

where

$$\begin{aligned} A &= \frac{-\gamma^2 \delta^2}{(1-\gamma^2)^2} - 1, \\ B &= \frac{\gamma^2 \delta}{1-\gamma^2} - \frac{2\gamma^2 \alpha^2 \delta}{(1-\gamma^2)^2} + \delta, \\ C &= \frac{\alpha^2 \gamma^2}{1-\gamma^2} - \frac{\gamma^2 \alpha^4}{(1-\gamma^2)^2}. \end{aligned}$$

For the discriminant of quadratic expression (3.26) we have

$$\begin{aligned} \Delta &= B^2 - 4AC \\ &= B^2 + 4\left(\frac{\gamma^2 \delta^2}{(1-\gamma^2)^2} + 1\right) \cdot \left(\frac{\alpha^2 \gamma^2}{1-\gamma^2} - \frac{\gamma^2 \alpha^4}{(1-\gamma^2)^2}\right). \end{aligned}$$

As $\beta \neq 0$ and $\|e\| = 1$, we see that $\alpha^2 + \gamma^2 < 1$. It implies that

$$\frac{\alpha^2 \gamma^2}{1-\gamma^2} - \frac{\alpha^4 \gamma^2}{(1-\gamma^2)^2} = \frac{\alpha^2 \gamma^2}{1-\gamma^2} \cdot \left(1 - \frac{\alpha^2}{1-\gamma^2}\right) \geq 0.$$

Hence, $\Delta \geq 0$ and the system of equations (3.23), (3.24) has a real solution u and v. This completes the proof. □

We now pass to the main result.

3.3.5. Theorem (Dorofeev-Sherstnev). *Any frame-type function on an infinite-dimensional inner product space is bounded.*

Proof: Seeking for a contradiction, suppose that there is an unbounded frame-type function on some inner product space of infinite dimension. By the previous lemmas we can find an unbounded frame-type function f on some infinite-dimensional inner product space S satisfying conditions (i) and (ii) in Lemma 3.3.2 and admitting an orthonormal set $\{e_1, e_2, e_3\}$ with $|f(e_i)| > 1$ for $i = 1, 2, 3$. Set

$$C = \max_{i=1,2,3} \{|f(e_i)|, \sup\{|f(x)| \mid x \in \{e_i\}^{\perp} \cap S\}\}.$$

By the unboundedness of f there is a unit vector h such that $|f(h)| > 3C$. Obviously, $(h, e_i) \neq 0$ for $i = 1, 2, 3$. By multiplying e_i with a suitable complex unit, we can arrange for (h, e_i) being real for all $i = 1, 2, 3$. Employing Lemma 3.3.4, we can write

$$h = x + y\,,$$

where x and y are nonzero and, for $z = y - \|y\|^2 h$,

$$(x, y) = (x, e_1) = (y, e_2) = (z, e_3) = 0\,. \tag{3.27}$$

Observe that

$$\begin{aligned}(h, z) &= (y - \|y\|^2 h, h) = (y - \|y\|^2 (x + y), x + y)\\ &= \|y\|^2 - \|y\|^2 (\|x\|^2 + \|y\|^2) = \|y\|^2 - \|y\|^2 = 0\,.\end{aligned}$$

Since the vectors x, y, h, z lie in a common two-dimensional subspace, we have

$$f(h) + f\left(\frac{z}{\|z\|}\right) = f\left(\frac{x}{\|x\|}\right) + f\left(\frac{y}{\|y\|}\right). \tag{3.28}$$

By construction and (3.27)

$$\left|f\left(\frac{x}{\|x\|}\right)\right|, \left|f\left(\frac{y}{\|y\|}\right)\right|, \left|f\left(\frac{z}{\|z\|}\right)\right| \leq C\,.$$

It gives us

$$\left|f(h) + f\left(\frac{z}{\|z\|}\right)\right| \leq 2C\,. \tag{3.29}$$

On the other hand,

$$\begin{aligned}\left|f(h) + f\left(\frac{z}{\|z\|}\right)\right| &\geq |f(h)| - \left|f\left(\frac{z}{\|z\|}\right)\right|\\ &> 3C - C = 2C\,,\end{aligned}$$

which is a contradiction. The proof is complete. □

The boundedness principle allows us to formulate the following improvement of Gleason Theorem in case of infinite-dimensional space.

3.3.6. Theorem. *Let f be a frame-type function on an infinite dimensional inner product space S. Then there is a trace class operator T on the completion H of S which represents f in the sense of the formula*

$$f(x) = (T\,x, x) \qquad \textit{for all } x \in S_1\,.$$

Proof: By Theorem 3.3.5 and Theorem 3.1.6 there is a bounded self-adjoint operator T on H such that

$$f(x) = (T\,x, x) \qquad \text{for all } x \in S_1\,.$$

We shall prove that T is a trace class operator. Looking for a contradiction suppose that this is not the case. There is no loss of generality in assuming that $\|T\| \leq 1$. Since T is not of trace class we can find by Proposition 2.2.1 an orthonormal sequence $x_1, \ldots, x_n \in H$ such that

$$\sum_{i=1}^{n} |(T\,x_i, x_i)| > 1\,.$$

Choose $\varepsilon > 0$ such that $\sum_{i=1}^{n} |(T\,x_i, x_i)| > 1 + \varepsilon$ and find an orthonormal sequence $h_1, \ldots, h_n \in S$ such that

$$\|h_i - x_i\| < \frac{\varepsilon}{2n}\,.$$

(This can be done by a standard approximation argument.) It is clear that

$$\begin{aligned} |(T\,h_i, h_i) - (T\,x_i, x_i)| &\leq |(T\,(h_i - x_i), h_i)| + |(T\,x_i, h_i - x_i)| \\ &\leq 2\,\frac{\varepsilon}{2\,n} = \frac{\varepsilon}{n}\,. \end{aligned}$$

Hence,

$$\begin{aligned} \sum_{i=1}^{n} |(T\,h_i, h_i)| &\geq \sum_{i=1}^{n} |(T\,x_i, x_i)| - \sum_{i=1}^{n} |(T\,x_i, x_i) - (T\,h_i, h_i)| \\ &\geq 1 + \varepsilon - \varepsilon = 1\,. \end{aligned}$$

For the orthogonal projection P of H onto $\mathrm{sp}\{h_1, \ldots, h_n\}$ one can write $T = P\,T + (1 - P)\,T$. As $P\,T$ is a trace class operator, $(1 - P)\,T$ is not of trace class. Since $(1 - P)\,(S)$ is dense in $(1 - P)(H)$ we can apply the same arguments as before and find an orthonormal sequence $h_{n+1}, h_{n+2}, \dots, h_{n+k}$ in $(1-P)(S) \subset S$ such that

$$\sum_{i=1}^{k} |(T h_{n+i}, h_{n+i}) \geq 1 .$$

Continuing this way, we can construct an orthonormal sequence (h_n) in S such that

$$\sum_{n=1}^{\infty} |f(h_n)| = \sum_{n=1}^{\infty} |(T h_n, h_n)| = \infty ,$$

which is a contradiction. □

In fact, the previous theorem says that any frame-type function on an infinite-dimensional Hilbert space is a frame function. The following result is a measure-theoretic version of Theorem 3.3.6.

3.3.7. Theorem. *Any completely additive measure on the projection lattice of the algebra $B(H)$ of all bounded operators on an infinite-dimensional Hilbert space H extends uniquely to a normal functional on $B(H)$.*

In Chapter 6 this result will be extended to nearly all projection lattices in von Neumann algebras.

3.4. Historical remarks and comments

Gleason Theorem for states on a Hilbert space structure is a cornerstone of the mathematical foundations of quantum mechanics and one of the most profound mathematical results of the 20th century. It was proved by Gleason in 1957 [140] as a respond to significant problems in the axiomatics of quantum mechanics. Gleason was motivated by Mackey's approach to mathematical formalism of quantum theory (see [224]) which is based on a natural concept of probability measure on the orthomodular structure referred today to quantum logic. Gleason succeeded in describing all (σ-additive) probability measures on the logic of all closed subspaces of a separable Hilbert space and in showing that, except for the obvious two-dimensional counterexamples, all probability measures can be identified with normal states in the sense of von Neumann approach [248]. Gleason's achievement confirmed von Neumann's original insight and put the calculus of Hilbert space quantum mechanics on natural physical grounds.

The original Gleason's proof combines intricate spherical geometric arguments with the theory of unitary representations of the rotation group. First, he reduces the problem to describing frame function on three dimensional real Hilbert space and then he proves the continuity of frame functions. Having established the continuity, Gleason appeals to the representation theory of rotation group G in $\mathbb{R}^3$ which says that every closed G-invariant subspace of the space of all continuous functions on the unit sphere endowed with the supremum norm is the closed linear span of certain spaces of the surface harmonics Q_n, where Q_n denotes the restriction to the unit sphere of the homogeneous polynomials of degree n satisfying Laplace's equation. Gleason then deduces that the G-invariant space F of all frame functions is of the form $\overline{\cup Q_{n_i}}$ for some increasing sequence (n_i) of integers. But a direct verification of the frame condition shows that one has to exclude all Q_{n_i}'s except $n_i = 0$ or $n_i = 2$. So $F = Q_0 + Q_2$, and every frame function has to be regular. Another analytic proof of this fact was given later by Paratharasy in [256]. He first proves that infinitely many differentiable frame functions on $\mathbb{R}^3$ are regular, which allows to show that the same holds for continuous frame functions by the convolution technique on the rotation group endowed with the Haar measure. Since the time of Gleason's breakthrough many mathematicians and physicists have tried to simplify the original proof of Gleason Theorem and to extend it to more general lattices and measures. In 1970 Piron [263] has presented a simpler proof for a physically relevant special case of a probability measures having value one at some one-dimensional projection. His reasoning is based on ingenious geometric arguments (see Piron's Geometrical Lemma). Another elementary proof of the regularity of simple frame functions has been given by Gudder [149]. Gudder's approach is unique in completely avoiding spherical geometry and in using only analytic arguments. An elementary proof of the regularity of general bounded frame function was published in 1985 by Cook, Keane, and Moran [86]. They used purely geometric and topological arguments and eliminated in a very interesting way hard analysis in the Gleason's proof concerning characterization of irreducible invariant subspaces associated with the rotation group in terms of harmonics. The prize they had to pay is the length of the paper. The proof presented here synthesizes both Gleason's argument and insight of [86]. We prove the continuity of frame function using Piron's Geometric Lemma and Proposition 3.2.3 borrowed from Gleason's paper [140]. Then we follow (and simplify) arguments of [86]. This organization enables us to eliminate both advanced harmonical analysis used by Gleason and fairly deep topological arguments (Tikhonov Theorem on compactness of product of uncountably many bounded intervals) used in [86]. We in turn reduce the proof to very basic calculations accessible for first grade students.

The arguments used in the proof of Gleason Theorem has also been examined recently from the point of view of constructive mathematics [190, 275]. In particular, a constructive proof (avoiding e.g. Bolzano-Weierstrass theorem) has been given in a paper by Richman [275].

The combinatorial nature of the Gleason's proof of continuity of frame function on three-dimensional sphere has been recognized by Pitowski, in [264, 265]. He proved that there is a subset Γ of the unit sphere which is large but nevertheless finite and such that the continuity of a frame function f can be deduced only from the restriction $f|\Gamma$.

Original Gleason Theorem deals with σ-additive probability measures on the projection lattice of a separable Hilbert space. Extensions of this result for signed measures was given by Sherstnev [293] and for nonseparable cases by Eilers and Horst and Drisch [104, 131] (under some set-theoretic assumptions). It is worth noting that eliminating the assumption of σ-additivity, the technique developed by Gleason cannot be used. That is why Gleason Theorem for finitely additive measures on the projection lattice of a Hilbert space has been established much later by Christensen [80] for positive measures and by Bunce and Wright for general case [58]. This line of the research will be dealt with in Chapter 5.

Gleason Theorem has also motivated an extensive study of measures on projection lattices of Hilbert spaces attaining infinite values. It has turned out that such measures may be described by bilinear forms. Many results in this direction obtained by Lugovaya, Sherstnev, Dvurečenskij, and others, may be found in the monograph [116].

The Gleason Theorem has many physically relevant consequences. One of the most famous applications of Gleason Theorem is the proof of nonexistence of dispersion-free state on the Hilbert space logic. The dispersion free state is a state (i.e. a probability measure) attaining only values 0 and 1. Such a state assigns to each projection probability zero or one and eliminates so the probability character of the model. Whether such a state can exist was a topic of a long discussion. Since any countably additive state φ on the projection lattice $P(H)$, where $\dim H \geq 3$, extends by Gleason Theorem to a linear functional $\hat{\varphi}$ on $B(H)$, the problem is transferred to a normal state $\hat{\varphi}$ on $B(H)$. Suppose that $\hat{\varphi}$ attains only values 0 and 1 on projections. Write $\hat{\varphi}$ in its canonical form as the σ-convex combination of vector states

$$\hat{\varphi} = \sum_n \alpha_n \, \omega_{x_n} ,$$

where (x_n) is an orthonormal sequence. Since $\hat{\varphi}(p_{x_n}) = \alpha_n$, for any projection p_{x_n} onto linear span $\mathrm{sp}\{x_n\}$, we infer that

$$\hat{\varphi} = \omega_{x_i} ,$$

for some integer i. However, any vector state attains on projections all values between zero and one, which is a contradiction. In this way, the nonexistence of dispersion-free states directly follows from Gleason Theorem. Nevertheless, in order to establish the nonexistence of dispersion-free state we need not apply Gleason Theorem in its full power. Moreover, we can easily extend the result to all finitely additive states, which is the case not covered by the original Gleason Theorem.

3.4.1. Theorem. *Let H be a Hilbert space with* $\dim H \geq 3$. *There is no 0-1 finitely additive state on the projection lattice $P(H)$.*

Proof: First we prove that the result is true for three-dimensional real Hilbert space $\mathbb{R}^3$. For a contradiction, suppose that μ is a 0-1 state on $P(\mathbb{R}^3)$. It induces a frame function $f\colon\ S \to \{0,1\}$ on the unit sphere S. The function f has to attain value 1 at some vector $p \in S$. Consider p to be the north pole. Then f is a simple frame function attaining its maximum at p and being zero on the equator E_p. In view of Lemma 3.2.5 f is nondecreasing with respect to the latitude. It is clear that for the weight $w(f)$, we have $w(f) = 1$. By (3.7) we can find a frame (r, s, t) such that $l(r) = l(s) = 1/2$ and $l(t) = 0$. As $w(f) = 1$, f has to attain value 0 either at r or s. By the monotonicity it means that $f(u) = 0$ whenever $l(u) < 1/2$. Take now a frame (x, y, z) such that $l(x) = l(y) = l(z) = 1/3$. Then f vanishes at all x, y, z, which is a contradiction.

Suppose now that $\dim H \geq 3$ and that μ is a 0-1 state on $P(H)$. There is no loss of generality involved in assuming that H is real (we can always pass to a completely real subspace). If $\infty > \dim H \geq 3$, then there has to be a one-dimensional projection such that μ has value one at it. Therefore μ is nonzero on some three-dimensional projection. But this was excluded by the previous part of the proof. It remains to consider possibility of $\dim H = \infty$. In this case $P(H)$ contains a unital sublattice L isomorphic to $P(\mathbb{R}^3)$ (see e.g. Chapter 1). Hence, μ restricts to a nonzero 0-1 state on L – a contradiction by the previous part of the proof. □

The nonexistence of 0-1 state on an orthomodular lattice is connected with a long dispute on hidden variables in quantum theory. The existence of sufficiently many hidden variables would mean that the probabilistic character of quantum mechanics disappears by adding certain new parameters. One of the consequences would be that the description of quantum world offered by the standard Hilbert space model is incomplete. In this connection Theorem 3.4.1 says that the projection lattice of a Hilbert space cannot be embedded (as a quantum logic) into a Boolean algebra because any such embedding would induce a 0-1 state. It is therefore an important consequence of Gleason Theorem that the Hilbert space model cannot be completed by considering auxiliary hidden variables to a classical model without violating its inner structure. The nonexistence of hidden variables demonstrates the essential difference between classical and quantum physics. The first proof that quantum mechanics is logically incompatible with hidden variables was given by John von Neumann. It was one of the first successes of his rigorous axiomatic approach to physics [248]. (For the history of von Neumann famous *impossibility proof* we refer to [70].) He proved that there is no dispersion-free (linear) state on the algebra of all bounded operators on an infinite-dimensional Hilbert space. Consequently, von Neumann assumed that given functional φ satisfied

$$\varphi(a+b) = \varphi(a) + \varphi(b)\,, \tag{3.30}$$

whenever a and b are elements of $B(H)$. Since neither $a+b$ nor the expectation value $\varphi(a+b)$ has any meaning for noncompatible (i.e. noncommuting) observables a and b, the assumption (3.30) has no physical sense. For this reason the von Neumann's arguments had been attacked in subsequent studies (see e.g [32]). It was desirable to prove the nonexistence of a dispersion-free state additive only with respect to commuting elements. This was achieved by Gleason Theorem, showing that even if we relax (3.30) to be valid only for a and b being commuting projections then φ cannot be two-valued on projections.

Because of its relevance to quantum mechanics many alternative proofs of Theorem 3.4.1 simpler then Gleason Theorem has been presented [32, 79, 232, 245, 246]. In [32] the argument is based on geometrical reasoning known as Bell's Geometric Lemma which provides quite different proof then one in Theorem 3.4.1 above.

Another approach to the problem of hidden variables consists in studying finite substructures in a Hilbert space lattice. It was proved by Kochen and Specker [215] that there is a finite sublattice of $P(\mathbb{R}^3)$ having no 0-1 state. Therefore, the nonexistence of hidden variables can be established in a finite constructive way. This result inspired many researchers and has led to so-called *Kochen-Specker type theorems* [16, 41, 232, 260, 261, 264, 265, 304]. Hitherto known most effective construction due to Conway and Kochen involves 31 vectors. The construction of [260] given by Peres involves 33 vectors, but it is highly symmetric. Peres has also provided very simple argument showing the nonexistence of 0-1 state on a four-dimensional space. Besides nonexistence of dispersion-free state there are many other approaches to the problem of hidden variables. Some of them will be discussed later in Chapter 6.

Gleason Theorem says that the probability structure is determined by quantum logic, i.e. by the ordered structure of projections in a Hilbert space. The fact that only linear restriction can qualify for being quantum state has, besides hidden variable theory, other interesting physical consequences. One of them is the *indeterminacy principle*: Let p and q be atomic nonorthogonal projections in a Hilbert space H. For every completely additive state μ on $P(H)$ the following equivalence holds

$$\mu(p), \mu(q) \in \{0,1\} \iff \mu(p) = \mu(q) = 0\,.$$

In other words, no quantum state assigns sharp probability zero or one to two atomic nonorthogonal projections unless they are both false. Indeed, if $\mu(p) = 1$ for an atomic projection p then μ has to be a vector state concentrated at p. Then the value of μ at any atomic projection nonorthogonal to p has to be strictly between 0 and 1. Similarly as the nonexistence of dispersion-free state, the indeterminacy principle has finite character. It was shown in [265] that its validity can be demonstrated on some finite subset of the projection lattice.

Gleason Theorem has also been extended to more general spaces. Its variant for Hilbert space over nonstandard field was obtained by Keller and can be found in [213], results on spaces over rationals are dealt with in [241]. Gleason-type theorems for structure of skew projections have been studied by Matvejchuk and Mushtari in [239, 240].

Nearly 45 years after Gleason's paper appeared the investigation of measures on Hilbert space logic is still an area of very active research. Also, the current interest in orthomodular structures and foundations of physics attracts experts from various fields.

4. COMPLETENESS CRITERIA

The aim of this chapter is to study the role of completeness of underlying inner product space in quantum measure theory. The research in this field has brought a rather surprising result: "Natural" subspace structures of an inner product space, S, admit a nonzero completely additive measure if, and only if, S is complete. As a consequence, the probability theory on incomplete spaces is virtually empty and so the Hilbert space is the only inner product space on which a reasonable measure and probability theory can be built. This once again advocates the role of the (topological) completeness assumption in mathematical foundations of quantum theory and illustrates the unique position of the Hilbert space lattice in the realm of various ordered structures of subspaces that describe quantum systems. At the beginning of the effort to build mathematical formalism of quantum physics, as witnessed by the work by J. von Neumann and others, the completeness of inner product space was adopted for purely mathematical and technical reasons. In contrast to this, the results in this chapter will clearly show that completeness can be justified entirely on physical grounds. In fact, upon adopting the probability structure of quantum theory we automatically accept completeness as a consequence of the intrinsic character of quantum world.

From purely mathematical standpoint this chapter will be concerned with newly emerged interplay between linear topology of inner product spaces, ordered properties of specific subspace structures, and continuity of measures on them. As a by-product it is seen that Gleason Theorem, as formulated for completely additive measures, is valid for all inner product spaces, because the Hilbert space lattice is the only subspace lattice supporting a nontrivial completely additive measure.

4.1. Functional-analytic completeness criteria

Incomplete inner product spaces frequently occur in mathematical analysis and its applications. Let us e.g. mention subspaces in various L^2-spaces consisting of smooth functions with compact supports. In another art, inner product spaces result naturally from the G.N.S. type constructions in the theory of operator algebras. It is therefore desirable to know to what extent the Hilbert space technique can be used without assuming completeness. The next theorem shows that basic principles of analysis on Hilbert spaces, like the Projection Theorem or the Riesz Representation Theorem, are valid solely for the Hilbert space structures.

Before formulating these results, let us introduce some notation. Let M be a set in an inner product space S. Let us write

$$M^{\perp} = \{x \in S \mid (x, y) = 0 \text{ for all } y \in M\}.$$

The set $M^{\perp}$ obviously becomes a normed closed subspace of S. Let us recall that an *orthonormal basis* of an inner product space is defined to be an orthonormal set the linear span of which is dense in S.

4.1.1. Theorem. *Let S be an inner product space. The following statements are equivalent.*

(i) *S is complete.*

(ii) *If M is a closed subspace of S, then*

$$S = M \oplus M^{\perp}$$

(The Projection Theorem).

(iii) *If M is a closed subspace of S, then $M = M^{\perp\perp} (= (M^{\perp})^{\perp})$.*

(iv) *If M is a proper closed subspace of S, then $M^{\perp} \neq \{0\}$.*

(v) *Every maximal orthonormal system in S is an orthonormal basis.*

(vi) *If f is a continuous linear function on S, then there is an element $y \in S$ such that*

$$f(x) = (x, y)$$

for all $x \in S$ (The Riesz Representation Theorem).

Proof: (i)⇒(ii) This is the well known Projection Theorem.

(ii)⇒(iii) Let M be a (norm) closed subspace of S. Obviously, $M \subset M^{\perp\perp}$. By (ii) we can write, for any $x \in M^{\perp\perp}$,

$$x = x_1 + x_2\,,$$

where $x_1 \in M$ and $x_2 \in M^{\perp}$. Therefore, $x_2 = x - x_1 \in M^{\perp\perp}$ and thus $x_2 = 0$. Hence, $M^{\perp\perp} \subset M$.

(iii)⇒(iv) If M is a closed subspace with $M^{\perp} = \{0\}$, then according to (iii) we have

$$M = M^{\perp\perp} = \{0\}^{\perp} = S\,.$$

(iv)⇒(v) Suppose that (x_α) is a maximal orthonormal system in S. Let M be a closed linear span of (x_α). Then $M^{\perp} = \{0\}$ by maximality of (x_α) and therefore, by (iv), $M = S$. Hence, (x_α) is an orthonormal basis.

(v)⇒ (vi) Consider a continuous nonzero linear form f on S and set $M = \{x \mid f(x) = 0\}$. Take a maximal orthonormal system $B = (x_\alpha)$ in M and extend it to a maximal orthonormal system, B', of S. Since the codimension of M in S is one, we have that either $B' = B$ or $B' = B \cup \{y'\}$. By (v), B' is an orthonormal basis of S, which excludes the former case because f is nonzero. Therefore

$$x = \sum_\alpha (x, x_\alpha)\, x_\alpha + (x, y')\, y' \qquad \text{for all } x \in S\,.$$

Consequently,

$$f(x) = (x, y')\, f(y')\,.$$

By denoting

$$y = \overline{f(y')}\, y'\,,$$

we obtain

$$f(x) = (x, y) \qquad \text{for all } x \in S\,.$$

(vi)⇒(i) Take any nonzero vector y in the completion H of S. Define a bounded linear form f on S by letting

$$f(x) = (x, y) \qquad \text{for all } x \in S\,.$$

If (vi) holds then there is $y' \in S$ such that $f(x) = (x, y')$ for all $x \in S$. Then $(y - y') \perp S$ and therefore $y = y'$ by the density of S in H. □

4.2. Algebraic completeness criteria

We introduce now two important classes of closed subspaces in an inner product space that replace naturally the lattice of (norm) closed subspaces in a Hilbert spaces and can, unlike the merely closed subspaces, carry an algebraic structure with orthocomplementation. Let S be an inner product space. A subspace M of S is called *orthogonally closed* if $M = M^{\perp\perp}$. The set of all orthogonally closed subspaces in S will be denoted by $F(S)$, i.e.

$$F(S) = \{M \text{ is a subspace of } S \mid M = M^{\perp\perp}\}.$$

Since for any subspace M of S we have $M^{\perp} = M^{\perp\perp\perp}$, a natural example of an orthogonally closed subspace is the space $X^{\perp}$, where X is an arbitrary nonempty set in S. Further, a subspace M of S is called the *splitting subspace* if $S = M \oplus M^{\perp}$. The set of all splitting subspaces in S will be denoted by the symbol $E(S)$, i.e.

$$E(S) = \{M \text{ is a subspace of } S \mid S = M \oplus M^{\perp}\}.$$

Natural examples of splitting subspaces are finite-dimensional and cofinite-dimensional subspaces of S. Also, any complete subspace of S has to be splitting. It is clear that

$$E(S) \subset F(S) \subset L(S),$$

where $L(S)$ denotes the set of all norm closed subspaces of S. By Theorem 4.1.1, $F(S)$ does not coincide with $L(S)$ for any incomplete S. Moreover, it turns out that $E(S)$ is strictly smaller then $F(S)$, whenever S is not a Hilbert space. Before proving this, let us reinforce our intuition by presenting a concrete example.

4.2.1. Example. Let H be a separable infinite-dimensional Hilbert space with an orthonormal basis, (e_n). Let S be a subspace linearly generated by the vectors $\{f, e_1, e_2, \dots\}$, where $f = \sum_{n=1}^{\infty} \frac{1}{n} e_n$. Let us set

$$F_1 = \text{sp}\{e_2, e_4, e_6, \dots\} \qquad F_2 = \text{sp}\{e_1, e_3, e_5, \dots\}.$$

We wont to show that $F_2 = F_1^{\perp}$ (in S). Suppose that $x \in F_1^{\perp}$. Expressing x in the linear basis we have

$$x = \alpha f + \lambda_1 e_1 + \lambda_2 e_2 + \dots + \lambda_n e_n .$$

For $2k$ large enough we obtain

$$0 = (x, e_{2k}) = \frac{\alpha}{2k}.$$

Hence, $\alpha = 0$ and so $x \in F_2$. Since, obviously, $F_2 \subset F_1^{\perp}$, we have $F_1^{\perp} = F_2$, and similarly $F_2^{\perp} = F_1$. Thus, $F_1 = F_1^{\perp\perp} \in F(S)$. On the other hand, $S \neq F_1 \oplus F_1^{\perp} = \mathrm{sp}\{e_1, e_2, \ldots\}$. As a result, F_1 is orthogonally closed but not splitting.

4.2.2. Theorem. *An inner product space S is complete if, and only if, $F(S) = E(S)$.*

We shall prove this theorem in a series of auxiliary lemmas.

4.2.3. Lemma. *Let S be a dense subspace in a Hilbert space H. Then $S \cap M$ is dense in M whenever M is a norm closed finite-codimensional subspace of H.*

Proof: Suppose first that $\operatorname{codim} M = 1$. Then $M = \{z\}^{\perp}$ for some nonzero $z \in H$. By the density of S there is $y \in S$ such that $(y, z) \neq 0$. Take arbitrary $x \in M$ and find $h \in S$ such that $\|h - x\| < \varepsilon$. Letting

$$u = h - \frac{(h, z)}{(y, z)}\, y,$$

we find an element of S with $(u, z) = 0$. As

$$|(h, z)| = |(h - x, z)| \leq \|h - x\| \cdot \|z\| \leq \varepsilon \|z\|,$$

we can estimate

$$\|u - x\| \leq \|u - h\| + \|h - x\| < \varepsilon \frac{\|z\| \cdot \|y\|}{|(y, z)|} + \varepsilon.$$

We see that we can make the norm $\|u - x\|$ arbitrarily small and so $M \cap S$ is dense in M. Suppose now that $\operatorname{codim} M = n$. Then we can write

$$H = M \oplus \mathrm{sp}\{z_1, \ldots, z_n\},$$

where $z_1, \ldots, z_n$ is an orthonormal set. By the previous argument

$$S \cap (M \oplus \mathrm{sp}\{z_1, \ldots, z_{n-1}\})$$

is dense in $M \oplus \mathrm{sp}\{z_1, \ldots, z_{n-1}\}$ and using the same reasoning successively we obtain that $S \cap M$ is dense in M. □

4.2.4. Lemma. *Let S be a dense subspace of a Hilbert space H. Then for all orthogonal vectors $x, y \in H$ there is a subspace $M \in F(S)$ such that*

$$x \in \overline{M^\perp \cap S} \qquad \textit{and} \qquad y \in \overline{M}.$$

(Here the closures are taken in H.)

Proof: By Lemma 4.2.3 we can take $x_1 \in S$ such that $\|x - x_1\| < 1$ and $x_1 \in \{y\}^\perp$. Using Lemma 4.2.3 again, we obtain $y_1 \in \{x, x_1\}^\perp \cap S$ such that $\|y - y_1\| < 1$. Then we find $x_2 \in \{y, y_1\}^\perp \cap S$ such that $\|x - x_2\| < 1/2$ and $y_2 \in \{x, x_1, x_2\}^\perp \cap S$ with $\|y - y_2\| < 1/2$. Continuing this way we construct sequences (x_n), (y_n) in S such that $x_n \to x$, $y_n \to y$, and $x_n \perp y_m$, for all m, n. Set $M = (x_n)^\perp \cap S$. Then $M \in F(S)$ and, obviously,

$$(y_n) \subset M, \qquad (x_n) \subset M^\perp \cap S.$$

This completes the proof. □

Proof of Theorem 4.2.2: Let us denote by H the completion of S. Take a unit vector $x \in H$. Then there exists $s \in S$ such that $(s, x) \neq 0$. Put

$$y = x - \frac{s}{(s, x)}.$$

Obviously, $x \perp y$ and $x - y \in S$. By Lemma 4.2.4 there is a subspace $M \in F(S)$ such that $x \in \overline{M}$ and $y \in \overline{M^\perp \cap S}$. By the hypothesis of Theorem 4.2.2, $M \in E(S)$, and so

$$x - y = s_1 + s_2, \tag{4.1}$$

where $s_1 \in M$ and $s_2 \in M^\perp \cap S$. Viewing decomposition (4.1) as the decomposition in H with respect to $\overline{M}$ and $\overline{M^\perp}$ we obtain, by uniqueness, that $x = s_1 \in S$. □

4.2.5. Remark. By studying the proof of Lemma 4.2.4 we see that even a stronger result is in force. Orthogonal vectors in the completion can be separated by specific elements of $F(S)$. Namely, if x and y are orthogonal vectors in a Hilbert space H then, for any dense subspace S of H, there is a sequence (x_n) in S such that $x \in (x_n)^{\perp\perp}$ and $y \in (x_n)^\perp$. Using the Gram-Schmidt orthogonalization process we can choose an element $M \in F(S)$

that is a supremum of an orthogonal sequence of one-dimensional subspaces and such that

$$x \in M \qquad \text{and} \qquad y \in M^{\perp}.$$

Using the above considerations, the proof of Theorem 4.2.2 yields that the inner product S is complete if, and only if, any subspace of S that is a union of countably many orthogonal atoms is splitting. We will introduce, in this connection, a substructure $D(S)$ of $F(S)$ consisting of all subspaces that are suprema (in $F(S)$) of orthogonal one-dimensional subspaces. Members of $D(S)$ are called the *Foulis-Randall* subspaces. The arguments above tell us that the inner product space S is complete if, and only if, $D(S) = E(S)$.

In the sequel we want to classify the structures $F(S)$ and $E(S)$ in the context of ordered sets with orthocomplementations. According to Theorem 4.2.2, $E(S)$ and $F(S)$ are different for all incomplete spaces. It will be seen that also orthocomplementation structures on these sets are substantially different.

4.2.6. Proposition. *Let S be an inner product space. The structure $F(S)$ endowed with $\{0\}$ and S as a least and a greatest elements, the ordering being given by set inclusion, and the orthocomplementation by the assignment $\perp : M \to M^{\perp}$, forms a complete orthocomplemented lattice in which*

$$\bigwedge_{\alpha\in I} M_\alpha = \bigcap_{\alpha\in I} M_\alpha \qquad \text{and} \qquad \bigvee_{\alpha\in I} M_\alpha = \Big(\bigcup_{\alpha\in I} M_\alpha\Big)^{\perp\perp}$$

for any system $(M_\alpha)_{\alpha\in I}$ in $F(S)$. Moreover, if $M \in F(S)$ and $v \in S$, then

$$M \vee \mathrm{sp}\{v\} = M + \mathrm{sp}\{v\}.$$

Proof: It is easy to verify that the operation $\perp : M \to M^{\perp}$ is an orthocomplementation on $F(S)$. As $(\bigcap_{\alpha\in I} M_\alpha)^{\perp} \supset M_\beta^{\perp}$ for each $\beta \in I$, we infer that

$$\Big(\bigcap_{\alpha\in I} M_\alpha\Big)^{\perp\perp} \subset M_\beta^{\perp\perp} = M_\beta$$

and so

$$\Big(\bigcap_{\alpha\in I} M_\alpha\Big)^{\perp\perp} \subset \bigcap_{\alpha\in I} M_\alpha .$$

Hence, $\bigcap_{\alpha\in I} M_\alpha$ is orthogonally closed and thereby it is the infimum $\bigwedge_{\alpha\in I} M_\alpha$. The space $(\bigcup_{\alpha\in I} M_\alpha)^{\perp\perp}$ is an element of $F(S)$ majorizing all M_α's. Suppose that $A \in F(S)$ is an upper bound of $(M_\alpha)_{\alpha\in I}$. Then we have,

$$A \supset \bigcup_{\alpha\in I} M_\alpha$$

and

$$A = A^{\perp\perp} \supset (\bigcup_{\alpha\in I} M_\alpha)^{\perp\perp}.$$

In other words,

$$\left(\bigcup_{\alpha\in I} M_\alpha\right)^{\perp\perp} = \bigvee_{\alpha\in I} M_\alpha .$$

Consider now $M \in F(S)$ and $v \in S$. We shall assume that $v \notin M\ (= M^{\perp\perp})$. Hence we can find $u \in M^\perp$ such that $(u, v) = 1$. If $y \in M^\perp$ then

$$y - (y, v)\, u \in (M + \mathrm{sp}\{v\})^\perp ,$$

as it can be directly verified. For any $x \in (M + \mathrm{sp}\{v\})^{\perp\perp}$ and $y \in M^\perp$ we have

$$\begin{aligned} (x - (x, u)\, v, y) &= (x, y) - (x, u)\, (v, y) \\ &= (x, y - (y, v)\, u) = 0 . \end{aligned}$$

It implies that,

$$x - (x, u)\, v \in M^{\perp\perp} = M ,$$

which gives $x \in M + \mathrm{sp}\{v\}$. Therefore $M + \mathrm{sp}\{v\} = M \vee \mathrm{sp}\{v\}$.

4.2.7. Proposition. *The structure $E(S)$ endowed with a least element, $\{0\}$, and a greatest element, S, the ordering being given by set inclusion, and the orthocomplementation by the assignment $\perp : M \to M^\perp$, forms an orthomodular poset. Moreover, if $M \in E(S)$ and $v \in S$ then $M \vee \mathrm{sp}\{v\}$ exists in $E(S)$ and*

$$M \vee \mathrm{sp}\{v\} = M + \mathrm{sp}\{v\} .$$

Proof: It follows from a simple linear algebra reasoning that for all $A, B \in E(S)$ with $A \subset B$ we have

$$B = A \oplus (B \wedge A^\perp) = A \vee (B \wedge A^\perp) .$$

Thus, the orthomodular law is satisfied. It is also straightforward to see that $E(S)$ is closed under sums of finitely many orthogonal subspaces. We conclude that the orthocomplementation makes $E(S)$ an orthomodular poset.

In order to prove the second statement it is enough to show that $M + \mathrm{sp}\{v\}$ is a splitting subspace whenever $M \in E(S)$. For this, write $v = v_1 + v_2$, where $v_1 \in M$ and $v_2 \in M^\perp$. Then

$$M + \mathrm{sp}\{v\} = M \oplus \mathrm{sp}\{v_2\} \in E(S)\,. \qquad \square$$

It will be clear later that the structure $E(S)$ need not be a complete lattice. However, if the supremum or infimum exits in $E(S)$, then it is obtained in the same way as in $F(S)$.

4.2.8. Proposition. *Suppose that $(M_\alpha)_{\alpha\in I}$ is a system of elements in $E(S)$. The following statements hold true:*

(i) *If $\bigwedge_{\alpha\in I} M_\alpha$ exists in $E(S)$, then*

$$\bigwedge_{\alpha\in I} M_\alpha = \bigcap_{\alpha\in I} M_\alpha\,.$$

(ii) *If $\bigvee_{\alpha\in I} M_\alpha$ exists in $E(S)$, then*

$$\bigvee_{\alpha\in I} M_\alpha = (\bigcup_{\alpha\in I} M_\alpha)^{\perp\perp}\,.$$

Proof: Suppose that $\bigwedge_{\alpha\in I} M_\alpha$ exists in $E(S)$ and let $M = \bigwedge_{\alpha\in I} M_\alpha$. Obviously, $M \subset \bigcap_{\alpha\in I} M_\alpha$. Since $M + \mathrm{sp}\{v\} \in E(S)$ for any $v \in \bigcap_{\alpha\in I} M_\alpha$ (see Proposition 4.2.7), $M + \mathrm{sp}\{v\}$ and M have to coincide. In other words,

$$\bigwedge_{\alpha\in I} M_\alpha = \bigcap_{\alpha\in I} M_\alpha\,.$$

(ii) Suppose that $\bigvee_{\alpha\in I} M_\alpha$ exists in $E(S)$. Using the obvious equality

$$\bigcap_{\alpha\in I} M_\alpha^\perp = (\ \mathrm{sp}\,(\bigcup_{\alpha\in I} M_\alpha))^\perp\,,$$

and the De Morgan's laws, we obtain from (i)

$$\begin{aligned} \bigvee_{\alpha\in I} M_\alpha &= (\bigwedge_{\alpha\in I} M_\alpha^\perp)^\perp = (\bigcap_{\alpha\in I} M_\alpha^\perp)^\perp \\ &= (\mathrm{sp}(\bigcup_{\alpha\in I} M_\alpha))^{\perp\perp} = (\bigcup_{\alpha\in I} M_\alpha)^{\perp\perp}. \end{aligned}$$ □

Now we pass to the algebraic completeness criterion, often called the Amemiya-Araki Theorem, which says that the orthomodular law is intimately connected with completeness of the respective inner product structure.

4.2.9. Theorem (Amemiya-Araki). *An inner product space S is complete if, and only if, $F(S)$ is an orthomodular lattice.*

Proof: Suppose that $F(S)$ is orthomodular. By Theorem 4.2.2, to prove the completeness of S it is enough to show that $F(S) = E(S)$. For this aim take $M \in F(S)$, $M \neq S$. We want to prove that $S = M \oplus M^\perp$. Choose $x \in S$ such that $x \notin M$. By the orthomodular law we infer that

$$(\,(\mathrm{sp}\{x\} \vee M\,) \wedge M^\perp) \vee M = \mathrm{sp}\{x\} \vee M = \mathrm{sp}\{x\} + M\,.$$

From this equality it is clear that $\dim((\mathrm{sp}\{x\} \vee M) \wedge M^\perp) \leq 1$ and therefore we can write $M + \mathrm{sp}\{x\} = M + \mathrm{sp}\{y\}$ for some $y \in M^\perp$. In other words, $x = \alpha\, y + m$, $\alpha \in \mathbb{C}$, $m \in M$ and so $x \in M \oplus M^\perp$. This means that M is splitting and the proof is completed. □

4.2.10. Remark. By inspecting the proof of Amemiya-Araki Theorem we can observe that the key condition for completeness of S is that the spaces $M + \mathrm{sp}\{x\}$ and M obey the orthomodular law. More precisely, an inner product space S is complete if, and only if,

$$M + \mathrm{sp}\{x\} = ((\mathrm{sp}\{x\} \vee M) \wedge M^\perp) \vee M\,, \tag{4.2}$$

whenever $M \in F(S)$ and $x \in S$. This property is called the *weak atomic orthomodularity* and it is, in the context of orthomodular structures, strictly stronger then orthomodularity. In view of Remark 4.2.5 we can even say that S is complete if, and only if, the condition (4.2) holds for all subspaces $M \in F(S)$ which are suprema of orthogonal sequences of one-dimensional subspaces.

Our discussion shows that the algebraic properties of subspace structures of S determine the functional-analytic properties of S. If S is an incomplete inner product space, then $F(S)$ is not orthomodular, but it is a complete lattice. On the other hand, $E(S)$ is always an orthomodular poset which, however, may be far from being complete. This is demonstrated by the following theorem which we have already proved on our way.

4.2.11. Theorem. *Let S be an inner product space. Then S is complete if, and only if, any sequence of mutually orthogonal atoms in $E(S)$ has a supremum in $E(S)$.*

Proof: It follows from Remark 4.2.5 and Proposition 4.2.8. □

4.3. Measure theoretic completeness criteria

The lattice, $L(H)$, of closed subspaces of a Hilbert space H which can be identified with the projection lattice, $P(H)$, of the von Neumann algebra, $B(H)$, of all bounded operators acting on H, has an abundance of completely additive measures. Surprisingly, if we remove the assumption of completeness and replace $L(H)$ by the structure of orthogonally closed or splitting subspaces, the situation changes dramatically. In fact, if S is an incomplete inner product space, then the structures $F(S)$ and $E(S)$ have no nonzero completely additive measure. As a result, completely additive measure and probability structures collapse to trivial ones if we do not work on a Hilbert space. The central result of this section is the following theorem:

4.3.1. Theorem. *An inner product space S is complete if, and only if, there is a nonzero completely additive measure on $F(S)$.*

We will prove this theorem in a series of statements. Some of them may have an independent bearing. First, let us recall that $\mu\colon F(S) \to \mathbb{C}$ is a completely additive measure (not necessarily bounded) if

$$\mu\left(\bigvee_{\alpha} M_{\alpha}\right) = \sum_{\alpha} \mu(M_{\alpha}),$$

whenever (M_α) is a collection of pairwise orthogonal elements of $F(S)$. By considering real and imaginary parts of μ, we can easily reduce the problem to a real measure. Suppose therefore that μ is real. In a standard way we associate to μ the frame type function $f_\mu\colon S_1 \to \mathbb{R}$ by letting

$$f_\mu(x) = \mu(\mathrm{sp}\{x\}) \qquad \text{for all } x \in S_1 .$$

(In case of μ being finitely additive the function f_μ is a weak frame function.) Since for any maximal orthonormal system (x_α) in S we have $S = \bigvee_\alpha \mathrm{sp}\{x_\alpha\}$, the complete additivity of μ ensures that f_μ is nonzero whenever μ is nonzero. By the Dorofeev-Sherstnev result, Theorem 3.3.5, we also know that f_μ is bounded if $\dim S = \infty$.

The following simple but useful observation, formulated in a more general context than it is needed here, tells us that the orthomodular law is a direct consequence of the richness of measure structure.

4.3.2. Proposition. *Let P be an orthoposet. Then for any finitely additive measure μ on P we have*

$$\mu(a \vee (b \wedge a^\perp)) = \mu(b) \tag{4.3}$$

for all $a, b \in P$ such that $a \le b$.

In particular, if for all $a, b \in P$ with $a < b$ there is a finitely additive measure μ on P satisfying $\mu(a) \neq \mu(b)$, then P is an orthomodular poset.

Proof: Take arbitrary elements $a, b \in P$ with $a \le b$. Suppose that μ is a finitely additive measure on P. By applying the De Morgan's laws we can compute

$$\begin{aligned} \mu(a \vee (b \wedge a^\perp)) &= \mu(a) + \mu(b \wedge a^\perp) = \mu(a) + \mu((b^\perp \vee a)^\perp) \\ &= \mu(a) + \mu(1) - \mu(b^\perp) - \mu(a) \\ &= \mu(a) + \mu(b) - \mu(a) = \mu(b) . \end{aligned}$$

Second statement follows immediately from (4.3) and the fact that $a \vee (b \wedge a^\perp) \le b$, whenever $a \le b$. □

4.3.3. Remark. In view of Remark 4.2.10 let us observe that the lattice $F(S)$ is orthomodular if, and only if, for each $M \in F(S)$ and $v \notin M$, there is a finitely additive measure μ on $F(S)$ separating subspaces M and $M + \text{sp}\{v\}$, i.e. satisfying

$$\mu(M + \text{sp}\{v\}) \neq \mu(M)\,. \tag{4.4}$$

Moreover, one can suppose only that M is a supremum of an orthogonal sequence of one-dimensional subspaces. Indeed, suppose that this statement holds. By Remark 4.2.10, $F(S)$ is orthomodular if $X = Y$, whenever $X = M + \text{sp}\{x\}$, and $Y = M \vee ((M + \text{sp}\{x\}) \wedge M^{\perp})$, where M is a supremum of a sequence of orthogonal atoms. By Proposition 4.3.2

$$\mu(X) = \mu(Y) \tag{4.5}$$

for any finitely additive measure μ on $F(S)$. It is clear that $Y \leq X$ and $\dim X/Y \leq 1$. If X and Y were different, then

$$X = M + \text{sp}\{x\}\,, \qquad Y = M\,,$$

and $x \notin M$. But this would lead to a contradiction of (4.5) and (4.4).

Given a unitary map U on an inner product space S and a finitely additive measure μ of $F(S)$, we shall often make use of the unitary transform μ_U of μ by defining

$$\mu_U(M) = \mu(U^{-1}(M))\,, \qquad M \in F(S)\,.$$

The following technical lemma is a key ingredient of the proof of Theorem 4.3.1.

4.3.4. Lemma. *Let v be a unit vector in the completion H of an inner product space S. Suppose that μ is a finitely additive measure on $F(S)$ for which the corresponding weak frame function f_μ is bounded. Then for every $\varepsilon > 0$ there exists $\delta > 0$ such that the following statement holds. If $w \in S$ is a unit vector such that $\|v - w\| < \delta$, then for each $A \in F(S)$ satisfying the properties $A \subset \{v\}^{\perp}$, $2 \leq \dim A < \infty$, we have the inequality*

$$|\mu_U(A \vee \text{sp}\{w\}) - \mu_U(A) - \mu_U(\text{sp}\{w\})| < \varepsilon$$

for any unitary U on S.

Proof: Let $\varepsilon > 0$ be given. The weak frame function f_μ corresponding to μ is bounded by the assumption and so it is uniformly continuous on the unit sphere S_1 of S (see Remark 3.2.7). For any unitary transform $U\colon S \to S$, the frame function f_{μ_U} is given by

$$f_{\mu_U}(x) = f(U^{-1}x), \qquad x \in S_1 .$$

Clearly, f_{μ_U} is also uniformly continuous with the same modulus of uniform continuity as f_μ. In other words, we can find $\delta_1 > 0$ such that for any unitary U

$$|f_{\mu_U}(x) - f_{\mu_U}(y)| < \varepsilon , \tag{4.6}$$

whenever $\|x - y\| < \delta_1$. By the continuity of the function

$$\varrho(t) = \sqrt{2 - 2\sqrt{1 - t^2}}$$

there exists $\delta \in (0,1)$ such that $\varrho(t) < \delta_1$ for any $t \in [0, \delta]$. Now if $\|v - w\| < \delta$, then

$$\|P_{\mathrm{sp}\{v\}^\perp}(w)\| = \|P_{\mathrm{sp}\{v\}^\perp}(v - w)\| \leq \|v - w\| < \delta .$$

Fix now a unit vector $w \in S$ with $\|w - v\| < \delta$ and assume that A is a finite-dimensional subspace of S fulfilling the assumptions of the theorem. Then, for orthogonal projection P_A of H onto A, we have

$$\|P_A(w)\| = \|P_A P_{\mathrm{sp}\{v\}^\perp}(w)\| \leq \|P_{\mathrm{sp}\{v\}^\perp}(w)\| < \delta.$$

We thus obtain

$$\left| \frac{(1 - P_A)w}{\|(1 - P_A)w\|} - w \right| = \varrho(\|P_A(w)\|) < \delta_1 .$$

Put

$$w' = \frac{(1 - P_A)w}{\|(1 - P_A)w\|} .$$

Then we have $\|w - w'\| < \delta_1$. It is also clear that $A \vee \mathrm{sp}\{w\} = A \vee \mathrm{sp}\{w'\}$ and $\mu_U(A \vee \mathrm{sp}\{w\}) = \mu_U(A) + \mu_U(\mathrm{sp}\{w'\})$. Making use of the latter observation and (4.6), we obtain

$$\begin{aligned} & |\mu_U(A \vee \mathrm{sp}\{w\}) - \mu_U(A) - \mu_U(\mathrm{sp}\{w\})| \\ = \; & |\mu_U(A) + \mu_U(\mathrm{sp}\{w'\}) - \mu_U(A) - \mu_U(\mathrm{sp}\{w\})| \\ = \; & |\mu_U(\mathrm{sp}\{w'\}) - \mu_U(\mathrm{sp}\{w\})| = |f_{\mu_U}(w') - f_{\mu_U}(w)| < \varepsilon . \; \square \end{aligned}$$

In the following Proposition we collect basic properties of completely additive measures on subspace structures.

4.3.5. Proposition. *Let μ be a completely additive measure on $F(S)$. The following statements hold true:*

(i) *Let (x_α) be a maximal orthonormal system in $M \in F(S)$. Then*

$$\mu(M) = \sum_\alpha \mu(\mathrm{sp}\{x_\alpha\}).$$

(ii) *(Regularity condition) For each $M \in F(S)$*

$$\lim_{n\to\infty} \mu(A_n) = \mu(M),$$

whenever A_n is an increasing sequence of finite dimensional subspaces with $\bigvee_{n=1}^{\infty} A_n = M$.

Proof: (i) Put $N = \{x_\alpha\}^{\perp\perp}$. Then $M \wedge N^\perp = \{0\}$ because of the maximality of (x_α) and therefore

$$\mu(M) = \mu(N \vee (M \wedge N^\perp)) = \mu(N).$$

(See Proposition 4.3.2). Hence, by the complete additivity of μ we see that

$$\mu(M) = \sum_\alpha \mu(\mathrm{sp}\{x_\alpha\}).$$

(ii) Making use of the standard orthogonalization process, we see that there is an orthonormal sequence (e_n) such that

$$\bigvee_{n=1}^{\infty} A_n = \bigvee_{n=1}^{\infty} \mathrm{sp}\{e_n\},$$

and each A_n has an orthonormal basis that is a finite subsequence of (e_n). The complete additivity of μ now gives us the equality

$$\mu(M) = \mu\Big(\bigvee_{n=1}^{\infty} A_n\Big) = \mu\Big(\bigvee_{n=1}^{\infty} \mathrm{sp}\{e_n\}\Big) = \sum_{n=1}^{\infty} \mu(\{\mathrm{sp}(e_n\}) = \lim_{n\to\infty} \mu(A_n).$$

□

4.3.6. Proposition. *Let μ be a finitely additive measure on $F(S)$. For all unit vectors $x, y \in S$ there is a transformed measure μ_U such that*

$$\mu_U(\mathrm{sp}\{y\}) = \mu(\mathrm{sp}\{x\}).$$

Proof: We will find a unitary transform $U\colon S \to S$ with $Ux = y$. This can be easily done by taking any unitary operator U on $\mathrm{sp}\{x, y\}$ with $Ux = y$ and extending it then by letting $Uz = z$ for all $z \in \mathrm{sp}\{x, y\}^{\perp}$. Then

$$\mu_U(\mathrm{sp}\{y\}) = \mu(U^{-1}(\mathrm{sp}\{y\})) = \mu(\mathrm{sp}\{x\}). \qquad \square$$

Finally, we are in the position to prove the main result.

Proof of Theorem 4.3.1: Suppose that $F(S)$ admits a nonzero completely additive measure μ. Assume now that $A, B \in F(S)$ with $\dim B/A = 1$ and A is a supremum of countably many orthogonal atoms. The proof will be completed when we construct a measure separating A and B (see Remark 4.3.3). As μ is completely additive it attains a nonzero value, say c, on some one-dimensional subspace. Since the completion of B contains a unit vector orthogonal to A (A is not dense in B), we can apply Lemma 4.3.4 and find a unit vector $w \in B \setminus A$ such that

$$|\mu_U(C + \mathrm{sp}\{w\}) - \mu_U(C) - \mu_U(\mathrm{sp}\{w\})| < \frac{|c|}{2}, \tag{4.7}$$

for all subspaces $C \subset A$ with $2 \leq \dim C < \infty$, and all unitaries U. According to Proposition 4.3.6 we can specify U to satisfy

$$\mu_U(\mathrm{sp}\{w\}) = c.$$

Since A is a countable union of atoms, we can find, by using the Gram-Schmidt orthogonalization process, an increasing sequence (A_n) of finite-dimensional subspaces, such that $\mu(A_n) \to \mu(A)$ and $\mu(A_n + \mathrm{sp}\{w\}) \to \mu(B)$. On substituting A_n for C in (4.7) and making then the limit $n \to \infty$, we obtain

$$|\mu(B) - \mu(A) - c| < \frac{|c|}{2}.$$

Then, obviously, $\mu(A) \neq \mu(B)$, for otherwise $0 < |c| < \frac{|c|}{2}$, which is absurd. The proof is completed. $\square$

As it is apparent from the proof of Theorem 4.3.1 various weakenings of the complete additivity are possible. For example, it is clear that an inner product space S is complete if and only if $F(S)$ admits a σ-additive measure that is nonzero on some atom in $F(S)$.

The previous proof of measure theoretic completeness criterion does not require to call upon the full strength of Gleason Theorem. We have only used the fact that bounded frame function is uniformly continuous, which is much easier to prove than Gleason Theorem. If we use Gleason Theorem in its full power, we obtain a shorter alternative proof based on purely functional-analytic arguments. Observe for a further use that the following theorem is also a quantitative expression of the fact that an inner product space is complete exactly when any maximal orthonormal system is an orthonormal basis.

4.3.7. Theorem. *An inner product space S is complete if, and only if, there is a unit vector x in its completion H and a positive number $c > 0$ such that*

$$0 < c \leq \sum_{\alpha \in I} |(x, x_\alpha)|^2 , \tag{4.8}$$

for any maximal orthonormal system $(x_\alpha)_{\alpha \in I}$ in S.

Proof: If S is complete, then for any maximal orthonormal system (x_α) and for any unit vector $x \in S$ we have

$$1 = \sum_{\alpha \in I} |(x, x_\alpha)|^2 ,$$

since any maximal orthonormal system is an orthonormal basis.

Suppose now that S is an inner product space admitting a vector $x \in H$ fulfilling (4.8). First we prove the statement under the condition that $x \in S$. Notice that the inequality (4.8) holds for any unit vector $y \in S$. For this, we construct a unitary map $U : S \to S$ with $Ux = y$ (see e.g. the proof of Proposition 4.3.6). Then

$$\sum_{\alpha \in I} |(y, x_\alpha)|^2 = \sum_{\alpha \in I} |(Ux, x_\alpha)|^2 = \sum_{\alpha \in I} |(x, U^{-1} x_\alpha)|^2 \geq c ,$$

because $(U^{-1} x_\alpha)$ is also a maximal orthonormal system. Having to reach a contradiction assume that S is not complete. According to Theorem 4.1.1

there is a maximal orthonormal system $(x_\alpha)_{\alpha\in I}$ of S that is not an orthonormal basis. Define a linear map

$$F : H \to l^2(I) \ : \ z \to ((z, x_\alpha))_\alpha .$$

Then $\|F(z)\|^2 = \sum_\alpha |(z, x_\alpha)|^2 \leq \|z\|^2$, meaning that F is norm decreasing. Since (x_α) is not an orthonormal basis, there is a unit vector $z \in H$ with $(z, x_\alpha) = 0$ for all $\alpha \in I$. In other words, $\|F(z)\|_{l^2} = 0$. Let (x_n) be a sequence of unit vectors in S converging to z. Then

$$0 = \|F(z)\|_{l^2}^2 = \lim_n \|F(x_n)\|_{l^2}^2 \geq c > 0 ,$$

which is a contradiction.

In case of a general $x \in H$ satisfying (4.8), we can approximate x by an element $y \in S$ such that, using continuity of $\|F(\cdot)\|$, we have

$$\|F(x)\|_{l^2}^2 \geq \frac{c}{2} > 0 .$$

We then complete the proof by the arguments used above. □

Recall that the frame-type function on an inner product space S is a function $f\colon S_1 \to \mathbb{R}$ that is a weak frame function and, moreover, the sum $\sum_\alpha f(x_\alpha)$ is convergent for each orthonormal set (x_α) in S (see section 3.3). In view of the results of Chapter 3 (Theorem 3.1.6) we know that, in case of $\dim S = \infty$, f is represented by a self-adjoint trace class operator T on the completion H of S in the standard way:

$$f(x) = (Tx, x) \qquad \text{for all } x \in S_1 .$$

4.3.8. Theorem. *An inner product space S is complete if, and only if, there is a frame-type function $f\colon S_1 \to \mathbb{R}$ such that there is a constant $c > 0$ for which*

$$\sum_\alpha |f(x_\alpha)| \geq c$$

for each maximal orthonormal system (x_α) in S.

Proof: We may and shall suppose that $\dim S = \infty$. Let T be a self-adjoint trace class operator on the completion H of S representing f. Then $T = T_+ - T_-$, where T_+ and T_- are the negative and the positive parts of

T, respectively. Then $|T| = T_+ + T_-$ is a nonnegative trace class operator. Moreover, for each maximal orthonormal system (x_α), we have

$$0 < c \le \sum_\alpha |f(x_\alpha)| \le \sum_\alpha (T^+ x_\alpha, x_\alpha) + (T^- x_\alpha, x_\alpha) = \sum_\alpha (|T|\, x_\alpha, x_\alpha) .$$

Let us represent $|T|$ in the canonical form

$$|T| = \sum_n \lambda_n P_n ,$$

where $\lambda_n \ge 0$ and P_n is a projection onto one-dimensional subspace $\mathrm{sp}\{e_n\}$, where (e_n) is an orthonormal sequence consisting of eigenvectors of $|T|$. Then, for any maximal orthonormal system (x_α) in S, we have

$$\begin{aligned} c &\le \sum_\alpha (|T|\, x_\alpha, x_\alpha) = \sum_n \lambda_n \sum_\alpha (P_n x_\alpha, x_\alpha) \\ &= \sum_n \lambda_n \sum_\alpha |(x_\alpha, e_n)|^2 . \end{aligned}$$

For at least one n, say n_0, we must have

$$\sum_\alpha |(x_\alpha, e_{n_0})|^2 > \frac{c/2}{\sum_n \lambda_n} ,$$

for otherwise

$$\sum_\alpha (|T|\, x_\alpha, x_\alpha) \le \sum_n \lambda_n \frac{c/2}{(\sum_n \lambda_n)} = c/2 ,$$

which is a contradiction. So

$$\sum_\alpha |(e_{n_0}, x_\alpha)|^2 \ge \frac{c/2}{\sum_n \lambda_n} > 0$$

for any maximal orthonormal systems (x_α) of S, and, by Theorem 4.3.7, S is complete. □

The previous theorem excludes the existence of nonzero frame functions on incomplete spaces.

4.3.9. Theorem. *An inner product space S is complete if, and only if, there is a nonzero frame function on S.*

Proof: If f is a nonzero frame function on S with a nonzero weight w, then for any maximal orthonormal system (x_α) in S we obtain

$$0 < |w| = \left|\sum_\alpha f(x_\alpha)\right| \leq \sum_\alpha |f(x_\alpha)|,$$

and one can apply Theorem 4.3.8. Suppose now that $w = 0$. Since there is a unit vector $x \in S$ with $f(x) \neq 0$, the restriction of f to $S \cap \{x\}^\perp$ gives the frame function with a nonzero weight equal to $-f(x)$. By the previous reasoning $S \cap \{x\}^\perp$ is complete and so is S. □

4.3.10. Corollary. *An inner product space S is complete if, and only if, $E(S)$ possesses at least one nonzero completely additive measure.*

Proof: Suppose that μ is a nonzero completely additive measure on $E(S)$. Let $f(x) = \mu(\mathrm{sp}\{x\})$, $x \in S_1$. Consider maximal orthonormal system $(x_\alpha)_{\alpha\in I}$ in S. Then $(x_\alpha)^{\perp\perp} = S \in E(S)$, and hence $S = \bigvee_{\alpha\in I} \mathrm{sp}\{x_\alpha\}$. Thus, $\mu(S) = \sum_{\alpha\in I} \mu(\mathrm{sp}\{x_\alpha\}) = \sum_{\alpha\in I} f(x_\alpha)$, and therefore S has to be complete by Theorem 4.3.9. □

4.3.11. Corollary. *Let μ be a finitely additive probability measure on $F(S)$. Suppose that there is $\varepsilon > 0$ such that for any maximal orthonormal system $(x_\alpha)_{\alpha\in I}$ of S there is a finite-dimensional subspace K of S generated by some members of (x_α) with*

$$\mu(K) > \varepsilon .$$

Then S is complete.

Proof: Let f be a weak frame function induced by μ. Then $\sum_\alpha f(x_\alpha) < \infty$ for each maximal orthonormal system (x_α) because any finite subsum is less then one. It is therefore apparent that f is a frame-type function satisfying assumptions of Theorem 4.3.8 with $c = \varepsilon$. Hence, S is complete, which completes the proof. □

The previous results tell us that the completely additive measure theories on $F(S)$ and $E(S)$ are either virtually empty or are equivalent to the theory of normal states on Type I factors. However, it is worth noting that finitely additive theories on these structures are quite different. In case of physically relevant 0-1 states it can be illustrated by the following result:

4.3.12. Proposition.
(i) *If* $\dim S \geq 3$, *then* $F(S)$ *admits no 0-1 state.*
(ii) *There is an inner product space* S *such that* $E(S)$ *admits a 0-1 state.*

Proof: (i) Suppose that $\dim S = \infty$. The nonexistence of a 0-1 state is based on the possibility to embed any finite-dimensional structure into $F(S)$. Let us organize a maximal orthonormal system B of S as a union of disjoint three-elements sets, $B = \bigcup_\beta B_\beta$. Let H_3 be a three-dimensional Hilbert space and $U_\beta\colon H_3 \to \mathrm{sp}\,\mathrm{B}_\beta$ be a unitary map of H_3 onto $\mathrm{sp}\,B_\beta$. For each subspace $M \subset H_3$ set

$$\tilde{U}(M) = \bigvee_\beta U_\beta(M)\,.$$

Then $\tilde{U}$ is an embedding of $L(H_3)$ into $F(S)$ that preserves the unit. Hence, any 0-1 state on $F(S)$ would induce a 0-1 state on $L(H_3)$, which is absurd by Theorem 3.4.1.

(ii) Let S be a dense linear hyperplane in a Hilbert space H. We show that for any $M \in E(S)$ either M is complete or $M^\perp$ is complete. To this end, $S = M \oplus M^\perp$ and passing to the completion $H = \overline{M} \oplus \overline{M^\perp}$. (The closures are taken in H.) Since $\dim H/S = 1$, we see that either $\dim \overline{M}/M = 0$, in which case M is complete, or $\dim \overline{M^\perp}/M^\perp = 0$, in which case $M^\perp$ is complete. Now, a 0-1 measure μ on $F(S)$ can be naturally defined by setting

$$\mu(M) = \begin{cases} 0 & \text{if } M \text{ is complete} \\ 1 & \text{if } M \text{ is not complete.} \end{cases}$$

□

Let us remark that it is difficult to find a nontrivial finitely additive measure on the structure $F(S)$ if S is incomplete. It was shown only recently by Chetcuti and Dvurečenskij [77] that such a measure may exist (see Section 4.4. for more details). On the other hand, there are many natural examples of nonzero finitely additive measures on $E(S)$. In fact, for any unit vector $x \in S$, the map $m_x \colon E(S) \to [0,1]$, defined by $m_x(M) = \|x_M\|^2$, where x_M denotes the orthogonal projection of x onto M, is a finitely additive probability measure on $E(S)$. Of course, none of these measures could be completely additive if S is incomplete.

4.4. Historical remarks and comments

The structure of the Hilbert space is one of the most important structures in mathematical and numerical analysis. For this reason many characterizations of Hilbert spaces have appeared not only in Banach space theory and the theory of topological linear spaces. Functional-analytic criteria characterizing Hilbert spaces among inner product spaces presented in Section 4.1 are due to Gudder and Gudder and Holland [150, 151, 152].

Subspace structures of inner product spaces are vital for the axiomatization program in quantum mechanics and have been intensely investigated since the advent of mathematical foundations of quantum theory. Unlike the closed subspaces, the orthogonally closed and splitting subspaces can be singled out by transparent algebraic conditions that reflect basic physical axioms. This makes it relevant to projective geometries describing the logic of quantum propositions, which was a motivation for many authors.

In 1967 Amemiya and Araki, inspired by the paper of Piron [262], proved in [22] that an inner product space is complete if and only if the structure of orthogonally closed subspaces is an orthomodular poset. This result, called Amemiya-Araki Theorem, has been a signal for further studying the interplay between algebraic and metric structure of the Hilbert space and it is one of the main tools in our exposition. The observation that only weak orthomodularity of $E(S)$ is sufficient for completeness is due to Pták and Weber [269]. Theorem 4.2.11 characterizing Hilbert spaces as spaces having σ-complete poset of splitting subspaces was obtained by Dvurečenskij [113], which completed the previous results in [73, 148]. A natural question arising in this connection was a question of whether $E(S)$ could be a lattice for S incomplete. This problem was answered by a careful analysis of lattice properties carried out by Pták and Weber in [271]. They first showed making use the technique of Keller [148] that the lattice structure of splitting spaces does not determine the metric structure of S. Indeed, $E(S)$ is not a lattice for any inner product space S with countable linear dimension. On the other hand, Pták and Weber proved that for any separable Hilbert space of infinite dimension there is a dense hyperplane S such that S contains no infinite-dimensional complete subspace [271]. In the latter case $E(S)$ reduces to the structure of subspaces that are either finite-dimensional or are orthogonal complements of finite-dimensional spaces, implying that $E(S)$ is even a modular lattice. This also implies the existence of 0-1 measure on $E(S)$. Although the property of $E(S)$ being a lattice does therefore not seem to have an explicit bearing on (topological) completeness, it was shown by Dvurečenskij that S is complete if, and only if, $E(S)$ has the *σ-Riesz property* [124]. (We say that a poset L satisfies the σ-Riesz property if for all sequences $(x_n), (y_n) \subset L$ with $x_n \leq y_n$ for all n, there is an element $x \in L$ such that $x_n \leq x \leq y_n$ for all n.) Except orthogonally closed and splitting subspaces many algebraic completeness criteria were obtained in terms of various distinguished subspace families [69, 72, 117, 119]. For example, it has been proved in [117] that an inner product space S is complete if, and only if, the structure of Foulis-Randall subspaces forms an orthomodular poset. The relevance of this result for test spaces in axiomatics of quantum mechanics is shown in [117]. Results on algebraic structures on Hilbert spaces over skew fields may be found in [119].

First measure-theoretic completeness criterion was obtained by Hamhalter and Pták in [157], where it was shown that a separable inner product space S is complete if and only if $F(S)$ possesses at least one nonzero σ-additive probability measure. This paper was a beginning of an intense investigation of inner product structures. The methods of [157] relays on uniform continuity of frame function and are fundamental in this exposition too. The work [157] has inspired many authors who, following the methods of [157], extended measure completeness criterion to general measures on various classes of subspaces in an inner product space. Credits extend to Dvurečenskij, de Lucia, Pulmannová, Cannetti, Marino, Mišík, Jr., Buhagiar, Chetcuti, and others (see original papers [69, 109, 111, 112, 113, 114, 115, 123, 125], and the monograph [116] which summarizes the development till early 90's). A new and quite elegant method of proving completeness

criteria, based on a direct application of Gleason Theorem that is presented in concluding part of Section 4.3, was discovered by Dvurečenskij [118].
It should be noted that several questions remain open, notably the one of whether every incomplete space S has nonzero measure on $F(S)$ [157, 267] and [116]. Very recent result of Chetcuti and Dvurečenskij [78] has shown that there is an inner product space admitting a nonzero measure on orthogonally closed subspaces. It has solved a long standing problem in this area. We shall outline their construction of nontrivial measure. Based on ideas of Ptak and Weber [271] it can be shown that for every separable Hilbert space H there is an incomplete dense hyperplane S in H such that $A \cap S \neq \{0\}$ whenever A is a closed infinite-dimensional subspace of H. Let φ be state on $P(H)$ vanishing on finite-dimensional subspaces. (Such states correspond to states on Calkin algebra $B(H)/K(H)$ – see Chapter 1.) Let us set $\mu(M) = \varphi(\overline{M})$ for each $M \in F(S)$. (Here the closure is taken in H.) We shall prove that μ is a state on $F(S)$. For this let us observe that for each $M \in F(S)$ with maximal orthonormal system (e_n)

$$\varphi\left(\overline{\bigvee_{n=1}^{\infty} \mathrm{sp}\{e_n\}}\right) = \varphi(\overline{M}).$$

To verify it let us consider the orthogonal decomposition (in H)

$$\overline{M} = F \oplus (F^{\perp} \cap M),$$

where $F = \overline{\bigvee_{n=1}^{\infty} \mathrm{sp}\{e_n\}}$. By the property of the hyperplane S, $\dim F^{\perp} \cap \overline{M} < \infty$ for otherwise we can find a nonzero vector in $F^{\perp} \cap \overline{M} \cap S = F^{\perp} \cap M$, contradicting the maximality of (e_n). As φ vanishes on finite-dimensional subspaces, one concludes that $\varphi(\overline{M}) = \varphi(\overline{\vee_{n=1}^{\infty} \mathrm{sp}\{e_n\}})$, and hence $\mu(M) = \mu(\vee_{n=1}^{\infty} \mathrm{sp}\{e_n\})$. Let now A and B be orthogonal subspaces in $F(S)$. Then, for orthonormal bases (e_n) and (f_n) of A and B, respectively, the union $(e_n) \cup (f_n)$ is a maximal orthonormal system in $A \vee B$. Hence,

$$\mu(A \vee B) = \varphi(\overline{\mathrm{sp}\,(e_n) \cup (f_n)}) = \varphi(\overline{\mathrm{sp}\{e_n\}}) + \varphi(\overline{\mathrm{sp}\{f_n\}}) = \mu(A) + \mu(B).$$

In summary, μ is a finitely additive state on $F(S)$.

On the other hand, by the recent result of Chetcuti and Dvurečenskij [78] an inner product space S is complete if, and only if, $F(S)$ possesses a state that is nonzero at some atom in $F(S)$. (Corollary 4.3.11 is a consequence of it.)

The relevance of classes of orthogonally closed spaces and measures on them is stressed by the fact that they represent subspace structures with orthocomplementation in a very general context. The following result has been proved by Willbur in [321]. Let X be an infinite-dimensional complex locally convex space whose Mackey topology is not weak. If the lattice $L(X)$ of all closed subspaces of X is orthocomplemented, then there is an inner product space S such that the lattices $L(X)$ and $F(S)$ are isomorphic. This result enabled Willbur to generalize the classical Mackey-Kakutani theorem [209, 210] by stating that complex Mackey space X whose topology is not weak is isomorphic to a Hilbert space whenever $L(X)$ is an orthomodular poset. Combined with the measure theoretic criterion we obtain the following general characterization of complex Hilbert spaces (see [159]).

4.4.1. Theorem. *Let X be an infinite-dimensional complex Mackey space whose topology is not weak. Then X is isomorphic to a Hilbert space if, and only if, the lattice $L(X)$ of all closed subspaces of X is an orthocomplemented lattice with at least one nonzero completely additive measure.*

Further application of completeness criteria can be anticipated in connection with Soler's Theorem characterizing Hilbert spaces among orthomodular spaces [297]. Many important consequences of this remarkable result are demonstrated in Holland's paper [192]. The question of whether completeness criteria can simplify the system of axioms of quantum mechanics given in [192] should be subject of further research (for partial results, see [122]).

Another line of development has been recently initiated by Sherstnev and Turilova [294]. Starting with a von Neumann algebra, M, acting on a Hilbert space, H, they consider the structure of subspaces of H affiliated with M (i.e. the subspaces invariant with respect to the commutant M' of M) and introduce an analogy to orthogonally closed and splitting subspaces in this context. In case of M being the algebra of all bounded operators on H these structures give exactly orthogonally closed and splitting subspaces investigated in this chapter. An interesting connections between Type of a von Neumann algebra, algebraic properties of these subspace structures and tracial properties of states in Tomita-Takesaki modular theory have been obtained. We expect this line of research to bring further interesting results in the future.

5. GENERALIZED GLEASON THEOREM

5.1. The Mackey-Gleason Problem

Gleason Theorem, which was a central subject of the previous chapters, says that any completely additive measure on the projection lattice of a Hilbert space extends to a linear functional on all bounded operators. The lattice of all projections on a Hilbert space H can be characterized among von Neumann projection lattices as being atomic and irreducible. Thus, Gleason Theorem covers only very special situation in this respect. Besides, it is important to describe *all* measures on projection lattices and not only completely additive ones. In this connection, a natural question arises of whether or not Gleason Theorem can be extended to finitely additive measures on projection lattices of general von Neumann algebras. This question was first posed by Mackey [224]. A more precise formulation reads as follows:

Does any bounded finitely additive measure on the projection lattice of a von Neumann algebra extend to a bounded linear functional on M?

Among others implications, this problem amounts to finding a common extension of two important principles of modern mathematics: the construction of abstract Lebesgue integral underlying measure and probability theory and Gleason Theorem that is behind the Hilbert space quantum mechanics. It has turned out that the answer is in the positive, except for obvious counterexamples arising from the 2x2 matrix case.

The Mackey's problem can be formulated easily, but the solution is enormously complex. The ideas developed have led to new methods, enriching the operator theory as well as foundations of quantum theory. The solution of Mackey's problem can be thought of as an extraordinary achievement of more then 30's years afford of many researches in quantum measure theory [1, 3, 58, 61, 59, 80, 258, 329, 330]. It was not before the 90's of the 20th century that the final form of the solution of the Mackey-Gleason problem for vector measures was found by Bunce and Wright as a culmination of many results obtained in this area [58, 61, 59]. The aim of this chapter is to make accessible this highly nontrivial result, called the *Generalized Gleason Theorem*, and to formulate it in in the most general case involving finitely additive Banach-space valued measures on projection lattices of von Neumann algebras. We shall try to simplify the arguments in some places and explain the history of this discovery.

Mathematical aspects of Generalized Gleason Theorem discussed here to detail should not prevent us from keeping in mind its physical meaning. Unlike the former von Neumann's formalism [248] which identified observables of a physical system with self-adjoint parts of von Neumann algebras and states with normal linear functionals, Mackey's model preassumes that the system of bounded observables corresponds to self-adjoint part of a Type I factor while states are described by probability measures on projections (Axiom VII of [224]). Mackey's approach seems to be more plausible from the physical point of view because it allows us to reconstruct all physical entities from probability distributions. Nevertheless, the restriction to Type I factors has no physical justification. The principal motivation for it was Gleason Theorem enabling us to identify probability measures on the propositional system of projections with linear functionals and in turn construct their expectation values (quantum integrals). As Mackey himself admitted, there was only limited mathematical knowledge behind this assumption. Also, Type I factor case does not cover all physically interesting situations. For example, models of quantum field theory often use Type III algebras. These circumstances stimulated the study of measures on general von Neumann algebras. The Generalized Gleason Theorem enables us to extend Mackey's axiom to all von Neumann algebras and thereby establish a link between quantum measure theory and linear functional analysis.

5.2. Reduction to scalar quasi-functionals

In this section we provide the basic formalism connected to the Generalized Gleason Theorem.

5.2.1. Definition. Let $P(M)$ be a projection lattice of a von Neumann algebra M, X a Banach space, and let $\mu: P(M) \to X$ be a map satisfying

$$\mu(e+f) = \mu(e) + \mu(f) \text{ whenever } ef = 0 .$$

Then μ is said to be a *(finitely additive) X-valued measure on $P(M)$.* If, moreover, $\sup\{\|\mu(e)\| \mid e \in P(M)\} < \infty$, we say that μ is *bounded.* Further, if X is the Banach space of all complex or real numbers, we call μ the complex or real measure on $P(M)$, respectively.

5.2.2. Example. Let $T: M \to X$ be a bounded linear map mapping a von Neumann algebra M into a Banach space X. Then the restriction of T to the projection lattice $P(M)$ is a bounded X-valued measure.

The Generalized Gleason Theorem says that, for almost all von Neumann algebras, the previous example covers all bounded vector measures on projection lattices. The heart of this theorem is the case of complex measures.

5.2.3. Theorem. *Let M be a von Neumann algebra with no direct summand of Type I_2. Then each bounded complex measure on $P(M)$ extends to a bounded linear functional on M.*

The proof of this theorem will be done in several steps. Since M can be decomposed into finite and properly infinite direct summands, it suffices to prove it for each type separately. In Section 5.3 it will be done for Type I homogeneous algebras, in Section 5.4 for properly infinite von Neumann algebras, and finally, in Section 5.5 for finite algebras. Having Theorem 5.2.3 at our disposal, we can easily obtain its generalization to vector measures.

5.2.4. Theorem (Generalized Gleason Theorem). *Let M be a von Neumann algebra with no direct summand of Type I_2 and let X be a Banach space. Then each bounded X-valued measure μ on $P(M)$ extends uniquely to a bounded linear operator T from M to X.*

Proof: We deduce this theorem from Theorem 5.2.3. By boundedness of μ there is a constant K such that

$$\|\mu(p)\| \leq K \qquad \text{for all } p \in P(M) .$$

Our strategy is to show that μ extends to a bounded linear map on the linear span $V(M)$ of $P(M)$. Suppose that

$$x = \sum_{i=1}^{n} \lambda_i \, p_i \tag{5.1}$$

is a linear combination of projections $p_1, \ldots, p_n$. We show that the value

$$\sum_{i=1}^{n} \lambda_i \, \mu(p_i)$$

depends on x only. For this, take a functional $\varphi \in X^*$. The composition $\varphi \circ \mu$ is a bounded complex measure on $P(M)$ that extends, by Theorem 5.2.3, to a linear functional $\hat{\varphi} \in M^*$. By (2.11) we can estimate

$$\|\hat{\varphi}\| \leq 4 \sup\{|\hat{\varphi}(p)| \mid p \in P(M)\}$$

and hence

$$\left|\varphi\Big(\sum_{i=1}^{n} \lambda_i \, \mu(p_i)\Big)\right| \leq 4 \, \|x\| \, \sup\{|\varphi(\mu(p))| \mid p \in P(M)\} \leq 4 \, \|x\| \, \|\varphi\| \, K .$$

By the Hahn-Banach Theorem, there exists $\psi \in X^*$ with $\|\psi\| = 1$ and such that

$$\left\|\sum_{i=1}^{n} \lambda_i \, \mu(p_i)\right\| = \left|\psi\Big(\sum_{i=1}^{n} \lambda_i \, \mu(p_i)\Big)\right| \leq 4K \, \|x\| . \tag{5.2}$$

This inequality implies immediately that if

$$x = \sum_{j=1}^{k} \alpha_j \, q_j$$

is another expression of x as a linear combination of projections $q_1, \ldots, q_k$, then

$$\left\|\sum_{j=1}^{k} \alpha_i \, \mu(q_i) - \sum_{i=1}^{n} \lambda_i \, \mu(p_i)\right\| \leq 4K \, \|x - x\| = 0 ,$$

and hence

$$\sum_{j=1}^{k} \alpha_j \, \mu(q_j) = \sum_{i=1}^{n} \lambda_i \, \mu(p_i) \, .$$

This allows us to define a map $T\colon V(M) \to X$ by setting

$$T\Big(\sum_{i=1}^{n} \lambda_i \, p_i\Big) = \sum_{i=1}^{n} \lambda_i \, \mu(p_i) \, .$$

It is obvious from (5.2) that $\|T\| \leq 4\,K$ on $V(M)$. Since $V(M)$ is dense in M, T extends uniquely to a bounded operator from M into X. □

Let us now return to scalar measures. By taking real and imaginary part, respectively, we see that it is enough to prove Theorem 5.2.3 for real measures. As an initial step we show that real measures are in a one-to-one correspondence with so-called quasi-linear functionals.

5.2.5. Definition. Let M be a von Neumann algebra. A mapping $\varrho\colon M \to \mathbb{C}$ is called a *quasi-linear functional* (also quasi-functional) if it satisfies the following conditions:

(i) ϱ is a linear functional on any abelian von Neumann subalgebra A of M,

(ii) $\varrho(x + i\,y) = \varrho(x) + i\,\varrho(y)$ for all self-adjoint elements $x, y \in M$,

(iii) ϱ is bounded on the unit sphere of M.

Moreover, we say that ϱ is *self-adjoint* if $\varrho(x) \in \mathbb{R}$ whenever x is a self-adjoint element of M.

5.2.6. Proposition. *Let $\mu\colon P(M) \to \mathbb{R}$ be a bounded measure on a von Neumann algebra M. Then μ extends uniquely to a quasi-linear functional $\hat{\mu}$ on M. If μ is real, then $\hat{\mu}$ is self-adjoint.*

Proof: Denote $K = \sup\{|\mu(p)| \mid p \in P(M)\}$. First suppose that M is abelian. Let B be a finite-dimensional von Neumann subalgebra of M. (Observe that B is a direct sum of one-dimensional algebras.) Then μ extends trivially from $P(B)$ to a unique linear functional μ_B on B. Moreover, if $x = \sum_{i=1}^{n} \lambda_i p_i$ is a linear combination of mutually orthogonal projections $p_1, \ldots, p_n \in B$, then

$$\begin{aligned} |\mu_B(x)| &= \left| \sum_{i=1}^{n} \lambda_i \, \mu(p_i) \right| \leq \max_{i=1,\ldots,n} |\lambda_i| \sum_{i=1}^{n} |\mu(p_i)| \\ &\leq 4 \max_{i=1,\ldots,n} |\lambda_i| \, K = 4 \, \|x\| \, K \, . \end{aligned} \tag{5.3}$$

Let us observe that any two elements from the linear span $V(M)$ of $P(M)$ are contained in a common finite-dimensional subalgebra of M. Therefore we obtain a well-defined linear map $\hat{\mu}$ on $V(M)$ by defining

$$\hat{\mu}(x) = \mu_B(x) : V(M) \to \mathbb{C},$$

Moreover,

$$|\hat{\mu}(x)| \leq 4 \, K \, \|x\| \, , \qquad \text{for all } x \in M \, .$$

by (5.3), and therefore $\hat{\mu}$ is bounded on $V(M)$. Since $V(M)$ is dense in M (the spectral theorem), we can extend $\hat{\mu}$ uniquely to a bounded linear map (denoted again by $\hat{\mu}$) on M.

Suppose now that M is an arbitrary von Neumann algebra. Since any self-adjoint element $x \in M$ is contained in some abelian von Neumann subalgebra A of M we can (unambiguously) define $\hat{\mu}(x) = \hat{\mu}_A(x)$, where $\hat{\mu}_A$ is a (unique) extension of μ to a linear functional on A. Finally, by setting $\hat{\mu}(x + i\, y) = \hat{\mu}(x) + i\, \hat{\mu}(y)$ we obtain the desired quasi-linear extension of μ. If μ is real then, from the construction, $\mu(x)$ is real for any self-adjoint x and so $\hat{\mu}$ is self-adjoint. □

Let us observe that if M is an abelian von Neumann algebra and thereby a copy of some measure space $L^\infty(X, \mathcal{P}, \omega)$, then the one-to-one correspondence between measures and quasi-linear functionals provided by Proposition 5.2.6 reduces to the well known correspondence between measure $\mu \colon \mathcal{P} \to \mathbb{R}$ and integral $\hat{\mu}$:

$$\hat{\mu} \colon L^\infty(X, \mathcal{P}, \omega) \to \mathbb{R} : \qquad \hat{\mu}(f) = \int_X f(x) \, d\mu(x) \, .$$

The crucial step in proving Theorem 5.2.3 is to show that any quasi-linear functional on a given von Neumann algebra is in fact linear. Usually we will denote the quasi-linear functional by the same letter as the measure that is extended. Given a self-adjoint quasi-linear functional μ on a von Neumann algebra M, let us introduce the following notation that will be useful in the sequel. Given $p \in P(M)$, we shall write

$$\alpha(p) = \sup\{\mu(e) \mid e \in P(M)\,,\, e \leq p\}$$

and, for the variation,

$$V(p) = \sup\{|\mu(e)| \mid e \in P(M)\,,\, e \leq p\}\,.$$

Obviously, $0 \leq \alpha(p) \leq V(p)$. Further, let us set

$$\|\mu\| = \sup\{|\mu(x)| \mid \|x\| \leq 1\}\,.$$

The following relations between the introduced quantities will be often used below.

5.2.7. Proposition. *Let μ be a self-adjoint quasi-linear functional on a von Neumann algebra M. Then the following statements hold true:*

(i) $\|\mu\| = \sup\{|\mu(x)| \mid x = x^*\,,\, \|x\| \leq 1\} = 2\,\alpha(1) - \mu(1)\,.$

(ii) $\sup\{\mu(x) \mid 0 \leq x \leq 1\} = \alpha(1) \leq V(1)\,.$

Proof: (i) First we show that $\|\mu\|$ is equal to the supremum taken over the self-adjoint part of the unit ball of M. Take $\varepsilon > 0$ and $x \in M$ with $\|x\| \leq 1$ such that

$$|\mu(x)| > \|\mu\| - \varepsilon\,.$$

On multiplying x by a suitable complex unit, we can assume that $\mu(x)$ is real. Let x_1 and x_2 be the real and imaginary part of x, respectively. Then

$$\mu(x) = \mu(x_1) + i\,\mu(x_2) \in \mathbb{R}\,,$$

implying immediately that $\mu(x_2) = 0$. Hence, for a self-adjoint element x_1 with $\|x_1\| \leq 1$, we get

$$|\mu(x_1)| = |\mu(x)| > \|\mu\| - \varepsilon\,.$$

Thus,

$$\|\mu\| = \sup\{|\mu(x)| \mid x = x^*\,,\, \|x\| \leq 1\}\,.$$

Since any self-adjoint element $x \in M$ generates an abelian subalgebra, $W^*(x)$, and the function

$$|\mu| : a \in W^*(x)_{sa} \to |\mu(a)|$$

is continuous and convex, the values of $|\mu|$ on the self-adjoint part of the unit ball of $W^*(x)$ are bounded from above by the supremum of the restriction of μ to the extreme points, i.e. to the symmetries. But every symmetry $s \in M$ is of the form $s = 2p - 1$, where $p \in P(M)$. It implies

$$\begin{aligned} \sup\{|\mu(x)| \mid x = x^*, \|x\| \leq 1\} &= \sup\{|\mu(2p-1)| \mid p \in P(M)\} \\ &= \sup\{|\mu(-2p+1)| \mid p \in P(M)\} \\ &= 2\alpha(1) - \mu(1). \end{aligned}$$

(We have made use of the fact that, for any projection $p \in M$, $-(2p-1) = 2(1-p) - 1$.)

(ii) Any element $0 \leq x \leq 1$ can be written as follows:

$$x = \sum_{n=1}^{\infty} \frac{1}{2^n} p_n ,$$

where (p_n) are projections from abelian algebra $W^*(x)$. Since $\mu(x) = \sum_{n=1}^{\infty} \frac{1}{2^n} \mu(p_n)$, we see that $\mu(p_n) \geq \mu(x)$ for at least one p_n. This establishes the condition (ii). □

5.3. Linear extensions of measures on Type I_n algebras

The goal of this section is to establish the Generalized Gleason Theorem for von Neumann algebras of Type I_n, $3 \leq n < \infty$. The algebras of that type are close to abelian in many respects. The key part of the proof is the uniform continuity of measures on the projection lattice. This fact can be proved by using Gleason Theorem for suitable substructures isomorphic to $M_3(\mathbb{C})$. An important step in this regard is to show that close projections p and q in a von Neumann algebra M admit a projection r such that both $W^*(p,r)$ and $W^*(q,r)$ can be embedded into a direct sum of matrix algebras $M_n(\mathbb{C})$, when $n = 3, 4$. The strategy is based on analyzing isoclinicity of projections in von Neumann algebra and will be isolated in following paragraph.

Isoclinic projections

5.3.1. Definition. Let p and q be nonzero projections in a C^*-algebra A. We say that p and q are *isoclinic with angle* α, $0 < \alpha < \pi/2$ if

$$pqp = \cos^2 \alpha \, p \,, \qquad qpq = \cos^2 \alpha \, q \,. \tag{5.4}$$

It can be easily verified that one-dimensional nonorthogonal distinct projections p and q acting on a Hilbert space H are isoclinic with angle α if, and only if,

$$|\,(\xi, \theta)\,|^2 = \cos^2 \alpha$$

for all unit vectors ξ and θ in the range of p and q, respectively. This explains the geometric content of isoclinicity.

5.3.2. Example. *Let* $0 \le a, b \le 1$ *and* $\varphi, \psi \in \mathbb{R}$. *Then projections*

$$p = \begin{pmatrix} a & \sqrt{a-a^2}\, e^{i\varphi} \\ \sqrt{a-a^2}\, e^{-i\varphi} & 1-a \end{pmatrix}, q = \begin{pmatrix} b & \sqrt{b-b^2}\, e^{i\psi} \\ \sqrt{b-b^2}\, e^{-i\psi} & 1-b \end{pmatrix}$$

in $M_2(\mathbb{C})$ *are either commuting or isoclinic projections in* $M_2(\mathbb{C})$ *with angle* α *given by*

$$\cos^2 \alpha = a\,b + (1-a)\,(1-b) + 2\,\sqrt{a-a^2}\,\sqrt{b-b^2}\,\cos(\varphi - \psi)\,. \tag{5.5}$$

Proof: The equation (5.5) can be verified by a straightforward computation. □

It turns out that all isoclinic projections are more or less one-dimensional projections in 2x2 matrix algebras.

5.3.3. Theorem. *If projections* p *and* q *in a* C^**-algebra* A *are isoclinic, then* $C^*(p, q)$ *is isomorphic to* $M_2(\mathbb{C})$.

Proof: First we observe that $C^*(p,q)$ is the linear span of the set $\{p, q, p\,q, q\,p\}$. (For example, $q\,p\,q\,p\,q = \cos^2 \alpha \, p\,q$, where α is the angle between p and q.) An easy algebraic argument also gives us that p and q are the the atomic projections in $C^*(p,q)$. Therefore $C^*(p,q)$ is a four-dimensional algebra with two nonorthogonal atoms and hence it has to be isomorphic to $M_2(\mathbb{C})$. □

5.3.4. Corollary. *Let p and q be isoclinic projections with angle α in a unital C^*-algebra A. Then*

(i) *p and q are unitarily equivalent in $C^*(p,q)$,*

(ii) $\|p-q\| = \sin\alpha$.

Proof: (i) It follows from Theorem 5.3.3 and the fact that atomic projections in $M_2(\mathbb{C})$ are unitarily equivalent.

(ii) By (i) there is no loss of generality in assuming that there is a unitary $u \in A$ with $u^* p u = q$. Since

$$(p-q)^2 = p(1-q)p + (1-p)q(1-p)$$

is a decomposition of $(p-q)^2$ into positive orthogonal elements, we have

$$\begin{aligned} \|p-q\|^2 &= \max\{\|p(1-q)p\|, \|(1-p)q(1-p)\|\} \\ &= \max\{\|p(1-q)\|^2, \|q(1-p)\|^2\}. \end{aligned}$$

As we also have

$$u^* p(1-q)u = q(1-p),$$

we see that $\|p(1-q)\| = \|q(1-p)\|$ and therefore

$$\|p-q\|^2 = \|p(1-q)p\| = \|p - pqp\| = 1 - \cos^2\alpha = \sin^2\alpha.$$

□

The following neat result sheds light on the link of general and isoclinic projections.

5.3.5. Theorem. *Let p and q be distinct projections in a von Neumann algebra M with $p \wedge q = 0$ and $\|p-q\| < 1$. Then there is a projection h in M such that h is isoclinic to both p and q with the angle*

$$\alpha = \frac{1}{2}\sin^{-1}\|p-q\|.$$

Proof: Since $\|p-q\| < 1$, we have $p \wedge (1-q) = (1-p) \wedge q = 0$. There is no loss of generality in assuming that $p \vee q = 1$. According to Chapter 2 (Tensor products of C^*-algebras) we can identify $W^*(p,q)$ with

$C(X, M_2(\mathbb{C}))$, where X is a Hausdorff compact space, in such a way that, for all $x \in X$,

$$p(x) = \begin{pmatrix} 1 & 0 \\ 0 & 0 \end{pmatrix} \quad q(x) = \begin{pmatrix} a(x) & \sqrt{a(x) - a^2(x)} \\ \sqrt{a(x) - a^2(x)} & 1 - a(x) \end{pmatrix},$$

where $a(x)$ is a continuous function on X with $0 \le a(x) \le 1$. The condition $0 < \|p - q\| < 1$ implies that $0 < \alpha < \pi/4$. Let us set $\lambda = \cos^2 \alpha$. Then

$$1 > \lambda > \frac{1}{2}. \tag{5.6}$$

Also, for all $x \in X$,

$$\|p(x) - q(x)\| \le \|p - q\| = \sin 2\alpha < 1.$$

Using Example 5.3.2, we see that $p(x)$ and $q(x)$ are either equal or isoclinic (in $M_2(\mathbb{C})$) with the angle $\beta(x)$ for which, by (5.5),

$$a(x) = \cos^2 \beta(x).$$

It implies that

$$\sin \beta(x) = \|p(x) - q(x)\| \le \sin 2\alpha,$$

and therefore

$$a(x) = \cos^2 \beta(x) \ge \cos^2 2\alpha = (2\cos^2 \alpha - 1)^2.$$

In other words,

$$a(x) \ge (2\lambda - 1)^2 > 0 \tag{5.7}$$

for all $x \in X$. We are going to find the required projection, $h \in C(X, M_2(\mathbb{C}))$, in the form

$$h(x) = \begin{pmatrix} \lambda & \sqrt{\lambda - \lambda^2}\, e^{i\psi(x)} \\ \sqrt{\lambda - \lambda^2}\, e^{-i\psi(x)} & 1 - \lambda \end{pmatrix}, \tag{5.8}$$

where $e^{i\psi(x)}$ is a continuous function on X with values in the unit circle. By virtue of (5.5), h is obviously isoclinic to e with angle α. It remains to find $e^{i\psi(x)}$ such that the same assertion holds for projections g and h. Making use of (5.5), this reduces to finding (real) $\psi(x)$ satisfying

$$\lambda = \lambda\, a(x) + (1 - \lambda)(1 - a(x)) + 2\sqrt{a(x) - a(x)^2}\,\sqrt{\lambda - \lambda^2}\,\cos(\psi(x)).$$

If $a(x) = 1$ we can set $\psi(x) = 0$. Suppose that $a(x) \neq 1$. Since $1 > a(x) > 0$ by (5.7), the previous equation reads

$$\cos(\psi(x)) = \frac{(1 - a(x))\,(2\lambda - 1)}{2\sqrt{a(x) - a(x)^2}\,\sqrt{\lambda - \lambda^2}} = \frac{\lambda - 1/2}{\sqrt{\lambda - \lambda^2}}\sqrt{\frac{1}{a(x)} - 1}\,.$$

The right-hand side of the previous equation is obviously nonnegative (see (5.6)). In order to guarantee the existence of $\psi(x)$, we have to verify that the right-hand side of the previous expression takes values in [0,1]. Employing (5.7), (5.6) and the monotonicity of the corresponding function we can estimate

$$\frac{\lambda - 1/2}{\sqrt{\lambda - \lambda^2}}\sqrt{\frac{1}{a(x)} - 1} \leq \frac{\lambda - 1/2}{\sqrt{\lambda - \lambda^2}}\sqrt{\frac{1}{(2\lambda - 1)^2} - 1} = 1\,.$$

(We have used inequality (5.7) for simplifying the resulting expressions.) Putting now

$$e^{i\,\psi(x)} = e^{\,i\,\cos^{-1}\left(\sqrt{\frac{1}{a(x)} - 1}\,\frac{\lambda - 1/2}{\sqrt{\lambda - \lambda^2}}\right)}$$

for $x \in X$ with $a(x) \neq 0$ and $e^{i\,\psi(x)} = 1$ for x with $a(x) = 1$, we obtain a continuous function on X. Hence, defining h by (5.8) with ψ specified as above, we obtain the projection in $W^*(p, q)$ isoclinic to both p and q with the same angle $\alpha = \frac{1}{2}\sin^{-1}\|p - q\|$. □

Uniform continuity of measures on projections

In this subsection we establish the uniform continuity of bounded real measures on the projection lattice $P(M)$, where M is a von Neumann algebra with no Type I_2 direct summand. For proving this, Gleason Theorem for the three-dimensional space will be used repeatedly.

5.3.6. Proposition. *Let M be a von Neumann algebra without Type I_2 direct summand. Any von Neumann subalgebra A of M which is isomorphic to $M_2(\mathbb{C})$ is contained in a subalgebra $C \oplus D$ of M satisfying the following properties: C is either zero or it is isomorphic to $M_4(\mathbb{C})$; D is either zero or it is a copy of $M_2(\mathbb{C})$ contained in a subalgebra of M isomorphic to $M_3(\mathbb{C})$.*

Proof: Since A is generated by two orthogonal equivalent projections, say p and q, we can write (see Chapter 2, Classification of von Neumann algebras)

$$\begin{aligned} p &= p_1 + p_2 + p_3 \\ q &= q_1 + q_2 + q_3 , \end{aligned}$$

where $p_1 \sim p_2 \sim q_1 \sim q_2$ are equivalent (orthogonal) projections and $p_3 \sim q_3$ are equivalent abelian projections. If $p_1 \neq 0$, then by choosing an appropriate matrix unit we can embed projections p_1, p_2, q_1, q_2 into algebra C isomorphic to $M_4(\mathbb{C})$. In the same way, provided that $p_3 \neq 0$, we can embed the projections p_3, q_3 into subalgebra D isomorphic to $M_2(\mathbb{C})$ and orthogonal to C. It remains to prove that C can be enlarged to a copy of $M_3(\mathbb{C})$. For this, since p_3 and q_3 are abelian, they have to live in a Type I part of M. Let $\sum_\alpha \oplus z_\alpha M$ be a decomposition of Type I part into Type I_α direct summands. By our assumption, $\alpha \neq 2$ for each $\alpha \in I$. Since p_3 and q_3 are equivalent and orthogonal, abelian components $z_1 p_3$ and $z_1 q_3$ have to be zero. So we can suppose that $\alpha \geq 3$ for each $\alpha \in I$. Now, by the structure theory of Type I_α von Neumann algebra (see Chapter 2, Classification of von Neumann algebras), there is an abelian projection, r_α, in each $z_\alpha M$ such that $z_\alpha p_3$, $z_\alpha q_3$, r_α are orthogonal and equivalent. By setting $r = \sum_\alpha r_\alpha$ we obtain a projection such that the triple p_3, q_3, r forms a system of orthogonal equivalent projections. In a standard way, this system can be embedded into algebra isomorphic to $M_3(\mathbb{C})$. □

5.3.7. Corollary. *Let M be a von Neumann algebra without Type I_2 direct summand and let μ be a bounded complex measure on $P(M)$. Then the quasi-linear extension of μ is linear on every subalgebra of M isomorphic to $M_2(\mathbb{C})$.*

Proof: By Proposition 5.3.6, every subalgebra A of M isomorphic to $M_2(\mathbb{C})$ can be embedded into the direct sum of algebras C and D, where C is either zero or isomorphic to $M_4(\mathbb{C})$, and D is either zero or it can be embedded into a copy of $M_3(\mathbb{C})$. By Gleason Theorem, the quasi-linear extension is linear on both C and D, and it is therefore linear on their direct sum $C \oplus D$. In particular, the quasi-linear functional extending μ is linear on A. □

Combining now the linearity of quasi-functional on all $M_2(\mathbb{C})$ substructures with the geometric arguments of Theorem 5.3.5, we establish the uniform continuity of a measure on the projection lattice (the metric is inherited from the given von Neumann algebra).

5.3.8. Theorem. *Any bounded complex measure on the projection lattice $P(M)$ of a von Neumann algebra M having no Type I_2 direct summand is uniformly continuous.*

Proof: Let $\hat{\mu}$ be a quasi-linear functional corresponding to a bounded complex measure μ on $P(M)$. Take distinct projections $p, q \in P(M)$ such that $\|p-q\| < 1$ and set $e = p - p \wedge q$, $f = q - p \wedge q$. Then $\|e - f\| < 1$ and $e \wedge f = 0$. Using Theorem 5.3.5 we can find a projection g in M which is isoclinic to both e and f with the angle

$$\alpha = \frac{1}{2} \sin^{-1} \|e - f\| < \frac{\pi}{4} \,.$$

By Theorem 5.3.3, $W^*(e,g) \simeq W^*(f,g) \simeq M_2(\mathbb{C})$, and so μ is linear on both $W^*(e,g)$ and $W^*(f,g)$. This, together with Corollary 5.3.4 (ii), allows us to estimate

$$\begin{aligned} |\mu(e) - \mu(g)| &\leq \|\hat{\mu}\| \cdot \|e - g\| = \|\hat{\mu}\| \cdot \sin\alpha \\ &= \|\hat{\mu}\| \cdot \sin\left(\frac{1}{2}\sin^{-1}\|e-f\|\right) \leq \|\hat{\mu}\| \cdot \sin\left(\sin^{-1}\|e-f\|\right) \\ &= \|\hat{\mu}\| \cdot \|e - f\| \,. \end{aligned}$$

Similarly,

$$|\mu(f) - \mu(g)| \leq \|\hat{\mu}\| \cdot \|e - f\| \,.$$

Finally, we can write

$$\begin{aligned} |\mu(p) - \mu(q)| &= |\mu(e) - \mu(f)| = |\mu(e) - \mu(g) + \mu(g) - \mu(f)| \\ &\leq |\mu(e) - \mu(g)| + |\mu(f) - \mu(g)| \leq 2\,\|\hat{\mu}\| \cdot \|p - q\| \,. \end{aligned}$$

□

Gleason Theorem for Type I_n algebras

An important step in proving Generalized Gleason Theorem is the establishing of the theorem for Type I homogeneous algebras. It will be done in this subsection by viewing Type I_n algebras as algebras of complex matrix-valued functions and by appealing to uniform continuity of bounded measures on projections.

5.3.9. Theorem. *Let M be a von Neumann algebra of Type I_n, where $n \neq 2$ and $n < \infty$. Then any bounded complex measure μ on $P(M)$ extends to a linear functional on M.*

Proof: We can identify M with $C(X, M_n(\mathbb{C}))$, where X is a hyperstonean compact space. The $*$ subalgebra A of M consisting of all locally constant functions is dense in $C(X, M_n(\mathbb{C}))$ (see Chapter 2, Classification of von Neumann algebras). Let $\hat{\mu}$ be the quasi-linear functional extending the measure μ. We show first that μ is linear on A. For this, let us take $a, b \in A$. We can find a partition $O_1, O_2, \ldots, O_k$ of X, consisting of clopen sets, such that both a and b are constant on each O_i $(i = 1, \ldots, k)$. Consider now a subalgebra

$$B = \{f \in C(X, M_n(\mathbb{C})) \mid f \text{ is constant on each } O_i\,,\ i = 1, \ldots, k\}\,.$$

Then B is isomorphic to a direct sum of k copies of $M_n(\mathbb{C})$. By Gleason theorem $\hat{\mu}$ is linear on B. Since $a, b \in B$ we obtain immediately

$$\hat{\mu}(a + b) = \hat{\mu}(a) + \hat{\mu}(b)$$

and so we have the linearity of $\hat{\mu}$ on A. Hence, by taking a continuous linear extension, we can find a bounded linear functional φ on M coinciding with $\hat{\mu}$ on A. Since $P(A)$ is dense in $P(M)$ and μ is continuous on the projection lattice $P(M)$, we see that φ coincides with μ on the whole projection lattice $P(M)$. By the unicity of the quasi-linear extension we see that $\hat{\mu} = \varphi$, and the proof is complete. □

Let us remark that the method of the previous proof extends quickly to subalgebras of the von Neumann algebra M not containing Type I_2 direct summand that are of the form $A \otimes B$, where A is abelian and B is a von Neumann algebra for which every quasi-linear functional on M is linear on B. Similarly, it gives the existence of linear extensions for completely additive measures on all homogeneous Type I algebras (not necessarily finite) that are not of Type I_2, for they can be organized as $A \otimes B$, where B is a Type I factor and A is abelian. The following consequences of Theorem 5.3.9 will often be used in the sequel.

5.3.10. Corollary. *Let M be a von Neumann algebra without Type I_2 direct summand and let ϱ be a quasi-linear functional on M. Then ϱ is linear on any subalgebra $W^*(p,q)$ generated by projections $p,q \in M$.*

Proof: This is a straightforward consequence of the previous reasoning and the fact that $W^*(p,q)$ is a direct sum of abelian von Neumann algebra and Type I_2 algebra which is of the form $A \otimes B$, where A is abelian and B is a subalgebra of a copy of $M_3(\mathbb{C})$ in M. □

5.3.11. Corollary. *Let M be a von Neumann algebra without Type I_2 direct summand and let ϱ be a quasi-linear functional on M. Then ϱ is linear on any subalgebra $W^*(1,p,q,z)$, where p,q and z are projections such that z commutes with both p and q.*

Proof: We can write

$$W^*(1,p,q,z) = W^*(z,pz,qz) \oplus W^*(1-z,(1-z)p,(1-z)q)$$

and apply Corollary 5.3.10 to each direct summand.

5.4. Linear extensions of measures on infinite algebras

In this section the Generalized Gleason Theorem for properly infinite von Neumann algebras will be proved. First we shall introduce a few conventions. Throughout this part M will always denote a von Neumann algebra and

$$\mu\colon P(M) \to \mathbb{R}$$

will be a bounded real measure. When a misunderstanding does not threaten, the self-adjoint quasi-linear extension of μ to M will be denoted by the same letter μ. We shall make the additional assumption that

$$\mu(1) = 0. \tag{5.9}$$

There is no loss of generality in assuming the latter condition because we can always replace μ by $\mu - \varrho$, where $\varrho \in M^*$ is self-adjoint and such that $\varrho(1) = \mu(1)$. Also, by a suitable scaling, we may suppose that

$$\|\mu\| = \sup\{|\mu(x)| \mid \|x\| \leq 1\} = 1 . \tag{5.10}$$

By Proposition 5.2.7, by writing $\|\mu\| = 2\,\alpha(1) - \mu(1)$, we have disguised

$$\begin{aligned} \alpha(1) &= \sup\{\mu(e) \mid e \in P(M)\} = -\inf\{\mu(e) \mid e \in P(M)\} \\ &= \frac{1}{2} \end{aligned} \tag{5.11}$$

and, also,

$$\begin{aligned} \alpha(1) &= \sup\{\mu(x) \mid 0 \leq x \leq 1\} = \sup\{-\mu(x) \mid 0 \leq x \leq 1\} \\ &= \sup\{|\mu(x)| \mid 0 \leq x \leq 1\} . \end{aligned} \tag{5.12}$$

At first we show that if e is a projection in M such that the value $\mu(e)$ is close to $\alpha(1)$, then μ "nearly diagonalizes" with respect to e in the sense that

$$|\mu(p) - \mu(e\,p\,e) - \mu((1-e)\,p\,(1-e))|$$

is close to zero for any projection $p \in M$. Observe that, in fact, this statement concerns linear functionals because μ is linear on $W^*(1, p, e)$ by the previous findings. The inequality can be derived by applying the Schwarz inequality to the positive and the negative part of a given linear functional. Alternative proof using direct calculations is shown in the following lemma.

5.4.1. Lemma. *Let $0 < \varepsilon < 1/2$ and let $e \in P(M)$ be such that $\mu(e) > 1/2 - \varepsilon$. Then*

$$|\mu(p) - \mu(e\,p\,e) - \mu((1-e)\,p\,(1-e))| < 4\,\varepsilon^{1/2} ,$$

for all $p \in P(M)$.

Proof: Set $x = (1-\varepsilon)^{1/2}\,e \pm \varepsilon^{1/2}\,(1-e)\,p$. Then

$$x^*\,x = (1-\varepsilon)\,e + \varepsilon\,p\,(1-e)\,p \leq (1-\varepsilon)\,e + \varepsilon 1 \leq 1$$

and so $0 \leq x\,x^* \leq 1$. Equivalently,

$$0 \leq (1-\varepsilon)\,e + \varepsilon\,(1-e)\,p\,(1-e) \pm \varepsilon^{1/2}\,(1-\varepsilon)^{1/2}\,(e\,p\,(1-e) + (1-e)\,p\,e) \leq 1 . \tag{5.13}$$

As μ is linear on $W^*(p,e)$ and $|\mu(y)| \le 1/2$ for all $0 \le y \le 1$, we obtain from (5.13) that

$$\begin{aligned}\frac{1}{2} &\ge (1-\varepsilon)\,\mu(e) + \varepsilon\,\mu((1-e)\,p\,(1-e)) \\ &\pm\ \varepsilon^{1/2}\,(1-\varepsilon)^{1/2}\mu(e\,p\,(1-e) + (1-e)\,p\,e) \\ &\ge (1-\varepsilon)\left(\frac{1}{2}-\varepsilon\right) - \frac{1}{2}\varepsilon \pm \varepsilon^{1/2}\,(1-\varepsilon)^{1/2}\mu(e\,p\,(1-p) + (1-e)\,p\,e)\,.\end{aligned}$$

Finally,

$$\begin{aligned}&|\mu(p) - \mu(e\,p\,e) - \mu((1-e)\,p\,(1-e))| \\ = \ &|\mu(e\,p\,(1-e) + (1-e)\,p\,e))| \le \frac{(2-\varepsilon)\,\varepsilon}{\varepsilon^{1/2}\,(1-\varepsilon)^{1/2}} < 4\,\varepsilon^{1/2}\,.\end{aligned}$$

□

5.4.2. Lemma. *Let e be a properly infinite projection in M and $\varepsilon > 0$. Then there is a projection $f \le e$ with $f \sim e \sim e - f$ and with the variation*

$$V(f) = \sup\{|\mu(p)| \mid p \le f,\, p \in P(M)\} \le \varepsilon.$$

Proof: Since e is properly infinite it can be decomposed,

$$e = e_1 + e_2 + \cdots + e_m$$

into the sum of equivalent projections $e_1 \sim e_2 \sim \cdots \sim e_m \sim e$, where m is an arbitrary integer. We will choose $m = 2^n$, where $2^n\,\varepsilon > 1$. Suppose, by contradiction, that

$$V(e_i) > \varepsilon \qquad \text{for all } i = 1, \ldots, m\,.$$

In that case there is, for each i, a projection $f_i \le e_i$ such that

$$|\mu(f_i)| > \varepsilon\,.$$

Suppose (without any loss of generality) that $\mu(f_i) > 0$ for at least half of the f_i's. Denote by p the sum of these projections. Then $\mu(p) > 2^{n-1}\,\varepsilon > 1/2$, a contradiction with $\alpha(1) = 1/2$. □

Next lemma shows that supremum, $\alpha(1)$, can be attained on very specific projections.

5.4.3. Lemma. *Let M be a properly infinite von Neumann algebra. Then*

$$\sup\{\mu(e) \mid e \in P(M),\ e \sim 1 \sim 1 - e\} = \frac{1}{2}.$$

Proof: For a prescribed $0 < \varepsilon < \frac{1}{8}$ we can find a projection $e \in P(M)$ satisfying $\mu(e) > 1/2 - \varepsilon$. Decompose e as

$$e = z\,e + (1 - z)\,e\,,$$

where $z \in Z(M)$, $z\,e$ is finite and $(1 - z)\,e$ is properly infinite. Since M is properly infinite the projection $f = z\,(1 - e) = z - z\,e$ is properly infinite too. Indeed, if $z' \in Z(M)$ were such that $z'\,z\,(1 - e)$ is nonzero and finite, then $z'\,z\,e + z'\,z\,(1 - e) = z'\,z$ would be a nonzero finite central projection in M - a contradiction. Also $e + f = z + (1 - z)\,e$ being a sum of two properly infinite projections is properly infinite. Employing Lemma 5.4.2, we can find a projection $p \le f$ so that $p \sim f \sim f - p$ and $V(p) < \varepsilon$. Then $q = e + p \sim e + f$ is properly infinite and $\mu(q) > 1/2 - \varepsilon - \varepsilon = 1/2 - 2\varepsilon$. So we have found properly infinite projection q with $\mu(q) > 1/2 - 2\,\varepsilon$. According to Lemma 5.4.2 again, we can pick up a projection $r \le q$, $r \sim q \sim q - r$ with $V(q - r) < \varepsilon$. Then

$$1 - r = (1 - q) + (q - r) \sim (1 - q) + q = 1$$

and

$$\begin{aligned} \mu(1 - r) &= \mu(1 - q) + \mu(q - r) = -\mu(q) + \mu(q - r) \\ &< -\frac{1}{2} + 2\,\varepsilon + \varepsilon = -\frac{1}{2} + 3\,\varepsilon. \end{aligned}$$

In other words,

$$\mu(r) > \frac{1}{2} - 3\,\varepsilon\,. \tag{5.14}$$

Finally, Lemma 5.4.2 gives us a projection $h \le 1 - r$ fulfilling $h \sim 1 - r \sim 1 - r - h$ and $V(1 - r - h) < \varepsilon$. Now $h \sim 1 - r \sim 1$, $1 - h = r + 1 - r - h \sim r + 1 - r = 1$, and

$$\begin{aligned} \mu(1 - h) &= \mu(1 - r - h + r) = \mu(1 - r - h) + \mu(r) \\ &> -\varepsilon + \frac{1}{2} - 3\,\varepsilon = \frac{1}{2} - 4\,\varepsilon. \end{aligned}$$

□

Now we pass to an important 4x4 matrix technique due to Christensen which allows us, together with the diagonalization technique, to establish the linearity of the quasi-functional on properly infinite algebras. Let e and f be orthogonal projections in a C^*-algebra A. Given another projection $p \in A$, the compressions $p\,e\,p$ and $p\,f\,p$ are nonnegative elements in the hereditary subalgebra $p\,A\,p$. The content of the following proposition is to prove the converse. Two nonnegative elements in a hereditary subalgebra are compressions of orthogonal projections provided that there is enough space in the C^*-algebra under consideration.

5.4.4. Proposition. (Christensen Lemma). *Let e, e_1, e_2, e_3 be equivalent, nonzero, orthogonal projections in a C^*-algebra A. Suppose that $0 \leq a, b \leq \frac{1}{2} e$. Then there are orthogonal projections p and q in A such that*

$$e\,p\,e = a \qquad \textit{and} \qquad e\,q\,e = b\,.$$

Proof: Taking suitable matrix unit we can, after harmless assumption that $e + e_1 + e_2 + e_3 = 1$, organize A into 4x4 matrices with entries from the hereditary subalgebra $e\,A\,e$ ($e\,A\,e$ will be identified with the upper left corner in $M_4(e\,A\,e)$.) Let us define the elements x, y as follows:

$$x = \begin{pmatrix} a^{1/2} & 0 & 0 & 0 \\ a^{1/2} & 0 & 0 & 0 \\ (e-2\,a)^{1/2} & 0 & 0 & 0 \\ 0 & 0 & 0 & 0 \end{pmatrix}, \qquad y = \begin{pmatrix} b^{1/2} & 0 & 0 & 0 \\ -b^{1/2} & 0 & 0 & 0 \\ 0 & 0 & 0 & 0 \\ (e-2\,b)^{1/2} & 0 & 0 & 0 \end{pmatrix}.$$

By an easy matrix calculation we can verify that

$$\begin{aligned} x^*\,x \;&=\; \begin{pmatrix} a^{1/2} & a^{1/2} & (e-2\,a)^{1/2} & 0 \\ 0 & 0 & 0 & 0 \\ 0 & 0 & 0 & 0 \\ 0 & 0 & 0 & 0 \end{pmatrix} \cdot \begin{pmatrix} a^{1/2} & 0 & 0 & 0 \\ a^{1/2} & 0 & 0 & 0 \\ (e-2\,a)^{1/2} & 0 & 0 & 0 \\ 0 & 0 & 0 & 0 \end{pmatrix} \\ &=\; \begin{pmatrix} e & 0 & 0 & 0 \\ 0 & 0 & 0 & 0 \\ 0 & 0 & 0 & 0 \\ 0 & 0 & 0 & 0 \end{pmatrix}. \end{aligned}$$

Similarly, $y^*y = e$. It means that x and y are partial isometries and we can define projections $p = x\,x^*$ and $q = y\,y^*$. Since,

$$
\begin{aligned}
x^*\,y &= \begin{pmatrix} a^{1/2} & a^{1/2} & (e-2\,a)^{1/2} & 0 \\ 0 & 0 & 0 & 0 \\ 0 & 0 & 0 & 0 \\ 0 & 0 & 0 & 0 \end{pmatrix} \cdot \begin{pmatrix} b^{1/2} & 0 & 0 & 0 \\ -b^{1/2} & 0 & 0 & 0 \\ 0 & 0 & 0 & 0 \\ (e-2\,b)^{1/2} & 0 & 0 & 0 \end{pmatrix} \\
&= \begin{pmatrix} 0 & 0 & 0 & 0 \\ 0 & 0 & 0 & 0 \\ 0 & 0 & 0 & 0 \\ 0 & 0 & 0 & 0 \end{pmatrix},
\end{aligned}
$$

we have $p\,q = x\,x^*\,y\,y^* = 0$, and so p and q are orthogonal. There is no difficulty involved in verifying that

$$
e\,x = \begin{pmatrix} e\,a^{1/2} & 0 & 0 & 0 \\ 0 & 0 & 0 & 0 \\ 0 & 0 & 0 & 0 \\ 0 & 0 & 0 & 0 \end{pmatrix},
$$

which gives us $e\,p\,e = e\,x\,x^*\,e = e\,a\,e = a$. In the same way $e\,q\,e = b$ and the proof is completed. □

5.4.5. Lemma. *Let ϱ be a quasi-functional on a von Neumann algebra M. Suppose that*

$$|\varrho\,(a+b) - \varrho(a) - \varrho(b)| < \delta\,,$$

whenever $0 \le a, b \le a+b \le 1$. Then

$$\left|\varrho\Big(\sum_{i=1}^{k} x_i\Big) - \sum_{i=1}^{k} \varrho(x_i)\right| < (k-1)\,\delta\,,$$

whenever $x_1, x_2, \ldots, x_k$ are positive, $\sum_{i=1}^{k} x_i \le 1$, and $k > 1$.

Proof: It follows immediately by induction on k.

5.4.6. Proposition. *Let M be a properly infinite von Neumann algebra and $\delta > 0$. Then there is a projection e in M such that*

(i) $|\mu(p) - \mu(e\,p\,e) - \mu((1-e)\,p\,(1-e))| < \delta$ *for all* $p \in P(M)$,

(ii) $|\mu(x+y) - \mu(x) - \mu(y)| < \delta$, *whenever* $0 \le x, y \le x+y \le e$,

(iii) $|\mu(x+y) - \mu(x) - \mu(y)| < \delta$, *whenever* $0 \le x, y \le x+y \le 1-e$

Proof: Let us fix $0 < \varepsilon < \frac{1}{4}$ and let us choose, by Lemma 5.4.3 a projection $e \sim 1 \sim 1-e$ with $\mu(e) > \frac{1}{2} - \varepsilon$. Then, by using Lemma 5.4.1, we have

$$|\mu(p) - \mu(e\,p\,e) - \mu((1-e)\,p\,(1-e))| < 4\,\varepsilon^{1/2}, \tag{5.15}$$

for all projections $p \in M$. Let $h = e + f$, where f is a projection such that $f \le 1-e$, $f \sim 1-e \sim 1$ and $V(f) < \varepsilon$, the existence of which is guaranteed by Lemma 5.4.2. Proper infinity of the algebra enables us to find three orthogonal projections equivalent to e that are underneath f. Suppose now that $0 \le x, y \le x+y \le e$. On applying Christensen's Lemma we can find orthogonal projections p, q majorized by h such that

$$x = 2\,e\,p\,e \qquad \text{and} \qquad y = 2\,e\,q\,e\,.$$

Since x is in $W^*(e,p)$ it follows from Corollary 5.3.10 that

$$|2\,\mu(p) - \mu(x)| = |\mu(2\,p - x)| = 2\,|\mu(p - e\,p\,e)|\,.$$

By Corollary 5.3.11, μ is linear on $W^*(h,p,q) = W^*(e,p,f)$. Making use of this fact, we have

$$\begin{aligned} 2\,|\mu(p - e\,p\,e)| &= 2\,|\mu(f\,p\,f + f\,p\,e + e\,p\,f)| \\ &\le 2\,|\mu(f\,p\,f)| + 2\,|\mu(f\,p\,e + e\,p\,f)|\,. \end{aligned}$$

By Proposition 5.2.7,

$$|\mu(f\,p\,f)| \le 2\,\alpha(f) - \mu(f) \le 3\,V(f) \le 3\,\varepsilon\,.$$

Hence,

$$|2\,\mu(p) - \mu(x)| < 6\varepsilon + 2\,|\mu(f\,p\,e + e\,p\,f)|\,,$$

as well as, by (5.15),

$$|2\,\mu(p) - \mu(x)| < 6\,\varepsilon + 8\,\varepsilon^{1/2} < 14\,\varepsilon^{1/2}\,.$$

In the same way we can establish the inequalities

$$\begin{aligned} |2\,\mu(q) - \mu(y)| &< 14\,\varepsilon^{1/2} \\ |2\,\mu(p+q) - \mu(x+y)| &< 14\,\varepsilon^{1/2}\,. \end{aligned}$$

In summary,

$$|\mu(x+y) - \mu(x) - \mu(y)| < 42\,\varepsilon^{1/2}\,.$$

Hence with $\varepsilon = (\delta/42)^2$ we have (i) and also (ii). For (iii), observe that

$$(-\mu)(1-e) = \mu(e) > \frac{1}{2} - \varepsilon\,.$$

Therefore, we get (iii) by applying the previous arguments to $-\mu$. □

The next lemma shows that a quasi-functional on a von Neumann algebra is approximately linear whenever it is linear on two orthogonal hereditary subalgebras with the "diagonalization property".

5.4.7. Lemma. *Let M be a von Neumann algebra without Type I_2 direct summand. Let $\varepsilon > 0$ and let m be a natural number such that $\sum_{n=m+1}^{\infty} 2^{-n} < \varepsilon$. Assume that p and q are orthogonal projections in M satisfying*

(i) $|\mu(e) - \mu(p\,e\,p) - \mu(q\,e\,q)| < \frac{\varepsilon}{m}$, *for each projection* $e \leq p+q$.

(ii) $|\mu(x+y) - \mu(x) - \mu(y)| < \frac{\varepsilon}{m}$, *whenever* $x, y \geq 0$ *and* $x+y \leq p$.

(iii) $|\mu(x+y) - \mu(x) - \mu(y)| < \frac{\varepsilon}{m}$, *whenever* $x, y \geq 0$ *and* $x+y \leq q$.

Then

$$|\mu(x+y) - \mu(x) - \mu(y)| < 20\,\varepsilon,$$

whenever $x, y \geq 0$ *and* $x+y \leq p+q$.

Proof: There is no loss in assuming that $p+q=1$. There are sequences of projections (p_n), (q_n) and (r_n) in $W^*(x)$, $W^*(y)$ and $W^*(x+y)$, correspondingly, such that

$$x = \sum_{n=1}^{\infty} 2^{-n}\,p_n\,, \qquad y = \sum_{n=1}^{\infty} 2^{-n}\,q_n\,, \qquad x+y = \sum_{n=1}^{\infty} 2^{-n}\,r_n\,.$$

By the hypotheses,

$$\left\| x - \sum_{n=1}^{m} 2^{-n}\, p_n \right\| = \left\| \sum_{n=m+1}^{\infty} 2^{-n}\, p_n \right\| \le \sum_{n=m+1}^{\infty} 2^{-n} < \varepsilon\,,$$

and, similarly, $\|y - \sum_{n=1}^{m} 2^{-n}\, q_n\| < \varepsilon$, $\|x + y - \sum_{n=1}^{m} 2^{-n}\, r_n\| < \varepsilon$. Using the fact that μ is linear on algebras generated by pairs of projections, we obtain, for $a = x - \sum_{n=1}^{m} 2^{-n}\, p_n$

$$\begin{aligned} \mu(x) &= \mu\left(a + \sum_{n=1}^{m} 2^{-n}\, p_n\right) = \mu(a) + \sum_{n=1}^{m} 2^{-n}\, \mu(p_n) \qquad (5.16) \\ &= \mu(a) + \sum_{n=1}^{m} 2^{-n}\, \mu(p_n - p\, p_n\, p - (1-p)\, p_n\, (1-p)) \\ &+ \sum_{n=1}^{m} 2^{-n}\, \mu(p\, p_n\, p) + \sum_{n=1}^{m} 2^{-n}\, \mu((1-p)\, p_n\, (1-p))\,. \end{aligned}$$

Further, by Lemma 5.4.5 applied to the hereditary subalgebra $p\,M\,p$ we have that

$$\left| \mu(p\,x\,p) - \mu(p\,a\,p) - \sum_{n=1}^{m} 2^{-n}\, \mu(p\, p_n\, p) \right| < m\,\frac{\varepsilon}{m} = \varepsilon\,. \qquad (5.17)$$

Similarly,

$$\left| \mu((1-p)\,x\,(1-p) - \mu((1-p)\,a\,(1-p) - \sum_{n=1}^{m} 2^{-n}\, \mu((1-p)\, p_n\, (1-p) \right| < \varepsilon\,. \qquad (5.18)$$

Now, by (5.16), (5.17) and (5.18) we see that

$$\begin{aligned} &|\, \mu(x) - \mu(p\,x\,p) - \mu((1-p)\,x\,(1-p))\,| \\ &\le \left| \mu(a) + \sum_{n=1}^{m} 2^{-n}\, \mu(p\, p_n\, p) + \sum_{n=1}^{m} 2^{-n} \mu((1-p)\, p_n\, (1-p)) \right. \\ &\left. -\mu(p\,x\,p) - \mu((1-p)\,x\,(1-p)) \right| \\ &+ \sum_{n=1}^{m} |\, 2^{-n}\, \mu(p_n - p\, p_n\, p - (1-p)\, p_n\, (1-p))\,| \end{aligned}$$

$$\leq | \mu(a) - \mu(p\,a\,p) - \mu((1-p)\,a\,(1-p))\,|$$
$$+2\,\varepsilon + \sum_{n=1}^{m} 2^{-n}\,|\mu(p_n - p\,p_n\,p - (1-p)\,p_n(1-p))\,|\,. \tag{5.19}$$

Whence,

$$|\,\mu(x) - \mu(p\,x\,p) - \mu((1-p)\,x(1-p))| \leq \|a\| + \|p\,a\,p\|$$
$$+\|(1-p)\,a\,(1-p)\| + 2\varepsilon + \frac{\varepsilon}{m} \leq 3\,\|a\| + 2\,\varepsilon + \frac{\varepsilon}{m} < 6\,\varepsilon\,.$$

In the same way we can show that the previous inequality holds if we replace x by y or $x+y$. Finally,

$$|\,\mu(x+y) - \mu(x) - \mu(y)\,| < |\mu(p\,(x+y)\,p) - \mu(p\,x\,p) - \mu(p\,y\,p)\,| +$$
$$|\mu((1-p)\,(x+y)\,(1-p)) - \mu((1-p)\,x(1-p)) - \mu((1-p)\,y\,(1-p)\,|$$
$$+18\,\varepsilon\,.$$

So,

$$|\,\mu(x+y) - \mu(x) - \mu(y)\,| < 2\,\frac{\varepsilon}{m} + 18\,\varepsilon < 20\,\varepsilon\,.$$

□

5.4.8. Theorem. *Let M be a properly infinite von Neumann algebra. Then any bounded complex measure μ on $P(M)$ extends to a linear functional on M.*

Proof: Combining Proposition 5.4.6 and Lemma 5.4.7 we can easily establish the inequality

$$|\,\mu(x+y) - \mu(x) - \mu(y)\,| < \varepsilon$$

whenever $0 \leq x, y \leq x+y \leq 1$ and ε is arbitrary. It immediately implies that the quasi-linear extension of μ is linear. □

5.5. Linear extensions of measures on finite algebras

The proof of Gleason Theorem for finite algebras requires, in addition to the techniques developed for infinite algebras some new and quite elaborate arguments. Roughly speaking, the basic idea is to associate to a given quasi-functional μ a suitable tracial state φ and to show first that μ is approximately linear on a hereditary subalgebra that is large with respect to φ. This will be achieved by careful analysis of values of μ at level sets of φ. Having done this, the technique of approximate diagonalization and structure arguments will complete the proof.

Unless specified otherwise, in this paragraph M will be a finite von Neumann algebra with no Type I_2 direct summand and $\mu\colon P(M) \to \mathbb{R}$ be a bounded real measure with the quasi-linear extension denoted again by μ. Let us recall the notation of the previous part:

$$\begin{aligned} V(p) &= \sup\{|\mu(e)| \mid e \le p,\, e \in P(M)\} \\ \alpha(p) &= \sup\{\mu(e) \mid e \le p,\, e \in P(M)\}\,. \end{aligned}$$

Let us observe that α is additive with respect to the orthogonal central projections. Indeed, let z_1 and z_2 be orthogonal central projections in M. If e is a projection less then $z_1 + z_2$ then

$$\mu(e) = \mu(e\,z_1 + e\,z_2) = \mu(e\,z_1) + \mu(e\,z_2) \le \alpha(z_1) + \alpha(z_2)$$

and so

$$\alpha(z_1 + z_2) \le \alpha(z_1) + \alpha(z_2)\,.$$

On the other hand, for projections $e_1 \le z_1$ and $e_2 \le z_1$ with $\mu(e_1) > \alpha(z_1) - \varepsilon$ and $\mu(e_2) > \alpha(z_2) - \varepsilon$ we have

$$\alpha(z_1) + \alpha(z_2) < 2\varepsilon + \mu(e_1 + e_2) \le 2\varepsilon + \alpha(z_1 + z_2)\,.$$

Since ε is arbitrary we see that

$$\alpha(z_1 + z_2) = \alpha(z_1) + \alpha(z_2)\,.$$

Now, $\mu|Z(M)$ is linear and we can find a linear functional $\varrho \in M^*$ coinciding with μ on the centre. By replacing μ by $\mu - \varrho$ if necessary, we can suppose that

$$\mu|Z(M) = 0\,. \tag{5.20}$$

It holds that α is a nonnegative measure on $Z(M)$, since $\alpha(z) \geq \mu(0) = 0$. By Proposition 5.2.7 we see that

$$\sup\{|\mu(x)| \mid \|x\| \leq 1\} = 2\,\alpha(1) - \mu(1) = 2\,\alpha(1)\,.$$

Suppose that z_n is a central projection in M such that $z_n M$ is zero or Type I_n, $n < \infty$, part of M. Employing Theorem 5.3.9, we infer that μ is linear on $z_n M$. By the inequality above applied to $z_n M$ we obtain, for $\|x\| \leq 1$,

$$|\mu(z_n\, x)| \leq 2\,\alpha(z_n)\,.$$

Hence,

$$\sum_{n=1}^{\infty} |\mu(x\, z_n)| \leq 2 \sum_{n=1}^{\infty} \alpha(z_n) \leq 2\,\alpha(1)\,.$$

In other words, $\sum_{n=1}^{\infty} \mu(x\, z_n)$ is absolutely convergent for each $x \in M$. Set $\psi(x) = \sum_{n=1}^{\infty} \mu(x\, z_n)$. Then $\psi \in M^*$ and by replacing μ by $\mu - \psi$ when needed, we can suppose that μ vanishes on each Type I_n finite direct summand. Let us denote by $T\colon\ M \to Z(M)$ the canonical centre-valued trace on M.

5.5.1. Lemma. *There is a positive tracial functional φ on M such that*

$$V(z) \leq \varphi(z) \leq 2\,V(z) \tag{5.21}$$

for each central projection z. Moreover, $\varphi(z_n) = 0$ for each central Type I_n projection z_n.

Proof: As α is a positive measure on the centre, it can be extended to a bounded positive linear functional $\hat{\alpha}$ on $Z(M)$. Set $\varphi_1 = \hat{\alpha} \circ T$. Starting with $-\mu$ we obtain, in the same way, a positive tracial functional φ_2. Let $\varphi = \varphi_1 + \varphi_2$. The inequality $V(z) \leq \varphi(z) \leq 2\,V(z)$ is clear for each central projection z. Since μ vanishes on all Type I_n, $n < \infty$, direct summands, we have $V(z_n) = 0$ for the corresponding central Type I_n projection z_n. This implies that

$$0 \leq \varphi(z_n) \leq 2\,V(z_n) = 0\,.$$

□

By scaling μ appropriately we may suppose that $\varphi(1) = 1$, i.e. we may suppose that φ is a tracial state. For the sake of clarity, we now summarize assumptions adopted in this section:

$$\mu | Z(M) = 0\,, \tag{5.22}$$

$$\mu \text{ vanishes on all Type } I_n\,, n < \infty\,, \text{ direct summands,} \tag{5.23}$$

$$\varphi(1) = 1\,. \tag{5.24}$$

Note that (5.24) yields

$$V(1) = \sup\{|\mu(p)| \mid p \in P(M)\} \le \varphi(1) = 1 \tag{5.25}$$

and, also, by Proposition 5.2.7 (ii) we have

$$\sup\{\mu(x) \mid 0 \le x \le 1\} = \alpha(1) \le \varphi(1) = 1\,. \tag{5.26}$$

Since φ is zero on all possible atoms in $P(M)$, the noncommutative version of the Lyapunov theorem on convexity of the range of nonatomic measures can be applied to show that the range of φ is an interval. We will prove this fact in our context, although the tracial property is not necessary for the validity of the following proposition.

5.5.2. Proposition. *Let $e \in P(M)$. For each $t \in [0, \varphi(e)]$ there is $q \in P(M)$ such that $q \le e$ and $\varphi(q) = t$.*

Proof: We shall suppose that $\varphi(e) > 0$. Let z be a Type II_1 central projection in the hereditary subalgebra $e\,M\,e$ and let W be the canonical centre-valued trace on the hereditary subalgebra $e\,M\,e$. Since W maps $P(z\,e\,M\,e)$ onto the positive part of the unit ball of $Z(z\,e\,M\,e)$, we can certainly find a projection $q_1 \le z$ with $W(q_1) = \frac{t}{\varphi(e)}\,z$. It means that

$$\varphi(q_1) = \varphi(W(q_1)) = \frac{t}{\varphi(e)}\,\varphi(z)\,.$$

Suppose now that z_n is a Type I_n projection in $e\,M\,e$. In that case we can identify $z_n\,e\,M\,e$ with the algebra $C(X_n)$, where X_n is a hyperstonean space. According to Chapter 2 (Classification of von Neumann algebras), the values of W on $P(z_n\,e\,M\,e)$ are exactly the continuous functions on X_n of the form

$$\frac{1}{n}\,h_1 + \frac{2}{n}\,h_2 + \cdots + \frac{n-1}{n}\,h_{n-1} + h_n\,, \tag{5.27}$$

where h_i $(i = 1, \ldots, n)$ are characteristic functions of clopen disjoint subsets of X_n. Consider now the function

$$g_n = \frac{t}{\varphi(e)} W(z_n) = \frac{t}{\varphi(e)} z_n \qquad \text{on } X_n .$$

It is clear that g_n can be approximated up to $\frac{1}{n} z_n$ by the functions of the type (5.27). More precisely, there is a projection $f_n \le z_n$ such that

$$W(f_n) \le \frac{t}{\varphi(e)} W(z_n) \le W(f_n) + \frac{1}{n} z_n .$$

Therefore,

$$W\Big(\sum_n f_n\Big) \le \frac{t}{\varphi(e)} \sum_n z_n \le W\Big(\sum_n f_n\Big) + \sum_n \frac{1}{n} z_n .$$

It implies,

$$\varphi\Big(\sum_{n=1}^{\infty} f_n\Big) \le \frac{t}{\varphi(e)} \varphi\Big(\sum_{n=1}^{\infty} z_n\Big) \le \varphi\Big(\sum_{n=1}^{\infty} f_n\Big) + \varphi\Big(\sum_{n=1}^{\infty} \frac{1}{n} z_n\Big) . \tag{5.28}$$

However, since $\varphi(z_n) = 0$ for each n (Lemma 5.5.1) and $\sum_{n=1}^{\infty} \frac{1}{n} z_n$ lies in the norm closure of the linear span of (z_n), we see that $\varphi(\sum_{n=1}^{\infty} \frac{1}{n} z_n) = 0$. Hence, (5.28) amounts to

$$\varphi\Big(\sum_{n=1}^{\infty} f_n\Big) = \frac{t}{\varphi(e)} \varphi(\sum_{n=1}^{\infty} z_n) .$$

Setting $q = q_1 + \sum_{n=1}^{\infty} f_n$ we obtain, taking into account that $z + \sum_{n=1}^{\infty} z_n = e$, that

$$\varphi(q) = \frac{\varphi(z)}{\varphi(e)} t + \frac{\varphi\Big(\sum_{n=1}^{\infty} z_n\Big)}{\varphi(e)} t = t .$$

□

For further study of measures on finite algebras, the following notation will be handy. For each $e \in P(M)$ and $0 \le t \le \varphi(e)$ let

$$P(e,t) = \{ f \in P(M) \mid f \le e ,\, \varphi(f) = t \} .$$

Proposition 5.5.2 guarantees that $P(e,t)$ is nonempty. Furthermore, upon setting

$$\begin{aligned} M(e,t) &= \sup\{\mu(f) \mid f \in P(e,t)\}, \\ m(e,t) &= \inf\{\mu(f) \mid f \in P(e,t)\}, \end{aligned}$$

we immediately see from the definition that

$$-V(e) \le m(e,t) \le M(e,t) \le V(e) \tag{5.29}$$

$$M(e,t) = \mu(e) - m(e, \varphi(e) - t). \tag{5.30}$$

Our first goal will be to exhibit the convexity properties of the functions $M(e,\cdot)$ and $m(e,\cdot)$. For this, the following lemma will be useful:

5.5.3. Lemma. *Let $e \in P(M)$ with $\varphi(e) \neq 0$ and let $0 < s \le t < \varphi(e)$.*

(i) *For a given $\varepsilon > 0$ and $f \in P(e,s)$ such that $\mu(f) > M(e,s) - \varepsilon$ there is a projection $g \ge f$ such that $g \in P(e,t)$ and $\mu(q) > M(e,t) - \varepsilon$.*

(ii) *For a given $\varepsilon > 0$ and a projection $f \in P(e,t)$ such that $\mu(f) > M(e,t) - \varepsilon$ there is a projection $g \le f$ such that $g \in P(e,s)$ and $\mu(g) > M(e,s) - \varepsilon$.*

Proof: (i) Let $h \in P(e,t)$ with $\mu(h) > M(e,t) - \delta$, where $\delta = \mu(f) - M(e,s) + \varepsilon > 0$. By the parallelogram law we have

$$h - h \wedge (1-f) \sim h \vee (1-f) - (1-f) \le 1 - (1-f) = f .$$

The tracial property of φ gives us

$$t - s = \varphi(h) - \varphi(f) \le \varphi(h \wedge (1-f)).$$

Using the Darboux property of φ (Proposition 5.5.2), we can find a projection h_0 in $P(e, t-s)$ satisfying $h_0 \le (1-f) \wedge h$. So $h - h_0 \in P(e,s)$. The inequality

$$M(e,s) \ge \mu(h - h_0) = \mu(h) - \mu(h_0) > M(e,t) - \delta - \mu(h_0)$$

implies

$$\mu(h_0) > M(e,t) - M(e,s) - \delta .$$

Let $g = f + h_0$. Of course, $g \geq f$ and $g \in P(e,t)$. Also,

$$\begin{aligned}\mu(g) &= \mu(f) + \mu(h_0) > \mu(f) + M(e,t) - M(e,s) - \delta \\ &= \mu(f) + M(e,t) - M(e,s) - (\mu(f) - M(e,s) + \varepsilon)) = M(e,t) - \varepsilon\,.\end{aligned}$$

(ii) The proof of (ii) uses a dual argument to the proof of (i). Let $\delta = \mu(f) - M(e,t) + \varepsilon$ and choose $h \in P(e,s)$ with $\mu(h) > M(e,s) - \delta$. Since

$$f - f \wedge (1-h) \sim f \vee (1-h) - (1-h) \leq 1 - (1-h) = h\,,$$

we infer that

$$t - s = \varphi(f) - \varphi(h) \leq \varphi(f \wedge (1-h))\,.$$

We can find a projection $h_0 \in P(e, t-s)$ with $h_0 \leq f \wedge (1-h)$. Then $\varphi(h + h_0) = t$ and

$$\begin{aligned}M(e,t) &\geq \mu(h+h_0) = \mu(h) + \mu(h_0) > M(e,s) - \delta + \mu(h_0) \\ &= M(e,s) + M(e,t) - \mu(f) - \varepsilon + \mu(h_0)\,.\end{aligned}$$

Put $g = f - h_0$. Then $g \in P(e,s)$ and

$$\begin{aligned}\mu(g) &= \mu(f) - \mu(h_0) > \mu(f) - M(e,t) + M(e,s) + M(e,t) - \mu(f) - \varepsilon \\ &= M(e,s) - \varepsilon\,.\end{aligned}$$

□

It turns out that the functions $t \to M(e,t) : t \to m(e,t)$ have nice convexity properties which are, in fact, a consequence of the tracial property of φ.

5.5.4. Proposition. *Let $\varphi(e) \neq 0$ for a projection $e \in P(M)$. Let $M(t) = M(e,t)$ and $m(t) = m(e,t)$ be functions defined on $(0, \varphi(e))$. Then the following statements hold true:*

(i) *$M(t)$ is a continuous concave function having proper one-sided limits at 0 and $\varphi(e)$.*

(ii) *$m(t)$ is a continuous convex function having proper one-sided limits at 0 and $\varphi(e)$.*

(iii) *$m(e,t_1) + m(e,t_2) \leq m(e, t_1 + t_2)$, whenever $0 \leq t_1, t_2, t_1 + t_2 \leq \varphi(e)$.*

(iv) $M(e,t_1)+M(e,t_2) \geq M(e,t_1+t_2)$, *whenever* $0 \leq t_1, t_2, t_1+t_2 \leq \varphi(e)$.

(v) $\lim_{t\to 0+} m(e,t) \leq 0$.

(vi) $\lim_{t\to 0+} M(e,t) \geq 0$.

Proof: (i) Let $0 < t_1 < \frac{t_1+t_2}{2} < t_2 < \varphi(e)$ and choose $\varepsilon > 0$. Employing Lemma 5.5.3 (ii), we can find projections $p \geq q \geq r$ in $P(e,t_2)$, $P(e, \frac{t_1+t_2}{2})$ and $P(e,t_1)$, respectively, with

$$\begin{aligned} \mu(p) &> M(e,t_2) - \varepsilon \\ \mu(q) &> M\left(e, \frac{t_1+t_2}{2}\right) - \varepsilon \\ \mu(r) &> M(e,t_1) - \varepsilon \,. \end{aligned}$$

It is clear that $r+p-q \in P(e, \frac{t_1+t_2}{2})$ Hence,

$$M\left(\frac{t_1+t_2}{2}\right) \geq \mu(r+(p-q)) = \mu(r) + \mu(p) - \mu(q)\,,$$

and therefore

$$M\left(\frac{t_1+t_2}{2}\right) + \mu(q) \geq M(t_1) + M(t_2) - 2\,\varepsilon\,.$$

By considering the supremum with respect to all possible $q \in P(e, \frac{t_1+t_2}{2})$ in the previous inequality and by realizing that ε can be arbitrary, we deduce that

$$2\,M\left(\frac{t_1+t_2}{2}\right) \geq M(t_1) + M(t_2)\,.$$

The function $M(t)$ is therefore continuous and concave and admits proper limits at boundary points the interval $[0,\varphi(e)]$.

The assertion (ii) follows from (i) and (5.30)

Let us prove assertions (iii) and (iv). Take $p \in P(e,t_1+t_2)$. By Lemma 5.5.3 there is a projection $q \in P(e,t_1)$, $q \leq p$. Then $p-q \in P(e,t_2)$. It follows that

$$m(e,t_1) + m(e,t_2) \leq \mu(q) + \mu(p-q) = \mu(p)\,.$$

Also,

$$\mu(p) = \mu(q) + \mu(p-q) \leq M(t_1) + M(t_2)\,.$$

The last two inequalities imply (by considering the corresponding suprema or infima) the inequalities (iii) and (iv).
Let us prove the statement (iv). For $0 < t < \varphi(e)$ we utilize (iii) to derive that $2\,m(e,t/2) \leq m(e,t)$. Using this rule repeatedly we obtain, for each integer n,

$$m(e, 2^{-n}\,t) \leq 2^{-n}\,m(e,t) \leq 2^{-n}\,V(e)\,.$$

Thus, $\lim_{t\to 0+} m(e,t) \leq 0\,.$
Finally, the proof of (vi) is same as the proof of (v). □

The properties of the functions $m(t)$ and $M(t)$ displayed in the previous lemma will be applied to estimating the values of μ on orthogonal hereditary algebras.

5.5.5. Lemma. *Let e and f be orthogonal projections in M such that $\varphi(e) \neq 0$, $\varphi(f) \neq 0$. Set $\delta = M(e+f, \varphi(e)) - \mu(e)$. Then there exists a real number λ such that*

$$\mu(e\,a\,e) \geq \lambda\,\varphi(a\,e) - \delta \qquad \textit{and} \qquad \mu(faf) \leq \lambda\,\varphi(af) + \delta\,,$$

whenever $0 \leq a \leq 1$.

Proof: Take $0 < t < \min(\varphi(e), \varphi(f))$. Let $p \in P(e,t)$ and let $q \in P(f,t)$. Obviously, $p\,q = 0$. Hence,

$$\varphi(e - p + q) = \varphi(e) - t + t = \varphi(e)$$

and so

$$\mu(e) - \mu(p) + \mu(q) = \mu(e - p + q) \leq M(e + f, \varphi(e)) = \mu(e) + \delta\,.$$

In other words, $\mu(q) \leq \mu(p) + \delta$, and this results in

$$\mu(q) \leq m(e,t) + \delta\,. \qquad \text{Thus,}\ \ M(f,t) \leq m(e,t) + \delta\,. \tag{5.31}$$

Now consider the following two sets in the plane:

$$\begin{aligned} O_1 &= \{(t,y) \mid 0 < t < \varphi(e),\ y > m(e,t) + \delta\} \\ O_2 &= \{(t,y) \mid 0 < t < \varphi(f),\ y < M(f,t)\}\,. \end{aligned}$$

By (5.31) O_1 and O_2 are disjoint. Moreover, by Proposition 5.5.4, both O_1 and O_2 are open convex sets. Therefore O_1 and O_2 can be separated by a line $y = \lambda t + \varepsilon$ in the sense that

$$\begin{aligned} M(f,t) &\le \lambda t + \varepsilon \qquad \text{for all } t \in [0, \varphi(f)), \\ \delta + m(e,t) &\ge \lambda t + \varepsilon \qquad \text{for all } t \in [0, \varphi(e)). \end{aligned}$$

(See Figure. 5.5.1)

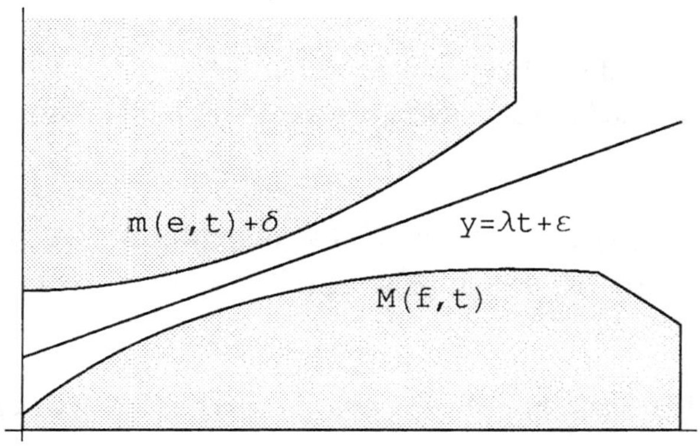

Figure 5.5.1

By letting $t \to 0+$ and employing Proposition 5.5.4 (v), (vi), we obtain $0 \le \varepsilon \le \delta$. Let us now take a projection $p \le e$. Then

$$\mu(p) \ge m(e, \varphi(p)) \ge \lambda \varphi(p) + \varepsilon - \delta \ge \lambda \varphi(p) - \delta\,.$$

For any $0 \le a \le 1$ we can write $e\,a\,e = \sum_{n=1}^{\infty} \frac{1}{2^n} h_n$, where (h_n) are commuting projections in $e\,M\,e$. It yields

$$\mu(e\,a\,e) \ge \lambda \varphi(e\,a) - \delta\,.$$

By the same argument we finally obtain

$$\mu(f\,a\,f) \le \lambda \varphi(a\,e) + \varepsilon \le \lambda \varphi(a\,f) + \delta\,. \qquad \square$$

The following lemma is from the C^*-theory and allows us to construct equivalent projections in an arbitrary C^*-algebra starting with a pair of orthogonal projections.

5.5.6. Lemma. *Let e, f be orthogonal projections in a C^* algebra A and let g be a projection dominated by $e+f$. Let us put $b = e\,g\,f$ and set, for $\lambda \in [-1,1]$,*

$$c_\lambda = \frac{1}{2}\,e + \left(\frac{1}{4}\,e - \lambda^2\,b\,b^*\right)^{1/2}, \qquad d_\lambda = \frac{1}{2}\,f - \left(\frac{1}{4}\,f - \lambda^2\,b^*\,b\right)^{1/2},$$

$$g_\lambda^+ = c_\lambda + d_\lambda + \lambda(b+b^*)\,, \qquad g_\lambda^- = c_\lambda + d_\lambda - \lambda(b+b^*)\,.$$

Then

$$0 \le e - c_\lambda \le \frac{1}{2}\,\lambda^2\,e\,, \qquad 0 \le d_\lambda \le \frac{1}{2}\,\lambda^2\,f\,. \tag{5.32}$$

Furthermore, g_+ and g_- are projections dominated by $e+f$ and satisfying $g_\lambda^+ \sim e \sim g_\lambda^-$, whenever $\lambda^2 + |\lambda| < 1$.

Proof: Obviously, $b\,b^* \in e\,A\,e$ and $b^*\,b \in f\,A\,f$. Further, $\|b\| \le 1/2$ by (2.6) and so $\lambda^2\,b\,b^* \le \frac{1}{4}\,e$ and $\lambda^2\,b^*\,b \le \frac{1}{4}\,f$. (Therefore c_λ and d_λ are correctly defined.) It is also clear that $c_\lambda \in e\,A\,e$, $d_\lambda \in f\,A\,f$, $e - c_\lambda = \frac{1}{2}\,e - (\frac{1}{4}\,e - \lambda^2\,b\,b^*)^{1/2} \ge \frac{1}{2}\,e - \frac{1}{2}\,e = 0$, and $d_\lambda \ge \frac{1}{2}\,f - \frac{1}{2}\,f = 0$. Define a continuous function

$$\psi(t) = \frac{1}{2} - \left(\frac{1}{4} - t^2\right)^{1/2} \qquad \text{for } t \in [0, 1/2]\,,$$

and observe that $\psi(t) \le 2\,t^2$. Using the function calculus (in the hereditary subalgebra $e\,A\,e$), we obtain $e - c_\lambda = \psi(|\lambda|\sqrt{b\,b^*})$. Therefore,

$$\begin{aligned} \|e - c_\lambda\| &= \sup\{|\psi(x)| \mid x \in \mathrm{Spec}(|\lambda|\,\sqrt{b\,b^*})\,\} \\ &\le \sup\{|\psi(x)| \mid x \in [0, |\lambda|\,\|\sqrt{b\,b^*}\|\,]\,\} \\ &\le \sup\{2\,x^2 \mid x \in [0, |\lambda|\,\|\sqrt{b\,b^*}\|\,]\,\} \\ &= 2\,|\lambda|^2\,\|\sqrt{b\,b^*}\|^2 \le 2\,\lambda^2\,\frac{1}{4} = \frac{1}{2}\,\lambda^2\,. \end{aligned}$$

This establishes the first inequality in (5.32). In exactly the same way one can prove that $d_\lambda \le \frac{1}{2}\,\lambda^2\,f$. An easy algebraic argument says that $p(b\,b^*)\,b = b\,p(b^*\,b)$ for any polynomial p. By the function calculus the above observation holds not only for polynomials but also for all continuous functions on $[0, \frac{1}{4}]$. Since c_λ and d_λ can be written in the form

$$c_\lambda = \frac{1}{2}\,e + \omega(b\,b^*) \qquad d_\lambda = \frac{1}{2}\,f - \omega(b^*\,b)\,,$$

where ω is a function defined by $\omega(x) = (1/4 - \lambda^2 x)^{1/2}$, $x \in [0, 1/4]$, we find out that

$$c_\lambda b + b d_\lambda = \frac{1}{2} b + \omega(b b^*) b + \frac{1}{2} b - b \omega(b^* b) = b. \tag{5.33}$$

An easy computations show that

$$\begin{aligned} c_\lambda d_\lambda = b c_\lambda = d_\lambda b &= 0 \\ c_\lambda^2 + \lambda^2 b b^* &= c_\lambda \\ d_\lambda^2 + \lambda^2 b^* b &= d_\lambda . \end{aligned} \tag{5.34}$$

Thanks to the orthogonality of e and f we also have

$$b^2 = b^{*2} = 0. \tag{5.35}$$

Making use of (5.33), (5.34) and (5.35) we obtain

$$\begin{aligned} g_\lambda^{+2} &= c_\lambda^2 + d_\lambda^2 + \lambda(c_\lambda b + d_\lambda b^* + b^* c_\lambda + b d_\lambda) + \lambda^2(b b^* + b^* b) \\ &= g_\lambda^+ \end{aligned}$$

Therefore, g_λ^+ is a projection. Evidently, $g_\lambda \le e + f$. Moreover,

$$\begin{aligned} \|e - g_\lambda^+\| &\le \|e - c_\lambda\| + \|d_\lambda\| + 2\,|\lambda|\,\|b\| \\ &\le \frac{1}{2}\lambda^2 + \frac{1}{2}\lambda^2 + |\lambda| = \lambda^2 + |\lambda|. \end{aligned}$$

Therefore, $\|e - g_\lambda\| < 1$, whenever $\lambda^2 + |\lambda| < 1$ and so $e \sim g_\lambda^+$ in this case (see Chapter 2, Projection structure of C^*-algebras). Similar arguments apply to g_λ^-. □

5.5.7. Lemma. *Let e and f be orthogonal projections in M and let $0 < \varepsilon < 1/2$. Suppose that*

$$\mu(e) > M(e + f, \varphi(e)) - \frac{1}{4}\varepsilon^4 .$$

Then

$$|\mu(g) - \mu(e g e) - \mu(f g f)| < \varepsilon \tag{5.36}$$

for all projections $g \le e+f$. Furthermore, suppose that L_1, L_2, L_3, L_4 are linear functionals on M such that

$$\begin{aligned} L_i(x) &= L_i(e\,x\,e) \qquad i=1,2 \\ L_i(x) &= L_i(f\,x\,f) \qquad i=3,4\,. \end{aligned}$$

Let $\gamma_1, \gamma_2, \gamma_3, \gamma_4$ be positive real numbers such that

$$\begin{aligned} L_1(a)-\gamma_1 \quad &\le \mu(e\,a\,e) \quad \le L_2(a)+\gamma_2 \\ L_3(a)-\gamma_3 \quad &\le \mu(f\,a\,f) \quad \le L_4(a)+\gamma_4\,. \end{aligned}$$

Then, for each $0 \le a \le 1$,

$$(L_1+L_3)(a)-(\gamma_1+\gamma_3+\varepsilon) \le \mu((e+f)\,a\,(e+f)) \le (L_2+L_4)(a)+(\gamma_2+\gamma_4+\varepsilon)\,. \tag{5.37}$$

Proof: First observe that $W^*(1,e,f,g) = W^*(1,e,g,e+f)$ and so we can apply Corollary 5.3.11 to obtain that μ is linear on $W^*(1,e,f,g)$. Therefore

$$\mu(g) = \mu((e+f)\,g\,(e+f)) = \mu(e\,g\,e)+\mu(f\,g\,f)+\mu(e\,g\,f)+\mu(f\,g\,e)\,. \tag{5.38}$$

Now we will use notation of Lemma 5.5.6 for $\lambda = \varepsilon^2$ and set $c = c_\lambda$, $d = d_\lambda$, $g_+ = {g_\lambda}^+$, $g_- = {g_\lambda}^-$. Using these symbols we can rewrite (5.38)

$$|\mu(g)-\mu(e\,g\,e)-\mu(f\,g\,f)| = |\mu(b+b^*)|\,.$$

Since $g_+ \sim e$, we have $\varphi(g_+) = \varphi(e)$, and hence

$$\mu(g_+) \le M(e+f, \varphi(e))\,. \tag{5.39}$$

As $g_+ - c - d = \varepsilon^2(b+b^*)$ we obtain

$$\mu(g_+) = \varepsilon^2\,\mu(b+b^*) + \mu(c) + \mu(d)\,.$$

By substituting into (5.39), we further have

$$\varepsilon^2\,\mu(b+b^*) + \mu(c) + \mu(d) \le M(e+f, \varphi(e))$$

and, in turn,

$$\begin{aligned} \varepsilon^2\,\mu(b+b^*) &\le M(e+f,\varphi(e)) - \mu(c) + \mu(d) \\ &= M(e+f,\varphi(e)) - \mu(e) + \mu(e-c) + \mu(d)\,. \end{aligned}$$

Thus, by our assumption,

$$\varepsilon^2 \mu(b+b^*) \leq \frac{1}{4}\varepsilon^4 + \mu(e-c) + \mu(d)\,. \tag{5.40}$$

By Lemma 5.5.6 ,

$$0 \leq e-c \leq \frac{1}{2}\varepsilon^4 e \qquad \text{and} \qquad 0 \leq d \leq \frac{1}{2}\varepsilon^4 f\,.$$

It implies (Proposition 5.2.7 (ii)) that

$$\frac{2}{\varepsilon^4}\mu(e-c) \leq V(e) \qquad \text{and} \qquad \frac{2}{\varepsilon^4}\mu(d) \leq V(f)\,.$$

Returning to (5.40) we have

$$\varepsilon^2 \mu(b+b^*) \leq \frac{1}{4}\varepsilon^4 + \frac{1}{2}\varepsilon^4 V(e) + \frac{1}{2}\varepsilon^4 V(f)\,.$$

By our convention, $V(e), V(f) \leq V(1) \leq \varphi(1) = 1$ and so

$$\varepsilon^2 \mu(b+b^*) \leq \frac{5}{4}\varepsilon^4\,.$$

It means that

$$\mu(b+b^*) \leq \frac{5}{4}\varepsilon^2 < \varepsilon\,.$$

Similarly, by replacing g_+ by g_- we can obtain

$$-\mu(b+b^*) < \varepsilon\,.$$

In other words, for any projection $g \leq e+f$,

$$|\mu(g) - \mu(e\,g\,e) - \mu(f\,g\,f)| < \varepsilon\,.$$

This concludes the proof of (5.36) and simultaneously establishes (5.37) for $a = g$. The general case of $0 \leq a \leq 1$ can be obtained from this special situation by writing

$$(e+f)\,a\,(e+f) = \sum_{n=1}^{\infty} \frac{1}{2^n}\, g_n\,,$$

where g_n are (commuting) projections in $(e+f)\, M\,(e+f)$, and by using linearity of μ on abelian algebra generated by (g_n). □

Next proposition is a cornerstone of the proof of linearity of quasi-functionals on finite algebras. It says that quasi-functional is nearly linear on a hereditary von Neumann algebra which is arbitrarily large when measured in terms of the tracial functional φ. This corresponds, in a way, to establishing approximate Radon-Nikodym derivative of μ with respect to φ.

5.5.8. Proposition. *For a given ε, $0 < \varepsilon < \frac{1}{2}$, there exist mutually orthogonal projections h_1, h_2, h_3 in M with $h_1 + h_2 + h_3 = 1$ and*

(i) $\varphi(h_1) = \varphi(h_3) < \varepsilon^2$

(ii) $|\mu(p) - \mu(h_i\, p\, h_i) - \mu((1-h_i)\, p\,(1-h_i))| < \varepsilon$ *for $i = 1, 3$ and for all $p \in P(M)$.*

(iii) $|\mu(x+y) - \mu(x) - \mu(y)| < 13\,\varepsilon$, *whenever x and y are positive elements of M with $x + y \leq h_2$.*

Proof: For

$$M_0 = \lim_{t\to 0+} M(1,t) \qquad \text{and} \qquad M_1 = \lim_{t\to 1-} M(1,t)\,,$$

let us fix n large enough to guarantee that $\frac{1}{n} < \varepsilon^2$ and, for $0 < s \leq \frac{1}{n}$,

$$|M_0 - M(1,s)| < \frac{1}{2}\,\varepsilon \qquad \text{and} \qquad |M_1 - M(1,1-s)| < \frac{1}{2}\,\varepsilon\,. \tag{5.41}$$

Let us put

$$\delta = \frac{1}{4}\left(\frac{\varepsilon}{2n}\right)^4 .$$

By repeated application of Lemma 5.5.3 we can construct a decreasing chain of projections

$$1 = e_n > e_{n-1} > e_{n-2} > \ldots > e_2 > e_1$$

such that $e_k \in P(1, \frac{k}{n})$ and $\mu(e_k) > M(1, \frac{k}{n}) - \delta$ for $k = 1, \ldots, n-1$.

The projections h_1, h_2, h_3 will now be defined as follows

$$h = e_1\,, \qquad h_2 = e_{n-1} - e_1\,, \qquad h_3 = 1 - e_{n-1}.$$

It is clear from the definition that

$$\varphi(h_1) = \varphi(h_3) = \frac{1}{n} < \varepsilon^2 .$$

So we have (i). By applying Lemma 5.5.7 to the pair of orthogonal projections e_1 and $1 - e_1$, we see from $\mu(h_1) > M(1, \frac{1}{n}) - \delta$ that

$$|\mu(p) - \mu(h_1\, p\, h_1) - \mu((1 - h_1)\, p\, (1 - h_1))| < \sqrt[4]{4\,\delta} = \frac{\varepsilon}{2n} < \varepsilon ,$$

for all $p \in P(M)$. Arguing in the same way for a pair of orthogonal projections $1 - e_{n-1}$ and e_{n-1} we obtain

$$|\mu(p) - \mu(h_2\, p\, h_2) - \mu((1 - h_2)\, p\, (1 - h_2))| < \varepsilon ,$$

for all $p \in P(M)$. Therefore, it remains to prove (iii). For $k = 1, \ldots, n-1$ we set

$$f_k = e_{k+1} - e_k .$$

Then,

$$M(e_k + f_k, \varphi(e_k)) - \mu(e_k) \leq M(1, \varphi(e_k)) - \mu(e_k) < \delta$$

and we can use Lemma 5.5.5 to obtain real numbers λ_k $(k = 1, \ldots, n-1)$ such that

$$\mu(e_k\, a\, e_k) \;\geq\; \lambda_k\, \varphi(a\, e_k) - \delta \qquad (5.42)$$

$$\mu(f_k\, a\, f_k) \;\leq\; \lambda_k\, \varphi(a\, f_k) + \delta \qquad (5.43)$$

for all $0 \leq a \leq 1$. On substituting $f_{k-1}\, a\, f_{k-1}$ for a in (5.42) (and making use of $f_{k-1} \leq e_k$) we see that

$$\mu(f_{k-1}\, a\, f_{k-1}) \geq \lambda_k\, \varphi(a\, f_{k-1}) - \delta ,$$

and hence

$$\lambda_{k+1}\, \varphi(a\, f_k) - \delta \leq \mu(f_k\, a\, f_k) \leq \lambda_k\, \varphi(a\, f_k) + \delta \qquad (5.44)$$

for all $0 \leq a \leq 1$ and $k = 1, \ldots, n-2$. Since

$$f_1 + f_2 + \cdots + f_k = e_{k+1} - e_1 ,$$

we have $\varphi(f_1 + f_2 + \cdots + f_k) = \varphi(e_{k+1}) - \varphi(e_1) = \frac{k+1}{n} - \frac{1}{n} = \frac{k}{n}$. Therefore we can find $g \in P(f_1 + f_2 + \cdots + f_k, \frac{k-1}{n})$. Then $g + e_1 \in P(1, \frac{k}{n})$. It gives us

$$\begin{aligned} M\left(f_1 + f_2 + \cdots + f_k, \frac{k-1}{n}\right) &\le M\left(1, \frac{k}{n}\right) - \mu(e_1) \\ &< \mu(e_k) + \delta - \mu(e_1)\,. \end{aligned}$$

Hence,

$$\begin{aligned} & M\left(f_1 + f_2 + \cdots + f_k, \frac{k-1}{n}\right) < \mu(e_k) - \mu(e_1) + \delta \qquad (5.45) \\ & = \mu(e_k - e_1) + \delta = \mu(f_1 + f_2 + \cdots + f_{k-1}) + \delta\,, \end{aligned}$$

for $k = 1, \ldots, n-1$. By repeated application of (5.37) in Lemma 5.5.7 (for $\frac{\varepsilon}{2n}$ instead of ε) we obtain from (5.44) and (5.45)

$$\begin{aligned} & \sum_{k=1}^{n-2} \lambda_{k+1}\, \varphi(a\, f_k) - \left((n-2)\,\delta + (n-3)\,\frac{\varepsilon}{2n}\right) \qquad (5.46) \\ & < \mu(h_2\, a\, h_2) < \sum_{k=1}^{n-2} \lambda_k \varphi(a\, f_k) + \left((n-2)\,\delta + (n-3)\,\frac{\varepsilon}{2n}\right) \end{aligned}$$

for all $0 \le a \le 1$. Let us take into account that

$$\begin{aligned} (n-2)\,\delta + (n-3)\,\frac{\varepsilon}{2n} &= (n-2)\,\frac{1}{4}\left(\frac{\varepsilon}{2n}\right)^4 + (n-3)\,\frac{\varepsilon}{2n} \\ &\le \frac{\varepsilon}{2} + \frac{\varepsilon}{2} = \varepsilon\,. \end{aligned}$$

(Recall that $\frac{1}{n} < \varepsilon^2 < \varepsilon < 1/2$.) Therefore, (5.46) can be simplified to

$$\sum_{k=1}^{n-2} \lambda_{k+1}\, \varphi(a\, f_k) - \varepsilon < \mu(h_2\, a\, h_2) < \sum_{k=1}^{n-2} \lambda_k\, \varphi(a\, f_k) + \varepsilon\,. \qquad (5.47)$$

Putting $a = 1$ in (5.44) we obtain

$$\lambda_{k+1}\,\frac{1}{n} - \delta \le \mu(f_k) \le \lambda_k\,\frac{1}{n} + \delta\,,$$

and hence

$$\lambda_{k+1} < \lambda_k + 2n\,\delta\,. \tag{5.48}$$

Take now positive x and y with $x+y \le h_2$. Replacing a by $x+y$ in (5.47) we obtain

$$L - 2\varepsilon < L - \varepsilon < \mu(x+y) < K + \varepsilon < K + 2\varepsilon\,,$$

where we have denoted

$$\begin{aligned} K &= \sum_{k=1}^{n-2} \lambda_k\,\varphi((x+y)\,f_k) \\ L &= \sum_{k=1}^{n-2} \lambda_{k+1}\,\varphi((x+y)\,f_k)\,. \end{aligned}$$

Similarly, replacing a by x and y in (5.47) and adding the resulting inequalities, we obtain

$$K - 2\varepsilon < \mu(x) + \mu(y) < K + 2\varepsilon\,.$$

Therefore,

$$|\mu(x+y) - \mu(x) - \mu(y)| < K - L + 4\varepsilon\,.$$

We will now try to estimate $K - L$. First observe that, by making use of the inequality

$$\varphi((x+y)f_k) = \varphi(f_k(x+y)f_k) \le \varphi(f_k) = \frac{1}{n}$$

(and using (5.48)), we deduce that

$$(\lambda_k - \lambda_{k+1})\,\varphi((x+y)\,f_k) \le (\lambda_k - \lambda_{k+1} + 2n\,\delta)\,\frac{1}{n} = \frac{1}{n}\,(\lambda_k - \lambda_{k+1}) + 2\,\delta\,.$$

So

$$\begin{aligned} |\mu(x+y) - \mu(x) - \mu(y)| &\le \frac{1}{n}(\lambda_1 - \lambda_{n-1}) + 2\,(n-2)\,\delta + 4\,\varepsilon \\ &< \frac{1}{n}\,(\lambda_1 - \lambda_{n-1}) + 5\,\varepsilon\,. \end{aligned} \tag{5.49}$$

It remains to analyze the value $\frac{1}{n}(\lambda_1 - \lambda_{n-1})$. Lemma 5.5.3 implies the existence of a projection $q \in P(1, 1 - \frac{1}{2n})$ with $q > e_{n-1}$ and $\mu(q) > M(1, 1 - \frac{1}{2n}) - \delta$. By (5.41) we see that

$$\begin{aligned} M\left(1, 1 - \frac{1}{2n}\right) &> M_1 - \frac{\varepsilon}{2} \\ -M\left(1, 1 - \frac{1}{n}\right) &> -M_1 - \frac{\varepsilon}{2}. \end{aligned}$$

Hence,

$$\begin{aligned} &\mu(q) - \mu(e_{n-1}) > M_1 - \frac{\varepsilon}{2} - \delta - \mu(e_{n-1}) \\ &= M_1 - \frac{\varepsilon}{2} - \delta - M\left(1, 1 - \frac{1}{n}\right) + M\left(1, 1 - \frac{1}{n}\right) - \mu(e_{n-1}) \\ &> M_1 - \frac{\varepsilon}{2} - \delta - M\left(1, 1 - \frac{1}{n}\right) > M_1 - \frac{\varepsilon}{2} - \delta - M_1 - \frac{\varepsilon}{2} \\ &= -(\varepsilon + \delta). \end{aligned} \tag{5.50}$$

Since

$$\begin{aligned} \varphi(q\, f_{n-1}) &= \varphi(q\,(1 - e_{n-1})) \\ &= \varphi(q - e_{n-1}) = \left(1 - \frac{1}{2n}\right) - \left(1 - \frac{1}{n}\right) = \frac{1}{2n}, \end{aligned}$$

we infer from (5.43) specialized to $k = n - 1$ and $a = q - e_{n-1}$ that

$$\mu(q - e_{n-1}) \leq \lambda_{n-1}\, \varphi(q - e_{n-1}) + \delta = \lambda_{n-1}\, \frac{1}{2n} + \delta.$$

Combining this with (5.50) we obtain

$$2n\,(-\varepsilon - \delta) < \lambda_{n-1} + 2\,n\,\delta.$$

In other words, $2n\,(\varepsilon + 2\,\delta) > -\lambda_{n-1}$ and so (using the relationship between ε and δ)

$$-\lambda_{n-1} < 4\,n\,\varepsilon. \tag{5.51}$$

In a similar way we now estimate λ_1. Again, by Lemma 5.5.3, there is a projection $r \in P(1, \frac{1}{2n})$ with $r < e_1$ and $\mu(r) > M(1, \frac{1}{2n}) - \delta$. So

$$\mu(e_1 - r) = \mu(e_1) - \mu(r) < M\left(1, \frac{1}{n}\right) - M\left(1, \frac{1}{2n}\right) + \delta.$$

But $M(1, \frac{1}{n}) < M_0 + \frac{\varepsilon}{2}$ and $M(1, \frac{1}{2n}) > M_0 - \frac{\varepsilon}{2}$. So

$$\mu(e_1 - r) \leq \varepsilon + \delta\,.$$

Put in (5.42) $a = e_1 - r$ and $k = 1$. Then we see that

$$\lambda_1\, \varphi(e_1 - r) - \delta \leq \mu(e_1 - r) \leq \varepsilon + \delta\,.$$

Hence

$$\frac{\lambda_1}{2n} \leq \varepsilon + 2\,\delta\,,$$

giving

$$\lambda_1 \leq 4\,n\,\varepsilon\,. \tag{5.52}$$

Now, (5.49), (5.52) and (5.51) force

$$|\mu(x + y) - \mu(x) - \mu(y)| < 13\,\varepsilon\,.$$

The proof is complete. □

We are now going to prove that the projections h_1, h_2 and h_3 in the previous proposition can be modified such that μ is almost linear on all subalgebras $h_i\, M\, h_i$ $(i = 1, 2, 3)$. For this we need the following auxiliary lemma.

5.5.9. Lemma. *Suppose that* $0 < \varepsilon < 1$ *and* $0 < \delta < 1$. *Let* n *be a natural number such that* $2^{n-1}\,\delta > 1$. *Let* $e \in P(A)$ *with*

(i) $|\mu(p) - \mu(e\,p\,e) - \mu((1 - e)\,p\,(1 - e))| < \varepsilon$ *for each* $p \in P(A)$.

(ii) $T(e) < \frac{1}{3 \cdot 2^n}\, T(1 - e)$.

Then

$$|\mu(a + b) - \mu(a) - \mu(b)| < 6\,(\delta + \varepsilon),$$

whenever $0 \leq a,\ b \leq a + b \leq e$.

Proof: The condition (ii) can be restated in terms of the existence of an orthogonal sequence of projections $f_1, \ldots, f_{2^n}$, with $f_i \leq 1 - e$ and $f_i = p_i + q_i + r_i$, for each i, where p_i, q_i, r_i are mutually orthogonal projections each of which is equivalent to e. For at least one j we must have $V(f_j) \leq \delta$. Indeed, suppose that for each j there is a projection $u_j \leq f_j$ with

$|\mu(u_j)| > \delta$. Since at least 2^{n-1} of the numbers $\mu(u_1), \ldots, \mu(u_{2^n})$ have the same sign, there is a projection q with $|\mu(q)| > 1$ – a contradiction. We will take f_j for which $V(f_j) < \delta$ and put $f = f_j$. By applying 4x4 matrix methods (Proposition 5.4.4), we can find projections p and q, with $p, q \le e + f$ and such that

$$a = 2\,epe \qquad b = 2\,eqe\,.$$

Working in the hereditary subalgebra $(e+f)\,M\,(e+f)$ and using linearity of μ on subalgebras generated by two projections, we can estimate

$$\begin{aligned} |2\,\mu(p) - \mu(a)| &= |2\,\mu(p) - 2\,\mu(epe) - 2\,\mu(fpf) + 2\,\mu(fpf)| \\ &\le 2\,|\mu(p - epe - fpf)| + 2\,|\mu(fpf)|\,. \end{aligned}$$

From (i) and the fact that $V(f) < \delta$ we obtain

$$|2\,\mu(p) - \mu(a)| \le 2\,\varepsilon + 2\,\delta\,.$$

In the same way we derive the inequalities

$$\begin{aligned} |2\,\mu(q) - \mu(b)| &\le 2\,(\varepsilon + \delta) \\ |2\,\mu(p+q) - \mu(p) - \mu(q)| &\le 2\,(\varepsilon + \delta)\,. \end{aligned}$$

The last three inequalities say that

$$|\mu(a+b) - \mu(a) - \mu(b)| \le 6\,(\varepsilon + \delta)\,.$$

□

5.5.10. Proposition. *For a given $0 < \gamma < 1$ there exist projections e_1, e_2, e_3 with $e_1 + e_2 + e_3 = 1$ and with*

(i) $|\mu(p) - \mu(e_i\,p\,e_i) - \mu((1-e_i)\,p\,(1-e_i))| < \gamma$ *for every $p \in P(M)$ and for $i = 1, 3$.*

(ii) $|\mu(a+b) - \mu(a) - \mu(b)| < \gamma$, *whenever $a, b \ge 0$ and $a + b \le e_i$ for some $i = 1, 2, 3$.*

Proof: We set $\delta = \frac{\gamma}{20}$. Choose a natural number n with $2^{n-1}\,\delta > 1$ and let $\varepsilon = \frac{1}{3\cdot 2^n}\,\delta$. Let us start with projections h_1, h_2, h_3 satisfying (i), (ii) and (iii) of Proposition 5.5.8. Let z be the spectral projection e_ε of $T(h_1 + h_3)$ in $Z(M)$. Then

$$z\,T(h_1 + h_3) \leq \varepsilon\, z \qquad \text{and} \qquad (1-z)\,T(h_1+h_3) \geq \varepsilon\,(1-z)\,. \tag{5.53}$$

As $\varphi = \varphi \circ T$ we see that

$$\varepsilon\,\varphi(1-z) \leq \varphi(h_1+h_3) < 2\,\varepsilon^2$$

and thus $V(1-z) \leq \varphi(1-z) < 2\,\varepsilon$, meaning that the variation of μ is "small" on $1-z$. Let us put

$$e_1 = h_1\,z, \qquad e_3 = h_3\,z \qquad \text{and} \qquad e_2 = 1 - e_1 - e_3\,.$$

We are going to prove that e_1, e_2, e_3 satisfy the properties required in the proposition. First, by decomposing the projection $p \in P(A)$ as $p = z\,p + (1-z)\,p$ we can write for $i = 1, 3$,

$$\begin{gathered} p - e_i\,p\,e_i - (1-e_i)\,p\,(1-e_i) = (1-z)\,p + z\,p - e_i\,z\,p\,e_i \\ -e_i\,(1-z)\,p\,e_i - (1-e_i)\,z\,p\,(1-e_i) - (1-e_i)(1-z)\,p\,(1-e_i)\,. \end{gathered}$$

Recalling that $1 - z \leq 1 - z\,h_i = 1 - e_i$, this equality simplifies to

$$\begin{aligned} & p - e_i\,p\,e_i - (1-e_i)\,p\,(1-e_i) \\ = \;& z\,p - h_i\,z\,p\,h_i - (1-e_i)\,z\,p\,(1-e_i) + (1-z)\,p - (1-z)\,p \\ = \;& z\,p - h_i\,z\,p\,h_i - (1-h_i)\,z\,p\,(1-h_i)\,. \end{aligned}$$

Therefore, by (ii) of Proposition 5.5.8,

$$|\mu(p) - \mu(e_i\,p\,e_i) - \mu((1-e_i)\,p\,(1-e_i))| < \varepsilon < \gamma$$

for each $p \in P(M)$ and for $i = 1, 3$. (We have again used the linearity of μ on the algebras generated by two projections.)
Since $\varepsilon = \frac{1}{3\cdot 2^n}\,\delta < \frac{1}{3\cdot 2^n\cdot 20} < \frac{1}{3\cdot 2^n + 1}$, we have, for $i = 1, 3$,

$$T(e_i) \leq \varepsilon\cdot 1 < (1-\varepsilon)\,\frac{1}{2^n\cdot 3}\cdot 1 \leq \frac{1}{2^n\cdot 3}\,T(1-e_i)\,,$$

by (5.53). By applying Lemma 5.5.9 we claim that

$$|\mu(a+b)-\mu(a)-\mu(b)| < 6\,(\delta+\varepsilon) < 7\,\delta$$

whenever $a,b \geq 0$ and $a+b \leq e_i$ for $i=1$ or $i=3$. It remains to establish "approximate linearity" on the hereditary subalgebra $e_2\,M\,e_2$. For this, consider now $a,b \geq 0$ with $a+b \leq e_2$. We can write $e_2 = 1-e_1-e_3 = 1-(1-h_2)\,z = h_2\,z+(1-z)$. So $z\,e_2 \leq h_2$ and, by the choice of h_2, we have

$$|\mu((a+b)\,z)-\mu(a\,z)-\mu(b\,z)| < 13\,\varepsilon.$$

This inequality, together with the fact that the variation of μ on $(1-z)\,M$ is small, gives the approximate linearity of μ on $e_2\,M\,e_2$. More precisely, for each positive x we obtain

$$|\mu(x)-\mu(z\,x)| = |\mu((1-z)\,x)| \leq \|x\|\cdot V(1-z) \leq 2\,\|x\|\,\varepsilon.$$

Therefore,

$$|\mu(a+b)-\mu(a)-\mu(b)| < 13\,\varepsilon+6\,\varepsilon < 19\,\varepsilon < 19\,\delta < \gamma\,.$$

□

Finally, we can establish the linearity of μ on M.

5.5.11. Theorem. *Let M be a finite von Neumann algebra without Type I_2 direct summand. Then μ is linear on M.*

Proof: Fix $0<\varepsilon<1$ and m such that $\sum_{n=m+1}^{\infty} 2^{-n} < \varepsilon$. Emplcying Proposition 5.5.10 we can find orthogonal projections e_1, e_2, e_3 with $e_1 + e_2 + e_3 = 1$ such that

$$|\mu(a+b)-\mu(a)-\mu(b)| < \frac{\varepsilon}{m^2}\,, \tag{5.54}$$

whenever $a,b \geq 0$ and $a+b \leq e_i$, $i=1,2,3$, and such that they satisfy

$$|\mu(p)-\mu(e_i\,p\,e_i)-\mu((1-e_i)\,p\,(1-e_i))| < \frac{\varepsilon}{m^2}\,, \tag{5.55}$$

whenever $p \in P(M)$ and $i=1,2,3$. By (5.55), we infer for any projection $p \leq e_2+e_3$,

$$\begin{aligned}&|\mu(p)-\mu(e_2\,p\,e_2)-\mu(e_3\,p\,e_3)|\\ &= |\mu(p)-\mu(e_3\,p\,e_3)-\mu((1-e_3)\,p\,(1-e_3))| < \frac{\varepsilon}{m^2}\,.\end{aligned}$$

Now we can apply Lemma 5.4.7 to obtain

$$|\mu(a+b) - \mu(a) - \mu(b)| < 20\,\frac{\varepsilon}{m} \tag{5.56}$$

for all $a, b \geq 0$ with $a + b \leq e_2 + e_3$. The inequality (5.56), together with (5.54) for $i = 1$, allows us to apply Lemma 5.4.7 again, to obtain

$$|\mu(a+b) - \mu(a) - \mu(b)| < 20^2\,\varepsilon = 400\,\varepsilon\,,$$

for all $a, b \geq 0$ with $a + b \leq 1$. Hence, μ is linear. □

5.6. Historical remarks and comments

The chef motivation for the study of linear extensions of measures on projections, discussed already in the introduction to this chapter, was the lack of mathematical background allowing to derive the linearity of quantum mechanical state from physically transparent assumptions. That was certainly one of the main stimulations for all effort ensuing Gleason's remarkable result. However, long time before Gleason's achievement the problem of extending measure to a linear functional appeared in the structure theory of von Neumann algebras. As early as in 1937, Murray and von Neumann showed that the dimension function on the projection lattice $P(M)$ of Type II_1 factor, which is manifestly a σ-additive unitarily invariant measure on the projection lattice extends uniquely to a normal state on M (the canonical trace). This result was a highly nontrivial "continuous" version of the well known fact that the function on projections of matrix algebra, assigning to each projection its dimension, extends uniquely to the usual matricial trace. It was a prototype for duality between dimension functions and traces on finite algebras established later (see e.g. [97, 206]). This is a cornerstone of general theory of operator algebras. It illustrates that Generalized Gleason Theorem, even when specified to distinguished classes of measures, gives deep results. The problem of the extension of centre-valued dimension function is precisely the problem of the additivity of the traces. It can be viewed as a special case of the problem of linearity of quasi-linear functional that is due to Kadison [202, 207].

The history of the proof of the Generalized Gleason Theorem is long and intriguing and we are not in the position to provide a complete historical analysis. Nevertheless, we would like to summarize the principal breakthroughs which, in our opinion, represent the most effective way of proving this nontrivial result and that have influenced our exposition. The first important steps in the treatment of quasi-linear functions were made by Aarnes [1, 3] who, inspired by Kadison's insight [202, 207] and interesting Varadarajan's discussion [314, 315], studied the connection between finitely additive probability measures on projections and their quasi-linear extensions. From the physical standpoint this correspondence reads as follows. Let μ be a probability measure on the projections (yes-no observables) in a Hilbert space H. For each observable, given by a self-adjoint operator a on H with the spectral resolution $(e_\lambda)_{\lambda\in\mathbb{R}}$, the value $\mu(e_\lambda)$ represents the probability that given observable has values less then λ. In other words, $\lambda \to \mu(e_\lambda)$ is the distribution function and we can assign to a the expectation value

$$\varrho(a) = \int_{\mathbb{R}} \lambda \, d\,\mu(e_\lambda)\,. \tag{5.57}$$

It is far from being clear that the resulting function ϱ on operators is linear. Studying the additivity of this expectation value on algebras of observables have led to the following concepts generalizing Definition 5.2.5.

5.6.1. Definition. The function $\varrho\colon A \to \mathbb{C}$ on a C^*-algebra A is called the *local quasi-linear functional* (in short local quasi-functional) if the following conditions are satisfied

(i) ϱ is a linear functional on each C^*-algebra generated be a self-adjoint element of A.

(ii) $\varrho(x + i\,y) = \varrho(x) + i\,\varrho(y)$ for all self-adjoint elements $x, y \in A$.

(iii) ϱ is bounded on the unit sphere of A.

Moreover, ϱ is called the *quasi-linear functional* if it is linear on each abelian subalgebra of A. The *local quasi-state* (resp. *quasi-state*) is defined as local quasi-functional (resp. *quasi-functional*) for which $\varrho(a) \geq 0$, whenever $a \geq 0$, and $\sup\{|\varrho(x)| \mid \|x\| \leq 1\} = 1$.

The natural idea, elaborated by Aarnes, was to extend the probability measure μ from projections to a local quasi-state on the algebra of observables. Although the theorem stated in [1] which said that any local quasi-state on an abelian C^*-algebra is linear turned out to be wrong in subsequent development, the arguments of Aarnes are an important part of the solution of the Mackey-Gleason problem. In the response to work of Aarnes [1],

Akemann and Newberger proved in [12] that if X is a totally disconnected compact Hausdorff space, then given functions $f, g \in C(X)$, we can find a function h such that f and g are in the C^*-algebra generated by h. It implies immediately that for algebras rich in projections local quasi-functionals coincide with quasi-functionals. This, of course, applies to von Neumann algebras. Aarnes [3] has also provided positive answer to Mackey-Gleason problem for algebras of compact operators having no quotient isomorphic to algebra of two-by-two matrices and studied the continuity of quasi-states and pure quasi-states (i.e. extreme points of the convex set of all quasi-states). The method of extending finitely additive probability measure to quasi-functional presented here is based only on simple linear algebra arguments and the density of elements with finite spectrum. We have borrowed this insight from [58].

While every positive linear form on C^*-algebra has to be automatically continuous, this is far from being true for quasi-states. In order to prove the continuity of probability measures on projections, that is clearly a necessary condition for the existence of linear extension, the technique of isoclinic projections was developed by Gunson and Christensen [80, 154]. Their methods have led to the following result.

5.6.2. Theorem. *Let e and f be nonzero projections in a unital C^*-algebra A such that $\|e - f\| < 1$, and such that there is a partial isometry $u \in A$ with $u^* u = e$ and $e u = f u = 0$. Then there exists a projection h in A isoclinic to both e and f with the angle $\alpha = \frac{1}{2} \sin^{-1} \|e - f\|$.*

In case of von Neumann algebra the assumption on the existence of u in the previous theorem can be relaxed (Theorem 5.3.5). It was shown by Christen [81] and it is documented in [225]. Our proof differs considerably from the original one by reducing the problem to easy matrix computations. Although Gunson and Aarnes used Theorem 5.6.2 for establishing continuity of measures on the projection lattice, the correct proof of continuity was obtained later [55, 56, 58, 225]. It was based on the diagonalization technique of Kadison and Fillomore [135, 208, 289]. However, since Theorem 5.3.5 has the power to avoid the deep results on diagonalizing normal elements, our approach simplifies arguments in this respect.

A prominent position in the history of Generalized Gleason Theorem presents Christensen's paper [80]. Using the representation of positive contractions by compressions of orthogonal projections (see Christensen

Lemma), Christensen succeeded in solving Mackey-Gleason problem for finitely additive positive measures on properly infinite von Neumann algebra and algebras of Type I_n, $n < \infty$. He also proved the linearity of a quasi-functional on a locally trivial field of algebras of compact operators on a locally compact Hausdorff space. By different methods, using Gleason's geometric arguments, the linear extensions of completely additive positive measures on von Neumann algebra of Type I were established by Tischer in [309]. The remaining case of finite algebras without Type I_n, $n < \infty$, turned out to be highly nontrivial. It required completely new ideas. The solution is due to Yeadon [329, 330] who developed ingenious methods for obtaining approximate derivations of probability measures with respect to a tracial functional. As a result, in the middle of eighties the Mackey-Gleason problem was established by conjugation of Christensen and Yeadon's papers for an important case of finitely additive positive measures on von Neumann algebras without Type I_2 direct summand (see a scholarly paper [225]). It should be noted that the problem of extending measures to linear functionals was also considered by the Kazan school, especially by Matvejchuk [226, 227].

The transition from positive measures to complex measures, which is in fact transition to vector measures as well, has proved to be extremely difficult. First result along this line is due to Paszkiewiz [258], who solved the Mackey-Gleason problem for σ-finite factors and signed measures. Finally, the final form of the Generalized Gleason Theorem was obtained by Bunce and Wright [58, 61, 59]. They used much of the insight and ingenious complex methods that are basically followed in this chapter. The crucial results of Bunce and Wright have opened the way to many important applications that will be discussed in next chapters.

The Generalized Gleason Theorem has many ramifications and local sharpenings. For example, it was proved by Bunce and Hamhalter that exclusion of Type I_2 direct summand in Generalized Gleason Theorem is not necessary for subadditive measures [51]. (A positive measure μ on the projection lattice $P(M)$ of a von Neumann algebra or a JBW algebra M is called *subadditive* if $\mu(e \vee f) \leq \mu(e) + \mu(f)$, whenever $e, f \in P(M)$.) It was proved in [51] that subadditive positive measure on *any* von Neumann algebra and JBW algebra extends uniquely to a trace.

It is also natural to consider the Mackey-Gleason problem in the model of quantum mechanics based on Jordan algebras. Bunce and Wright proved in [55, 56] the following theorem.

5.6.3. Theorem. *Let M be a JBW algebra without Type I_2 direct summand. Then any positive finitely additive measure μ on the projection lattice $P(M)$ extends to a positive functional on M.*

The new phenomenon here lies in this fact: There are measures not having corresponding "integral" not only for algebras resulting from two-by-two matrices but also from infinite-dimensional spin factors.

The Gleason type theorems can also be obtained for measures on other orthoposets associated with the projection structure in von Neumann algebras. Let $\mathcal{P}(M)$ be a system of all idempotents (not necessarily self-adjoint) in a von Neumann algebra M. Then $\mathcal{P}(M)$ forms a quantum logic with the order inherited from M and with orthocomplementation $p \to 1 - p$. It has been proved by Matvejchuk and Mushtari that, among others, any completely additive measure on $\mathcal{P}(M)$, where M is an infinite-dimensional Type I factor, can be described by the usual tracial formula [227, 239, 240]. It discovers a new feature of $\mathcal{P}(M)$: Although $\mathcal{P}(M)$ has plenty of completely additive measures it has no positive one. Also, measures on algebras of operators acting on spaces with indefinite metric have been pursued by Matvejchuk [228, 229, 230].

The Mackey-Gleason problem can also be formulated for C^*-algebras. If A is a unital C^*-algebra, then its projection structure, $P(A)$, forms an orthomodular poset. It is therefore natural to ask in what cases given complex measure μ on $P(A)$ extends to a linear functional on A. If the linear span of $P(A)$ is dense in A, then linear extension of a measure μ has to be unique as in the case of von Neumann algebras. The natural obstacle for the existence of linear extension is the presence of a representation on a two-dimensional space. Indeed, suppose that $\pi\colon A \to M_2(\mathbb{C})$ is a representation of A onto $M_2(\mathbb{C})$ and μ be a measure on $P(M_2(\mathbb{C}))$ having no extension to a linear functional. Then $m = \mu \circ \pi$ is a finitely additive measure on $P(A)$. If m had a linear extension, f, then f would be zero on $\operatorname{Ker} \pi$ and so it would induce a linear functional on $M_2(\mathbb{C})$ extending μ, which was excluded. Consequently, the problem of constructing the expectation value in case of C^*-algebras should be formulated as follows.

5.6.4. Problem. *Let A be a C^*-algebra having no quotient isomorphic to $M_2(\mathbb{C})$. Is any bounded finitely additive complex measure on $P(A)$ a restriction of a linear functional on A ?*

The Problem 5.6.4 is very difficult and it has been settled only for special cases. For example, having the positive solution of the Mackey-Gleason conjecture for von Neumann algebras, we can extend it quickly to all quotients of von Neumann algebras. This can be verified as follows. Let M be a von Neumann algebra without Type I_2 direct summand and I be a closed two-sided ideal in M. Suppose that μ is a finitely additive bounded measure on the projection structure $P(M/I)$. Denote by $\pi\colon M \to M/I$ the canonical quotient map. It is clear that $\mu\circ\pi$ is a bounded measure on $P(M)$ having, by Generalized Gleason's Theorem, a bounded linear extension f on $P(M)$ vanishing on I. It induces a functional ϱ on M/I by

$$\varrho(\pi(x)) = f(x) \tag{5.58}$$

for all $x \in M$. Suppose that p is a projection in M/I. We shall show that p lifts to a projection in M. For this, using the function calculus we can find an element $0 \leq x \leq 1$ with $\pi(x) = p$. Then $x = \sum_n \frac{1}{2^n} q_n$, where q_n are projections lying in $W^*(x, 1)$. Then $p = \pi(x) = \sum_n \frac{1}{2^n} \pi(q_n)$ is an expression of p as σ-convex combination of projections $\pi(q_n)$ sitting in some abelian C^*-subalgebra B containing p. However, since p is an extreme point of the positive part of unit ball of M/I (Chapter 2, Abstract C^*-algebras), we see that $p = \pi(q_n)$ for all n. Using now (5.58) we obtain that ϱ is a functional extending μ. That is why the Generalized Gleason Theorem can be established for quotients of von Neumann algebras. The following generalization of this fact was proved by Bunce and Wright in [62].

5.6.5. Theorem. *Let C^*-algebra A be equal to a quotient of I, where I is a norm closed two-sided ideal of a von Neumann algebra M without Type I_2 direct summand, and let X be a Banach space. Let $\mu\colon P(A) \to X$ be a bounded finitely additive measure. Then μ extends uniquely to a bounded linear map $\varrho\colon A \to X$.*

So, the Gleason Theorem holds, for example, for the Calkin algebra, i.e for the quotient of algebra of all bounded operators on a Hilbert space by the ideal of compact operators. Despite this progress the solution of

Problem 5.6.4 is not known even if we restrict ourselves to AW^*-algebras. If the C^*-algebra A is projectionless then Problem 5.6.4 has trivial solution because $P(A) = \{0\}$. However, Gleason-Mackey problem can be formulated in the following form which is nontrivial for all C^*-algebras.

5.6.6. Problem. *Is any quasi-linear function $\varrho\colon A \to \mathbb{C}$, where A is a C^*-algebra without quotient isomorphic to $M_2(\mathbb{C})$, linear?*

As seen above this problem, first recognized by Kadison, predates the Mackey-Gleason problem and contains it as a very special case. Any C^*-algebra can be viewed as pasting of abelian subalgebras, each isomorphic to some algebra of continuous functions. From this point of view the Problem 5.6.6 asks whether also the dual space of a C^*-algebra is determined by the system of duals of abelian subalgebras. Specialization of the Problem 5.6.6 to quasi-linear traces on AW^*-algebras is equivalent to Kaplansky problem, which is one of the most famous open problems in operator theory. It is therefore evident that Problem 5.6.6 remains a big challenge for functional analysis and its application to quantum theory. Various partial solutions of quasi-linearity problem for special algebras and special types of quasi-functionals can be found in [1, 3, 62, 63, 65, 80, 211]. In the previous sections we have seen that one of the essential ingredients of the proof of Mackey-Gleason conjecture was the establishing of the continuity of quasi-functionals. In this connection it was shown by Bunce and Wright in [64] that a quasi-functional on a C^*-algebra A without quotient isomorphic to $M_2(\mathbb{C})$ is linear, if and only if, it is uniformly continuous on the unit sphere of A endowed with the weak topology. Indeed, in this case one can extend the quasi-linear functional to a quasi-linear functional on the enveloping von Neumann algebra and use then the Generalized Gleason Theorem. Another result was obtained in [63] for so called approximately additive quasi-states. A quasi-linear function on a C^*-algebra A is called *approximately additive* if, for every $\varepsilon > 0$, there is a $\delta > 0$ such that $\|x\,y - y\,x\| < \delta$, for $x, y \in A$, implies $|\varrho(x+y) - \varrho(x) - \varrho(y)| < \varepsilon$. (This concept was introduced by Kaplan in [211]). Bunce and Write proved that any approximately additive quasi-state on Type I C^*-algebra with no quotient isomorphic to $M_2(\mathbb{C})$ is linear [64].

Another interesting line of the research, carried on so far for abelian C^*-algebras only, concerns the relationship between local and general quasi-functionals. It was shown in 1991 by Aarnes [4] that the algebra $C(X)$, where X is a unit square, admits a local quasi-functional that is not linear. This surprising counterexample was a beginning of the study of nonlinear local quasi-states on abelian algebras. This theory, developed mainly by Aarnes [4, 5, 6] shows that nonlinear quasi states give rise to a new kind of set functions on compact spaces, called quasi-measures. Quasi-measures are functions defined only on union of closed and open sets in a compact Hausdorff space enjoying some, but not all, properties of regular Borel measures. It turns out that quasi-measures are typically nonsubadditive. There is a one-to-one integral correspondence between local quasi-functionals and quasi-measures parallel to the correspondence between regular Borel measures and states via Riesz Theorem. In this connection many interesting results, including nonlinear Gelfand transform, general frame for the study of invariant measures on attractors, etc; lying at the interface of measure theory, topology and operator theory, have been obtained.

Our discussion on Generalized Gleason Theorem in the realm of operators algebras has no ambition to cover all results existing in this field. Many other references may be found in the bibliography. In Chapter 9 we return to Gleason Theorem in the context of general ordered structures with orthocomplementations.

6. BASIC PRINCIPLES OF QUANTUM MEASURE THEORY

In the present chapter we shall discuss noncommutative analogies of basic principles of measure theory. In every textbook on measure theory one can certainly encounter some of the basic measure-theoretic tools like the Hahn-Jordan and Yosida-Hewitt decompositions of measures, the Boundedness Principle, the Vitali-Hahn-Saks Theorem, the Radon-Nikodým Theorem, the Egoroff Theorem, the Lusin Theorem, and the Lyapunov Theorem. We shall address these principles in the framework of projection lattices instead of Boolean algebras. In the first section we derive the boundedness principle. It says that any completely additive measure on the projection lattice without finite-dimensional component is bounded. This deep result strengthens the Generalized Gleason Theorem for completely additive measures by eliminating the assumption of boundedness. A few consequences, notably the Nikodým boundedness principle for a system of measures, are then discussed. Section 6.2 is devoted to the analysis of continuous and discontinuous parts of finitely additive measures on projections, especially to the Yosida-Hewitt decompositions. The theorems on convergence of sequences of measures, in particular the Vitali-Hahn-Saks theorem, are treated in Section 6.3. In Section 6.4. the historical background is commented on and notes on remaining basic principles (Egoroff Theorem, Lusin Theorem, Radon-Nikodým Theorem, Lyapunov Theorem) are provided.

6.1. Boundedness of completely additive measures

In the classical measure theory based upon a measure space $(X, \mathcal{P}, \mu)$, where X is a nonempty set, $\mathcal{P}$ is a σ-field of subsets of X, and μ is a real measure on $\mathcal{P}$, the following important principle holds:

If μ is completely additive, then μ is bounded. Moreover, there are disjoint sets A_+ and A_- in $\mathcal{P}$ and positive completely additive measures μ_+ and μ_- living on A_+ and A_-, respectively, such that

$$\mu = \mu_+ - \mu_- \tag{6.1}$$

(see e.g. [105, 278]). The former statement is referred to as the *boundedness principle*, while the decomposition (6.1) is known as the *Hahn-Jordan decomposition* of μ. In the theory of normal states on operator algebras there is the following analogy of Hahn-Jordan decomposition. Any self-adjoint normal functional on a von Neumann algebra can be expressed as a difference of two positive normal functionals having orthogonal support projections (see Chapter 2, Normal states and normal homomorphisms). However, in the quantum measure theory the boundedness principle, and thereby the Hahn-Jordan decomposition, does not hold in general. Indeed, we have seen in Chapter 3 that there are many unbounded (completely additive) measures on projections in finite-dimensional algebras. On the other hand, a deep result of Dorofeev and Sherstnev, dealt with in Chapter 3, says that the boundedness principle does hold for all Type I factors of infinite dimensions. Having established both positive and negative results, we shall then address in this section a question of validity of the boundedness principle for the projection lattices of general von Neumann algebras. We will show that finite matrix algebras are, in fact, the only obstacles for the boundedness principle to hold. More precisely, it will be shown that any complex completely additive measure on the projection lattice $P(M)$ of a von Neumann algebra M having no finite-dimensional direct summand is automatically bounded. This result is intimately connected with the Generalized Gleason Theorem proved in Chapter 6. In the complicated proof of Gleason Theorem we have seen that boundedness of the measure was essential for the arguments. From this point of view the present section continues the study of linear extensions of measures by demonstrating that the assumption of boundedness can be eliminated in Generalized Gleason Theorem in the case of completely additive measures on nearly all von Neumann algebras. Having completed the programme of the linear extension of completely additive measure, we can obtain its Hahn-Jordan decomposition. The plan of our arguments will be as follows. First we show that the boundedness principle can be established in a straightforward manner for a wide class of von Neumann algebras having atomless centre. This allows us to reduce the problem to factors. Since the answer for Type I factor is in the positive (Chapter 3), the problem lies in studying factors of Type II and III. However, the arguments concerning this remaining case are fairly nontrivial. Finally, we conclude the discussion by proving a noncommutative generalization of the Nikodým boundedness theorem.

Let us fix the notation. In the sequel, μ will be a finitely additive real measure on the projection lattice $P(M)$ of a von Neumann algebra M. We shall continue to use the following symbols for $p \in P(M)$:

$$\begin{aligned} \alpha(p) &= \sup\{\mu(q) \mid q \in P(M),\, q \le p\} \\ V(p) &= \sup\{|\mu(q)| \mid q \in P(M),\, q \le p\}\,. \end{aligned}$$

It is clear that both α and V are nonnegative functions on $P(M)$. Since we do not assume boundedness of μ, it can happen that $\alpha(p) = \infty$ or $V(p) = \infty$. The measure μ is called *locally bounded* if, whenever Q is a family of mutually commuting projections, the collection $\{\mu(q) \mid q \in Q\}$ is a bounded set of real numbers. For the convenience of the reader, let us prove that any σ-additive measure on the projection lattice of an abelian von Neumann algebra is bounded, which corresponds to the classical case.

6.1.1. Proposition. *Let μ be a real σ-additive measure on an abelian von Neumann algebra M. Then μ is bounded.*

Proof: Suppose on the contrary that μ is unbounded. Let us say that a projection p in M is unbounded if $V(p) = \infty$. Let us assume first that any unbounded projection contains an unbounded subprojection of arbitrarily large measure. In that case we can find a decreasing sequence of unbounded projection (p_n) such that $|\mu(p_n)| > n$ for all n. Then

$$\mu(p_n) = \mu\left(\bigwedge_{i=1}^{\infty} p_i\right) + \sum_{i=n}^{\infty} \mu(p_i - p_{i+1}) \tag{6.2}$$

for all n. As $\lim_{n\to\infty} \sum_{i=n}^{\infty} \mu(p_i - p_{i+1}) = 0$ we obtain, by letting $n \to \infty$ in (6.2), that $|\mu(\wedge_{n=1}^{\infty} p_n)| = \infty$. This is a contradiction.

Consider now the second possibility. Suppose there exists an unbounded projection $p \in M$ and $K > 0$ such that $V(q) < \infty$ whenever $q \le p$ and $|\mu(q)| \ge K$. Take a subprojection $q_1 \le p$ such that $|\mu(q_1)| > K$. Then

$V(q_1) < \infty$ and so $p - q_1$ is unbounded. (Observe that the commutativity yields $V(e + f) < \infty$ whenever e, f are orthogonal projections with $V(e), V(f) < \infty$.) Applying the same reasoning as above to $p - q_1$ we can find a projection $q_2 \leq p - q_1$ with $|\mu(q_2)| > K$ and with $V(p - q_1 - q_2) = \infty$. Therefore, we are able to construct a sequence (q_n) of pairwise orthogonal projections with $|\mu(q_n)| > K$ for each n. Then $\sum_n |\mu(q_n)| = \infty$ which contradicts the σ-additivity of μ. □

By this result, any σ-additive measure on $P(M)$ is locally bounded, since any system of mutually commuting projections lives in some abelian subalgebra of M. The following lemma explores basic properties of locally bounded measures.

6.1.2. Lemma. *Let $(p_i)_{i \in I}$ be a family of nonzero, pairwise orthogonal projections in a von Neumann algebra M and let μ be a locally bounded measure on $P(M)$. The following statements hold true:*

(i) *$\alpha(p_i) < \infty$ for all p_i $(i \in I)$ with a possible exception for at most finitely many p_i's.*

(ii) *If $\alpha(p_i) < \infty$ for all $i \in I$ then $\sum_{i \in I} \alpha(p_i) < \infty$.*

Proof: (i) Suppose on the contrary that there is a sequence $p_{i_1}, p_{i_2}, \ldots$ such that $\alpha(p_{i_j}) = \infty$ for all j. Passing to appropriate subprojections we can find a sequence of pairwise orthogonal projections $q_1, q_2, \ldots$, such that $\mu(q_n) > n$ for all $n = 1, 2, \ldots$, which contradicts the local boundedness of μ.

(ii) Suppose that $\alpha(p_i) < \infty$ for all i. By definition of α we can take projections $q_i \leq p_i$ such that $\mu(q_i) \geq \frac{\alpha(p_i)}{2} \geq 0$. It follows from the local boundedness that the partial sums of the series $\sum \mu(q_i)$ are bounded and so $\sum \mu(q_i) < \infty$. Then, immediately,

$$\sum_i \alpha(p_i) \leq 2 \sum_i \mu(q_i) < \infty .$$

□

6.1.3. Proposition. *Let $(z_i)_{i\in I}$ be a collection of nonzero, pairwise orthogonal, central projections in M with $\sum_{i\in I} z_i = 1$. Suppose that μ is completely additive and upper bounded on each sublattice $P(z_i\, M)$. Then μ is upper bounded on $P(M)$.*

Proof: By the hypothesis, $\alpha(z_i) < \infty$ for each i and μ is locally bounded. By Lemma 6.1.2 (ii) we have $\sum_{i\in I} \alpha(z_i) < \infty$. Given a projection $p \in P(M)$ it follows from the complete additivity that

$$\mu(p) = \sum_{i\in I} \mu(z_i\, p) \leq \sum_{i\in I} \alpha(z_i)\,.$$

In other words, μ is upper bounded on the whole of $P(M)$. □

6.1.4. Theorem. *Let the centre $Z(M)$ of M have no minimal projection. Then any completely additive real measure μ on $P(M)$ is bounded.*

Proof: First we prove that there is a system of nonzero pairwise orthogonal central projections (z_i) with $\sum z_i = 1$ and such that $\alpha(z_i) < \infty$ for all i. In order to follow standard Zorn's lemma argument we will show that any nonzero central projection z in M dominates a nonzero central projection q such that $\alpha(q) < \infty$. Indeed, since $Z(M)$ has no atoms we can decompose z into an infinite sum $z = \sum_{j=1}^{\infty} q_j$ of nonzero central projections q_j's. By Lemma 6.1.2 $\alpha(q_j) < \infty$ for at least one q_j. By this and Zorn's lemma there is a maximal system (z_i) of nonzero pairwise orthogonal central projections in M with $\alpha(z_i) < \infty$. Using the previous argument, $\sum z_i = 1$. By Proposition 6.1.3 μ is upper bounded. Since the same holds for $-\mu$, we obtain the boundedness of μ. □

The previous theorem covers without any special additional effort a surprisingly large class of von Neumann algebras and has the power to reduce the problem to factors. Indeed, for any von Neumann algebra M there is a central projection z such that z is a sum of projections atomic in the center (if not zero) and $(1-z)\,M$ is a von Neumann algebra with atomless centre. Any completely additive real measure μ on $(1-z)\,M$ is bounded by the previous Theorem. Since the hereditary subalgebras corresponding to atomic central projections are factors, we see that $z\,M$ is a direct sum of factors. So, if the measure μ is bounded on each factor, then it is bounded on $z M$ by Proposition 6.1.3. As a result, μ has to be bounded.

Suppose now that M is of Type I and does not contain any finite-dimensional direct summand. Then the largest central summand of M with the atomic centre is either zero or it is is a direct sum of infinite-dimensional Type I factors. By applying the previous argument and Dorofeev-Sherstnev Theorem we immediately obtain an extension of the boundedness principle to the Type I case.

6.1.5. Corollary. *Let M be a Type I von Neumann algebra not containing any finite-dimensional direct summand. Any real completely additive measure on $P(M)$ is bounded.*

Let us further observe that the problem of boundedness can also be reduced to σ-finite algebras. This can be justified by the following reasoning. Let μ be completely additive measure on M. If μ is unbounded, then there is a sequence of projections (p_n) with $|\mu(p_n)| > n$ for each $n \in \mathbb{N}$. Without loss of generality we can assume that $\mu(p_n) > n$ for each $n \in \mathbb{N}$. Any projection p_n is a sum of orthogonal cyclic projections (Chapter 2, Projection structures of von Neumann algebras), i.e.

$$p_n = \sum_{\alpha \in I_n} f_\alpha^n , \tag{6.3}$$

where f_α^n are cyclic projections. By complete additivity we obtain, by taking an appropriate finite subsum of the series on the right-hand side of (6.3), that there is a cyclic projection q_n with

$$\mu(q_n) > n .$$

Since the supremum of countably many cyclic projections is a σ-finite projection, we obtain that μ is unbounded on the projection lattice of the hereditary σ-finite subalgebra $e M e$ where $e = \bigvee_n q_n$. As a result, the Boundedness Principle will be proved when it is established for σ-finite factors.

The following geometric arguments will be useful in what follows.

6.1.6. Proposition. *Let p and q be projections acting on a Hilbert space H. Write*

$$d(p,q) = \inf\{\|x-y\| \mid x \in p(H),\ \|x\| = 1,\ y \in q(H), \|y\| = 1\}.$$

Suppose that $d(p,q) = \delta > 0$. Then

$$p \vee q \leq \frac{4}{\delta^2}\,(p+q). \tag{6.4}$$

Proof: At first we prove that the sum $p(H) + q(H)$ is a closed subspace in H. Consider sequences $(x_n) \subset p(H)$, $(y_n) \subset q(H)$ with $x_n + y_n \to z$. Then, for $m, k \to \infty$ we have $\|x_k + y_k - x_m - y_m\| \to 0$. Using the lower bound δ for distances of unit vectors from the corresponding subspaces and an easy geometric argument, we deduce that

$$\|u - v\| \geq c \cdot \delta, \tag{6.5}$$

whenever $\|u\|, \|v\| \geq c \geq 0$, $u \in p(H)$, $v \in q(H)$. So from (6.5) we obtain that $\|x_k - x_m\| \to 0$ or $\|y_k - y_m\| \to 0$, for $k, m \to \infty$. Suppose, without loss of generality, that $\|x_k - x_m\| \to 0$ as $k, m \to \infty$. In other words, (x_n) is a Cauchy sequence and has a limit $x \in p(H)$. Thus, (y_n) converges to $y \in q(H)$ and so $z = x + y \in p(H) + q(H)$. Expressed in terms of projections,

$$(p \vee q)(H) = p(H) + q(H).$$

Assume now that $f \in p(H)$ is a unit vector. Since $(f, q\,f) = \|q\,f\|^2 \geq 0$ the vectors f and $q\,f$ can be embedded into at most two-dimensional real Hilbert space. Suppose that $q\,f \neq 0$ and put

$$d = \left\| f - \frac{q\,f}{\|q\,f\|} \right\|.$$

A trivial trigonometric calculation implies that $\sin\alpha = d\sqrt{1 - \frac{d^2}{4}}$, where α is the angel between f and $q\,f$. Since $\sqrt{2} \geq d \geq \delta$, we infer that

$$\sin\alpha \geq \delta \cdot \frac{1}{\sqrt{2}}.$$

Hence, $\|f - q\,f\| = \sin\alpha \geq \frac{\delta}{\sqrt{2}}$. Let us now take an arbitrary unit vector h in $(p \vee q)(H)$. Then $h = \lambda\,f + \mu\,g$, where f and g are unit vectors in $p(H)$ and $q(H)$, respectively, and $\lambda, \mu \in \mathbb{C}$. Then the inequality

$$\begin{aligned} 1 \geq \|h - q\,h\| &= \|\lambda\,f + \mu\,g - \lambda\,q\,f - \mu\,g\| = |\lambda|\,\|f - q\,f\| \\ &\geq |\lambda|\,\frac{\delta}{\sqrt{2}} \end{aligned}$$

yields $|\lambda| \leq \frac{\sqrt{2}}{\delta}$. Similarly, $|\mu| \leq \frac{\sqrt{2}}{\delta}$. It implies that

$$\begin{aligned} 1 = (h,h) &= \lambda\,(f,h) + \mu\,(g,h) \\ &\leq |\lambda|\,|(f,p\,h)| + |\mu|\,|(g,q\,h)| \leq \frac{\sqrt{2}}{\delta}\,(\|p\,h\| + \|q\,h\|)\,. \end{aligned}$$

Hence, $\|p\,h\| + \|q\,h\| \geq \frac{\delta}{\sqrt{2}}$. As $\|p\,h\|^2 + \|q\,h\|^2 \geq \frac{(\|p\,h\| + \|q\,h\|)^2}{2}$, we obtain

$$((p+q)\,h,h) = \|p\,h\|^2 + \|q\,h\|^2 \geq \frac{1}{2}\cdot(\frac{\delta}{\sqrt{2}})^2 = \frac{\delta^2}{4}\,.$$

Alternatively,

$$p + q \geq \frac{\delta^2}{4}\,(p \vee q)\,,$$

giving immediately (6.4). □

The following lemma says that a σ-additive measure is bounded whenever it is uniformly bounded on some increasing sequence of hereditary subalgebras "covering" the whole algebra.

6.1.7. Lemma. *Let μ be a σ-additive measure on a von Neumann algebra M. Suppose that there is a constant C and a sequence of projections (q_n) such that $q_n \nearrow 1$ and*

$$V(q_n) \leq C \qquad \textit{for all } n\,.$$

Then μ is bounded on $P(M)$.

Proof: Fix a projection $p \in M$ and set $p_n = 1 - q_n$. Then $p_n \searrow 0$ and so $p_n \wedge p^\perp \searrow 0$, $p_n \wedge p \searrow 0$. By passing to appropriate subsequences we can arrange for

$$|\mu(p_n)|,\ |\mu(p_n \wedge p^\perp)|,\ |\mu(p_n \wedge p)| \le 1\,,$$

for all n. As for the range projection $r(p\,p_1\,p)$ we have $r(p\,p_1\,p) = p - p \wedge p_1^\perp$, we can estimate

$$|\mu(p)| \le |\mu(r(p\,p_1\,p))| + |\mu(p \wedge p_1^\perp)|\,.$$

Since $p \wedge p_1^\perp \le q_1$, we conclude that

$$|\mu(p)| \le C + |\mu(r(p\,p_1\,p))|\,. \tag{6.6}$$

Our strategy is now to estimate $|\mu(r(p\,p_1\,p))|$. For this we shall, using the function calculus, consider projections

$$g_n = 1_{(0,1-1/n)}(p\,p_1\,p) + 1_{\{1\}}(p\,p_1\,p) = 1_{(0,1-1/n)}(p\,p_1\,p) + p \wedge p_1\,.$$

It is clear that $g_n \nearrow r(p\,p_1\,p)$. Thanks to the countable additivity of μ $\mu(g_n) \to \mu(r(p\,p_1\,p))$ and we can take m large enough to satisfy

$$|\mu(g_m) - \mu(r(p\,p_1\,p))| < 1\,.$$

Then (6.6) can be rewritten in the form

$$|\mu(p)| \le C + 1 + |\mu(g_m)|\,. \tag{6.7}$$

Set

$$g = 1_{(0,1-1/m)}(p\,p_1\,p)\,.$$

According to (6.7)

$$|\mu(p)| \le C + 1 + |\mu(p \wedge p_1)| + |\mu(g)| \le C + 2 + |\mu(g)|\,.$$

The crucial property for estimating $|\mu(g)|$ is the fact that projections g and p_n are always separated in the sense of Proposition 6.1.6. Indeed, by definition of g, for any unit vector $v \in g(H)$, we have

$$\|p_1\,v\|^2 = (p_1\,v, v) = (p\,p_1\,p\,v, v) \le 1 - \frac{1}{m}\,.$$

As $\|v - p_1 v\|^2 = 1 - \|p_1 v\|^2 \geq \frac{1}{m}$ we see that $d(g, p_1) \geq \frac{1}{\sqrt{m}}$. Then immediately $d(g, p_n) \geq \frac{1}{\sqrt{m}}$ because $p_n \leq p_1$ for all n. By virtue of Lemma 6.1.6

$$g \vee p_n \leq K\,(g + p_n)\,, \tag{6.8}$$

where $K = 4\,m$. Since $p_n \searrow 0$, by taking limits $n \to \infty$ in (6.8), we derive, for a projection h with $g \vee p_n \searrow h$, that

$$g \leq h \leq K\,g\,,$$

which immediately implies that $h = g$. So $g \vee p_n \searrow g$. Again, using the countable additivity of μ we can find k large enough to satisfy

$$|\mu(g)| \leq |\mu(g \vee p_k)| + 1\,. \tag{6.9}$$

However,

$$\begin{aligned} |\mu(g \vee p_k)| &= |\mu(g \vee p_k - p_k) + \mu(p_k)| \\ &\leq |\mu(g \vee p_k - p_k)| + |\mu(p_k)| \leq C + 1\,. \end{aligned}$$

(Observe that $g \vee p_k - p_k \leq q_k$.) Returning to (6.6) we can write

$$|\mu(p)| \leq C + 2 + C + 2 = 2\,C + 4\,. \qquad \square$$

6.1.8. Lemma. *Let μ be an unbounded finitely additive measure on a von Neumann algebra M. Suppose that p_0 is a projection in M such that $V(p_0), V(1 - p_0) < \infty$. Then, for any positive integer n, there exists a projection $q \in M$ such that $|\mu(q)| > n$ and $d(q, p_0) \geq 1/8$.*

Proof: Put $\sup\{|\mu(p)| \mid p \leq p_0 \text{ or } p \leq 1 - p_0\} = C$ and find a projection p with

$$|\mu(p)| > 2\,n + 5\,C\,.$$

Like in the proof of Lemma 6.1.7, let us consider the range projection $r(p\,p_0\,p) = p - p \wedge p_0^{\perp}$. Then

$$\begin{aligned} |\mu(r(p\,p_0\,p))| &\geq |\mu(p)| - |\mu(p \wedge p_0^{\perp})| \\ &\geq 2\,n + 5\,C - C = 2\,n + 4\,C\,. \end{aligned} \tag{6.10}$$

Let us now consider the spectral projections

$$q_1 = 1_{(0,1/2]}(p\,p_0\,p) \qquad q_2 = 1_{(1/2,\,1]}(p\,p_0\,p)\,.$$

As in the proof of Lemma 6.1.7, we have that $d(q_1, p_0) \geq \frac{1}{\sqrt{2}}$. Similarly, it holds that $d(q_2, p_0^\perp) \geq \frac{1}{\sqrt{2}}$. To see this, consider a unit vector v in the range of q_2. Then

$$\|p_0 \, v\|^2 = (p_0 \, v, v) = (p \, p_0 \, p \, v, v) > \frac{1}{2}$$

and hence $\|v - (1 - p_0) \, v\| = \|p_0 \, v\| > \frac{1}{\sqrt{2}}$. As $r(p \, p_0 \, p) = q_1 + q_2$, the inequality (6.10) implies that either $|\mu(q_1)| > n + 2\,C$ or $|\mu(q_2)| > n + 2\,C$. In the former case we can set $q = q_1$ and the proof is complete. Therefore, assume that $|\mu(q_2)| > n + 2\,C$. Let us set

$$q_3 = q_2 \vee p_0^\perp - q_2 \, .$$

Take a unit vector v in the range of q_3. Then $q_2 \, v = 0$, $(q_2 \vee p_0^\perp) \, v = v$, and therefore, using Lemma 6.1.6,

$$\|p_0^\perp \, v\|^2 = ((p_0^\perp + q_2) \, v, v) \geq \frac{1}{8} \, ((q_2 \vee p_0^\perp) \, v, v) = \frac{1}{8} \, .$$

Equivalently, $\|v - p_0 \, v\|^2 = \|p_0^\perp \, v\|^2 \geq \frac{1}{8}$ and thus

$$d(q_2 \vee p_0^\perp - q_2, p_0) \geq \frac{1}{\sqrt{8}} \, .$$

Now we can set $q = q_3$. For this observe that

$$|\mu(q_2 \vee p_0^\perp)| \leq |\mu(p_0^\perp)| + |\mu(q_2 \vee p_0^\perp) - \mu(p_0^\perp)| \leq 2\,C$$

and thus

$$|\mu(q_3)| \geq |\mu(q_2)| - |\mu(q_2 \vee p_0^\perp)| \geq n + 2\,C - 2\,C = n \, . \quad \square$$

The following result is a variant of Lemma 3.3.2 in Chapter 3.

6.1.9. Lemma. *Suppose that there is a von Neumann algebra of Type II or III with unbounded completely additive measure μ on its projection lattice. Then there exists a σ-finite factor M of Type II or III and a completely additive unbounded measure μ on $P(M)$ with the following property*

$$\text{If } |\mu(p)| > 1 \text{ for } p \in P(M)\,, \text{ then } \mu \text{ is bounded on } P((1 - p)\, M \,(1 - p))\,. \tag{6.11}$$

Proof: Since the problem of boundedness can be reduced to σ-finite factors, it follows from the previous discussion that the assumption made implies the existence of a σ-finite factor M, not of Type I, with an unbounded real completely additive measure μ on $P(M)$. If the property (6.11) is satisfied for a pair (M, μ), then we are done. If not, we can find a projection p_1 with $|\mu(p_1)| > 1$ and such that μ is unbounded on the projection lattice of the subfactor $M_1 = (1 - p_1)\, M\, (1 - p_1)$. If the pair $(M_1, \mu|M_1)$ satisfies the property in the Lemma 6.1.9, we are done. If not, we can find a projection $p_2 \leq 1 - p_1$ with $|\mu(p_2)| > 1$ and such that μ is unbounded on the projection lattice of the subfactor $(1 - p_2)\, M\, (1 - p_2)$. Continuing this way we either complete the proof or produce a sequence of orthogonal projections (p_n) satisfying

$$|\mu(p_n)| > 1 \qquad \text{for all } n\,.$$

But the latter contradicts the local boundedness of μ. □

The following lemma concerning properties of normal states will frequently be used in the sequel.

6.1.10. Lemma. *Let M be a von Neumann algebra containing no minimal projection, φ a normal state on M and $0 < \varepsilon < 1$. For any projection $p \in M$ there exist projections $p_1, p_2, \ldots, p_n$ such that*

(i) $p = \sum_{i=1}^{n} p_i\,,$

(ii) $\varphi(p_i) \leq \varepsilon$ *for all* $i = 1, 2, \ldots, n,$

(iii) $n \leq \frac{2}{\varepsilon}$.

Proof: Assume that p is nonzero. By the assumption we can decompose p into an infinite sum of nonzero projections (g_α), $p = \sum_\alpha g_\alpha$. By normality of φ we have $\varphi(p) = \sum_\alpha \varphi(g_\alpha)$. In particular, there exists a nonzero subprojection g_j of p with $\varphi(g_j) \leq \varepsilon$. Now, by applying the standard Zorn's lemma argument, we can express p as an infinite sum, $p = \sum_\alpha r_\alpha$, of nonzero projections r_α with $\varphi(r_\alpha) \leq \varepsilon$ for all α. By approximating the sum by the partial sum we obtain a finite decomposition of p satisfying (ii).

Suppose now that p admits a decomposition with the property (ii). If $n \leq \frac{2}{\varepsilon}$, the proof is complete. In the opposite case, at least one number among the collection

$$\varphi(p_1 + p_2),\ \varphi(p_2 + p_3),\ \ldots,\ \varphi(p_n + p_1)$$

has to be less then ε, for otherwise by summing them we would obtain $2 \geq 2\varphi(p) \geq n\varepsilon$ – a contradiction with $n > \frac{2}{\varepsilon}$. Therefore we can reduce the number of summands in the decomposition (i). One can continue this process till the number of summands is less then $\frac{2}{\varepsilon}$. □

Let us remark that any factor M is either finite or properly infinite. If it is properly infinite, then it contains, as a unital subalgebra, Type I subfactor. Thus, we can always find a unital subfactor that is of Type II_1 in any infinite-dimensional factor. For solving the problem of boundedness we shall often work in this special subfactor in the manner of the following proposition.

6.1.11. Proposition. *Let μ be an unbounded σ-additive measure on $P(M)$, where M is a von Neumann factor satisfying the property* (6.11) *in Lemma* 6.1.9. *Let N be a unital von Neumann subfactor of M of Type II_1 with a normal faithful tracial state τ. Then there exist constants $K > 0$ and $0 < \delta < 1$ with the following property: If $q \in P(N)$ and $\tau(q) \leq \delta$, then* $\sup\{|\mu(p)| \mid p \leq q,\, p \in M\} \leq K$.

Proof: Suppose on the contrary that the property asserted in the proposition does not hold. Then we can find a sequence (q_n) of projections in N such that $\tau(q_n) \leq \frac{1}{2^n}$ and $\sup\{|\mu(p)| \mid p \leq q_n,\, p \in P(M)\} > n$ for all n. Let us put

$$p_n = \bigvee_{k=n}^{\infty} q_k \,.$$

Employing the subadditivity of the trace τ we have

$$\tau(p_n) \leq \sum_{k=n}^{\infty} \tau(q_k) \leq \sum_{k=n}^{\infty} 2^{-k} \to 0$$

as $n \to \infty$. So $p_n \searrow 0$, since τ is a faithful normal state. Obviously, μ is unbounded on each hereditary subalgebra $p_n\, M\, p_n$ of M. By the property (6.11)

$$\sup\{|\mu(p)| \mid p \leq 1 - p_n\} \leq 1\,.$$

Since $1 - p_n \nearrow 1$ we obtain that μ is bounded by Proposition 6.1.7, which is absurd. □

Let us observe that at this stage the boundedness principle is established for finite Type II algebras.

6.1.12. Corollary. *Any real completely additive measure on the projection lattice of Type II_1 algebra is bounded.*

Proof: As we have seen in the previous considerations, the existence of unbounded completely additive measure on Type II finite algebra would imply the existence of a real unbounded measure, say μ, on a σ-finite Type II_1 factor M satisfying the property (6.11). Let τ be a faithful normal tracial state on M. Let $\delta > 0$ and $K > 0$ be as described in Lemma 6.1.11 (for $N = M$). Take any projection $p \in M$ and decompose it by Lemma 6.1.10 into the sum $p = \sum_{i=1}^{n} p_i$, where p_i are projections with $\tau(p_i) \leq \delta$ and $n \leq \frac{2}{\delta}$. It gives

$$|\mu(p)| \leq \sum_{i=1}^{n} |\mu(p_i)| < K \cdot \frac{2}{\delta}.$$

So μ is bounded, which is a contradiction. □

6.1.13. Lemma. *Let μ be a real unbounded σ-additive measure on a σ-finite factor M of Type II or III such that the property* (6.11) *in Lemma* 6.1.9 *is satisfied. Let N be a unital finite Type II_1 subfactor of M with the canonical tracial state τ. Then there is a constant C_0 such that*

$$\sup\{|\mu(p)| \mid p \in q\,M\,q\} \leq C_0\,,$$

whenever μ is bounded on the projection lattice of the subalgebra $q\,M\,q$, where $q \in N$.

Proof: Let K and δ be as described in Proposition 6.1.11. Suppose that μ is bounded on $P(q\,M\,q)$ for $q \in N$. Since $q\,M\,q$ is a factor of Type II or III we can apply the Generalized Gleason Theorem and extend μ to a bounded linear functional on $q\,M\,q$, denoted again by μ. Applying Lemma 6.1.10, we can find pairwise orthogonal projections $q_1, q_2, \ldots, q_n$ in N such that $\tau(q_i) \leq \delta/2$, $q = \sum_{i=1}^{n} q_i$ and $n \leq 4/\delta$. Take now a projection $p \in M$ dominated by q. Then $p = \sum_{i,j=1}^{n} q_i\,p\,q_j$ and so

$$\mu(p) = \sum_{i,j=1}^{n} \mu(q_i\,p\,q_j)\,. \tag{6.12}$$

As $q_i\,p\,q_j$ is an element in the algebra $A_{ij} = (q_i + q_j)\,M\,(q_i + q_j)$ and $\tau(q_i + q_j) \leq \delta$, we infer that

$$\sup\{|\mu(r)|\mid r \in P(A_{ij})\}| \leq K\,.$$

According to the inequality (2.11) we obtain that

$$|\mu(q_i\,p\,q_j)| \leq 4\,K$$

for all $i, j = 1, \ldots, n$. Finally, by (6.12) we can estimate

$$|\mu(p)| \leq n^2 \cdot 4\,K \leq \frac{16}{\delta^2}\,4\,K = \frac{64\,K}{\delta^2}\,.$$

Hence, by setting $C_0 = \dfrac{64\,K}{\delta^2}$ we obtain the desired constant. □

6.1.14. Lemma. *Suppose that M is a σ-finite von Neumann algebra and that the unbounded real completely additive measure μ on $P(M)$ satisfies condition* (6.11) *in Lemma* 6.1.9. *Let N be a unital Type II_1 subalgebra of M. Then there exists a projection g_0 in N with the following properties*

(i) *$\mu|P(g_0\,M\,g_0)$ is bounded*

(ii) *If p is a projection in N with $p > g_0$, then $\mu|P(p\,M\,p)$ is unbounded.*

Proof: Consider the following system of subsets

$$\begin{aligned}\mathcal{P} \;=\; & \{(q_i)_{i\in I} \subset P(N)\setminus\{0\}\mid q_i \perp q_j \text{ if } i \neq j\,, \text{ and for any finite subset} \\ & S \subset I \text{ we have that } \mu \text{ is bounded on } P(q_S\,M\,q_S)\,, \text{ where} \\ & q_S = \sum_{i\in S} q_i\}\end{aligned}$$

in $P(M)$. Then $\mathcal{P}$ is nonempty, and using Zorn's lemma we can find a maximal element of $\mathcal{P}$ with respect to the ordering given by the set inclusion. Denote this maximal system by $(f_i)_{i\in I}$. Since M is σ-finite, I is at most countable and so we can assume that $I = \mathbb{N}$. Put $g_0 = \sum_{n=1}^{\infty} f_n$ and $p_n = \sum_{i=1}^{n} f_i$. By the assumption μ is bounded on each hereditary subalgebra $p_n\,M\,p_n$ and so by Lemma 6.1.13 the sequence $(V(p_n))$ is bounded. By virtue of Proposition 6.1.7 μ is bounded on $P(g_0\,M\,g_0)$ because $p_n \nearrow g_0$.

It is clear from the construction that μ is unbounded on each hereditary subalgebra induced by projection from N strictly dominating g_0. The proof is complete. □

In the previous arguments based on Proposition 6.1.11, we have used essentially the fact that the trace is subadditive. As it is known, only the trace has this exclusive property (see e.g. [51]). Since Type II_∞ and Type III factors have no finite trace, the method of the proof cannot be extended directly to infinite algebras. However, we shall try to get around by replacing the trace by faithful normal functional and by showing that it is "nearly subadditive" on projections separated in the sense of Lemma 6.1.6. At first we prove a variant of Lemma 6.1.11.

6.1.15. Lemma. *Let M be a σ-finite von Neumann algebra and μ a real unbounded completely additive measure on $P(M)$ satisfying the property* (6.11) *in Lemma* 6.1.9. *Let φ be a faithful normal state on M. Then there is a projection, h_0, and $\delta > 0$ such that μ is unbounded on the hereditary subalgebra $h_0\, M\, h_0$ and, moreover, the following property holds:*

$$\text{If } p \leq h_0 \text{ is a projection with } \varphi(p) \leq \delta\,, \\ \text{then } \mu \text{ is bounded on } P(p\,M\,p). \tag{6.13}$$

Proof: If the statement of the lemma holds for $h_0 = 1$, we are done. Suppose that it does not. So there is a projection $p_1 \in M$ for which $\varphi(p_1) \leq 1/2$ and $\mu | P(p_1\, M\, p_1)$ is unbounded. Similarly, if p_1 does not satisfy the statement of the lemma, we can find a projection $p_2 \leq p_1$ with $\varphi(p_2) \leq 1/3$, and such that μ is unbounded on $P(p_2\, M\, p_2)$. Continuing in this way we either prove the lemma or construct a sequence of projections $p_n \searrow 0$ (thanks to faithfulness of φ) such that $\mu | P(p_n\, M\, p_n)$ is unbounded for each n. By the property (6.11)

$$\sup\{|\mu(p)| \mid p \leq 1 - p_n\} \leq 1$$

for all n. Since $1 - p_n \nearrow 1$, we have by Lemma 6.1.7 that μ is bounded. This contradiction concludes the proof. □

Now we are in the position to prove the main result.

6.1.16. Theorem (Dorofeev). *Any complex completely additive measure on the projection lattice of a von Neumann algebra M with no direct summand isomorphic to algebra of complex matrices is bounded.*

Proof: It follows from the preceeding results that it suffices to show the nonexistence of unbounded real completely additive measure μ on $P(M)$, where M is a σ-finite factor of Type II or III and such that μ satisfies condition (6.11) of Lemma 6.1.9. Take such a measure μ and a normal faithful state φ on M and look for a contradiction. By Lemma 6.1.15 we can suppose that the property (6.13) is satisfied with $h_0 = 1$ and δ. (If not, we can restrict ourselves to a suitable hereditary subalgebra.) Let N be a unital Type II_1 subalgebra of M and g_0 be a projection satisfying the assertions of Lemma 6.1.14. Since $g_0 \neq 1$, we can find a sequence of pairwise orthogonal nonzero projections $(g_n) \subset N$ such that $g_n \leq 1 - g_0$ for all n. Then μ is unbounded on each hereditary subalgebra $(g_0 + g_n)\, M\, (g_0 + g_n)$. We claim that μ is bounded on $(1 - g_0)\, M\, (1 - g_0)$, too. To this end, take any sequence $(q_n) \subset N$ of nonzero projections dominated by $1 - g_0$ such that $q_n \searrow 0$. Then μ is unbounded on $P((g_0 + q_n)\, M\, (g_0 + q_n))$ and so

$$\sup\{|\mu(p)| \mid p \leq 1 - g_0 - q_n\} \leq 1\,.$$

As $1 - g_0 - q_n \nearrow 1 - g_0$, we obtain the boundedness of μ on $(1 - g_0)\, M\, (1 - g_0)$ from Lemma 6.1.7. We can now apply Lemma 6.1.8 and find a sequence of projections (p_n) such that $p_n \leq g_0 + g_n$,

$$|\mu(p_n)| > \frac{n \cdot 2^{10n}}{\delta}\,, \tag{6.14}$$

and $d(p_n, g_0) \geq 1/8$. We show that

$$d(p_n, \bigvee_{i \neq n} p_i) \geq 1/8\,,$$

for each n. For example, let x and y be unit vectors in the ranges of projections p_1 and $\bigvee_{i \geq 2} p_i$, respectively. Then $y = \alpha\, u_1 + \beta\, u_2$, $\alpha, \beta \in \mathbb{C}$, where u_1 is a unit vector in the range of g_0 and u_2 is a unit vector in the range of the projection $\sum_{i \geq 2} g_i$. Moreover, the vectors u_1 and u_2 can be chosen in such a manner that $\alpha, \beta \geq 0$. Therefore,

$$\begin{aligned} \|x - y\|^2 &= \|x - \alpha\, u_1\|^2 + \|\beta\, u_2\|^2 \\ &\geq (\|x - u_1\| - \|(1 - \alpha)\, u_1\|)^2 + \beta^2 \\ &= (\|x - u_1\| - 1 + \alpha)^2 + 1 - \alpha^2\,. \end{aligned}$$

The right-hand side of the last inequality, viewed as a function of variable $\alpha \in [0,1]$, attains its minimum either for $\alpha = 0$ or for $\alpha = 1$. If $\alpha = 0$, then y is in the range of g_0 and we have $\|x - y\| \geq \frac{1}{8^2}$. If $\alpha = 1$, then

$$\|x-y\|^2 \geq \|x-u_1\|^2 \geq \frac{1}{8^2}\,.$$

In other words, $d(p_1, \bigvee_{i\geq 2} p_i) \geq 1/8$. Using Lemma 6.1.10 we can, for each p_n, find a pairwise orthogonal system of projections $p_1^1, \dots, p_n^{k_n}$ such that $p_n = \sum_{i=1}^{k_n} p_n^i$, $\varphi(p_n^i) \leq \frac{\delta}{2^{9n}}$ for all admissible n and i, and $k_n \leq \dfrac{2\cdot 2^{9n}}{\delta}$. By (6.14) for each n there must be an index $i_n \in \{1,\dots,k_n\}$ with

$$|\mu(p_n^{i_n})| > n\,.$$

By replacing, if necessary, p_n by $p_n^{i_n}$, we may suppose that $\varphi(p_n) \leq \frac{\delta}{2^{9n}}$ and $|\mu(p_n)| \to \infty$ for $n \to \infty$. Put now $K = \frac{4}{(1/8)^2} = 2^8$. By successive applications of Proposition 6.1.6 we obtain

$$\begin{aligned} p_1 \vee p_2 \vee \cdots \vee p_n &\leq K\,(p_1 + p_2 \vee p_3 \vee \cdots \vee p_n) \\ &\leq K\,p_1 + K^2\,(p_2 + p_3 \vee \cdots \vee p_n) \\ &\leq \cdots \leq K\,p_1 + \cdots K^n\,p_n\,. \end{aligned}$$

Hence,

$$\varphi(p_1 \vee p_2 \vee \cdots \vee p_n) \leq K\,\varphi(p_1) + \cdots + K^n\,\varphi(p_n) \leq \sum_{i=1}^{\infty} \frac{\delta}{2^{9i}} \cdot 2^{8i} = \delta\,.$$

Hence, for $p = \bigvee_{n=1}^{\infty} p_n$ we have $\varphi(p) \leq \delta$ and thus μ should be bounded on $P(p\,M\,p)$ by the property (6.13). This contradiction concludes the proof. □

The results of Dorofeev and Sherstnev enable us to obtain Generalized Gleason Theorem for completely additive measures without assuming boundedness on a wide class of von Neumann algebras. The Hahn-Jordan decomposition then follows as an easy consequence.

6.1.17. Theorem. *Let μ be a real completely additive measure on the projection lattice $P(M)$ of a von Neumann algebra M not containing either Type I_2 direct summand or direct summand isomorphic to the algebra $M_n(\mathbb{C})$ of all $n \times n$ matrices $(n > 1)$. Then μ extends uniquely to a normal functional on M. Moreover, there are positive completely additive measures μ_+ and μ_- such that*

$$\mu = \mu_+ - \mu_-$$

and orthogonal projections p_+ and p_- in M with

$$\mu_+(1 - p_+) = \mu_-(1 - p_-) = 0\,.$$

Proof: It follows immediately from Generalized Gleason Theorem, Theorem 6.1.16 and from the decomposition of a self-adjoint normal functional into a difference of orthogonal normal positive functionals. □

Let us now add that once the boundedness principle is established for complex measures, it can be easily extended to vector-valued measures.

6.1.18. Theorem. *Let M be a von Neumann algebra not containing any direct summand isomorphic to the algebra $M_n(\mathbb{C})$ of all $n \times n$ complex matrices, where $n > 1$. Let*

$$\mu \colon P(M) \to X$$

be a completely additive measure with values in a normed space X. Then μ is bounded.

Proof: Let ψ be any bounded functional on X. Then $\psi \circ \mu$ is a completely additive measure on M and thereby it has to be bounded. Equivalently, $\mu(P(M)) \subset X$ is a bounded set in the weak topology of X and thereby it is bounded. □

One of the most important principles of classical measure theory, known as the *Nikodým Boundedness Theorem* says that if a system of σ-additive measures on a σ-algebra $\mathcal{P}$ of subsets of a nonempty set X is pointwise bounded then it is uniformly bounded (see e.g. [105, Theorem IV 9.8]). Based on the previous results we now present the noncommutative versions of Nikodým Boundedness Theorem that simultaneously extend the main Theorem 6.1.16 from individual measure to a collection of measures.

6.1.19. Theorem. *Let $(\mu_\alpha)_{\alpha\in I}$ be a collection of σ-additive measures on the projection lattice of a von Neumann algebra M such that*

$$\sup\{|\mu_\alpha(p)| \mid \alpha \in I\} < \infty$$

for each $p \in P(M)$. Assume that M does not contain any Type I_2 direct summand and that at least one of the following conditions is satisfied:

(i) *Each measure μ_α is bounded.*

(ii) *Each measure μ_α is completely additive and M does not contain any direct summand isomorphic to $M_n(\mathbb{C})$ $(n > 1)$.*

Then

$$\sup\{|\mu_\alpha(p)| \mid p \in P(M), \alpha \in I\} < \infty .$$

Proof: Suppose that (i) is satisfied. By Generalized Gleason Theorem there is a system $(f_\alpha)_{\alpha\in I}$ of bounded functionals on M extending (μ_α). We will show that the system $(f_\alpha)_{\alpha\in I}$ is pointwise bounded, i.e. that $\sup\{|f_\alpha(x)| \mid \alpha \in I\}$ is finite for every $x \in M$. To verify that it is sufficient to consider $x \in M$ with $0 \le x \le 1$. Then $x = \sum_n \frac{1}{2^n} p_n$ for a system of commuting projections (p_n). The subalgebra $A = W^*(p_n)$ is an abelian von Neumann algebra. By the Nikodým Boundedness Theorem the system of measures $(\mu_\alpha | P(A))$ is uniformly bounded. In other words, $\sup\{|f_\alpha(p_n)| \mid \alpha \in I, n \in \mathbb{N}\} = K < \infty$. It means that $\sup\{|f_\alpha(x)| \mid \alpha \in I\} \le \sum_{n=1}^{\infty} \frac{1}{2^n} K = K$. By the Banach-Steinhaus Uniform Boundedness Principle the system (f_α) are uniformly bounded. So there is a constant $C > 0$ with

$$\|f_\alpha\| \le C \text{ for each } \alpha \in I .$$

It yields

$$\sup\{|\mu_\alpha(p)| \mid p \in P(M), \alpha \in I\} < C .$$

By Theorem 6.1.16 the condition (ii) implies the condition (i). □

6.2. Yosida-Hewitt decompositions of quantum measures

In the previous section the decomposition of a real completely additive measure into a difference of two positive completely ones was proved. Another type of decomposition, referred to as the Yosida-Hewitt decomposition, is the expression of a given finitely additive measure as a sum of a continuous measure (completely additive, σ-additive, etc.) and a discontinuous remaining component. The Yosida-Hewitt decomposition plays an important role in the classical measure and probability theory because it enables one to analyze how large the continuous parts of probability distributions are and how the properties of "pathological" discontinuous parts look like. The aim of the present section is to review new recent results, extending classical decomposition theorems obtained formerly for measures on Boolean algebras. First we prove the standard result of the theory of operator algebras saying that any positive functional on a von Neumann algebra is a unique sum of a normal positive functional and a singular positive functional. A possible analogy to σ-additive functionals and measures, which is more natural from the point of view of quantum probability theory, seems to be more difficult and will be the main theme of this section. All results in the present part will be formulated for linear states. Nevertheless, using the Generalized Gleason Theorem they can be translated quickly to "proper" measure theory by replacing the word functional by measure and by assuming, in addition, that the von Neumann algebra in question does not contain Type I_2 part. The conclusions of the present discussion will be applied later to establishing continuity properties of specific states arising in mathematical foundations of quantum mechanics.

Our approach to decompositions of states will be geometric and will be based on the analysis of the facial structure of the dual ball of the algebra under consideration. For this reason we recall first some concepts of convexity theory. Let M be a von Neumann algebra. The set $S(M)$ of all states on M is a convex set compact in the weak*-topology. A convex subset $F \subset S(M)$ is called a *face* if the assumption $\frac{\varrho_1 + \varrho_2}{2} \in F$, for $\varrho_1, \varrho_2 \in S(M)$, implies that both ϱ_1 and ϱ_2 lie in F. In other words, elements of F cannot be expressed as convex combinations of states lying outside F. Let $R \subset S(M)$ and let $co(R)$ be the convex hull of R. The face, $F(R)$, of $S(M)$ *generated by* R is the set of all $\tau \in S(M)$, such that, for some $0 < \lambda \leq 1$, $\lambda\tau$ is majorized by an element of $co(R)$. $F(R)$ is the smallest face of $S(M)$ containing R. We say that $S(M)$ is a *convex direct sum* of the convex subsets S_1 and S_2 of $S(M)$ if for any $\varrho \in S(M)$ there is a unique convex combination $\alpha\,\varrho_1 + (1-\alpha)\,\varrho_2$, $0 \leq \alpha \leq 1$, such that $\varrho_1 \in S_1$, $\varrho_2 \in S_2$ and

$$\varrho = \alpha\,\varrho_1 + (1-\alpha)\,\varrho_2\,.$$

A face F of $S(M)$ is said to be a *split face* if there is a face, $F^{\#}$, of $S(M)$ for which $S(M)$ is the direct convex sum of F and $F^{\#}$. The face $F^{\#}$ is uniquely determined by F and is called the *(facial) complement* of F. The structure of norm closed faces is isomorphic to the projection lattice of the second dual. Indeed, the norm closed faces are precisely the sets $F_p = \{\varrho \in S(M) \mid \varrho(p) = 1\}$, where p is a projection in M^{**} uniquely determined by the given face. On the other hand, the split faces are precisely the sets $F_z = \{\varrho \in S(M) \mid \varrho(z) = 1\}$, where z is a central projection in M^{**}, and $F_z^{\#} = F_{1-z}$ (see [19, 129]). The correspondence between split faces and central projections in the double dual is one-to-one. As a consequence, split faces are always norm closed. Moreover, split faces can be characterized among all normed closed faces as being invariant under transformations. Let $\varrho \in S(M)$ and $a \in M$ such that $\varrho(a^*\,a) \neq 0$. Then the *transformed state*, ϱ_a, of ϱ is defined by

$$\varrho_a(x) = \frac{\varrho(a^*\,x\,a)}{\varrho(a^*\,a)} \qquad \text{for all } x \in M\,.$$

The symbol E_ϱ will denote the norm closure of the set $\{\varrho_a \mid \varrho(a^*\,a) \neq 0, a \in M\}$ of all transformed states. Let us recall that E_ϱ is the set of all states $\omega_h \circ \pi_\varrho$, where h is a unit vector in the G.N.S. Hilbert space H_ϱ (see Chapter 2, The Gelfand-Naimark-Segal construction). From the point of view of physics, the transformed states correspond to some manipulation with the system initially prepared in the state ϱ under measurement modelled by a completely positive mapping (filter) $x \to a^*\,x\,a$.

6.2.1. Proposition. *Let F be a norm closed face of $S(M)$. Then the following conditions are equivalent:*

(i) *F is a split face,*

(ii) *$E_\varrho \subset F$ for each $\varrho \in F$,*

(iii) *$\varrho_u \in F$ for each $\varrho \in F$ and each unitary $u \in M$.*

Proof: The face $F = \{\varrho \in S(M) \mid \varrho(e) = 1\}$ for some unique projection e of M^{**}. If F is a split face of $S(M)$, then e is a central projection of M^{**} so that, for a in M, $\varrho(a^* a)\, \varrho_a(1-e) = \varrho((a^*a)(1-e)) \leq (\varrho(1-e))^{1/2}\, (\varrho(a^* a))^{1/2} = 0$ by the Cauchy–Schwarz inequality. Therefore $\varrho_a(e) = 1$ and hence $\varrho_a \in F$. We have proved that (i) implies (ii). On the other hand, if the condition (ii) holds then so does the condition (iii) which, by the uniqueness of e, implies $u^* e\, u = e$ for all unitaries u of M. By the Kaplansky Density Theorem, this gives $u^* e\, u = e$ for all unitaries u of M^{**} so that e is central in M^{**}. We see that F is a split face. □

Let $S_n(M)$ denote the set of all normal states on a von Neumann algebra M. As it was stated in Chapter 2 (Normal states and normal homomorphisms), the set $S_n(M)$ is norm closed and weak* dense in $S(M)$. Let us denote by $S_\sigma(M)$ the face of all σ-additive states on M. It is clear that $S_\sigma(M) = S_n(M)$ provided that M is σ-finite. The set $S_\sigma(M)$ is normed closed. Indeed, let $\varrho \in S(M)$ be such that $\varrho_n \to \varrho$ in norm for a sequence $(\varrho_n) \subset S_\sigma(M)$. Let $e = \sum_{n=1}^{\infty} e_n$, where $e,\, e_n \in P(M)$ for all n. Since $A = W^*(e_n) \simeq l^\infty$ is a σ-finite algebra, we get that ϱ is normal on A and so $\varrho(e) = \sum_{n=1}^{\infty} \varrho(e_n)$. In other words, ϱ is σ-additive. Since both $S_n(M)$ and $S_\sigma(M)$ are unitarily invariant they are split faces. We shall now present the standard result of theory of states characterizing the facial complement $S_n^{\#}(M)$ of the normal state space. First we recall the terminology.

6.2.2. Definition. A positive functional ϱ on a von Neumann algebra M is called *singular* if $\varrho = \lambda\varphi$, where $\lambda \geq 0$ and φ is a state such that $\phi \in S_n(M)^{\#}$.

It is clear that every state on M can be uniquely expressed as a convex combination of a normal state and a singular state. In particular, a pure state is either normal or singular.

6.2.3. Theorem. *Let ϱ be a nonnegative linear functional on a von Neumann algebra M. Then ϱ is singular if, and only if, the following statement holds. For every nonzero projection $e \in M$ there is a nonzero projection $f \leq e$ such that $\varrho(f) = 0$.*

Proof: Suppose that the statement holds. Let $\varrho = \varrho_1 + \varrho_2$, where ϱ_1 is a normal positive functional and ϱ_2 is a singular positive functional. We shall prove that $\varrho_1 = 0$. Looking for a contradiction suppose that ϱ_1 is nonzero. Then ϱ_1 has a nonzero support projection e. It means, that ϱ is nonzero on every subprojection of e - a contradiction. Hence ϱ has to be singular.

For proving the reverse implication suppose now that ϱ is singular and $e \in M$ is a nonzero projection. If $\varrho(e) = 0$ we are done. Assume that $\varrho(e) > 0$. We can fix a normal positive functional φ such that $\varphi(e) > \varrho(e)$. Let us define the set $\mathcal{F} = \{p \in P(e M e) \mid \varphi(p) \leq \varrho(p)\}$. Let (p_α) be an increasing net of elements in $\mathcal{F}$. For $p \in P(M)$ with $p_\alpha \nearrow p$ we have $\varrho(p) \geq \varrho(p_\alpha) \geq \varphi(p_\alpha)$ and so, by normality of φ, $\varrho(p) \geq \varphi(p)$. Hence, $\mathcal{F}$ is closed under suprema of increasing nets. Thus, by the standard Zorn's lemma argument, $\mathcal{F}$ has a maximal element, q. Set $f = e - q$. By the assumption, f is nonzero. By the maximality of q we have $\varrho(h) \leq \varphi(h)$ for all projections $h \leq f$. The spectral theorem gives us that $\varrho \leq \varphi$ on the hereditary subalgebra $f M f$. In other words, $\varrho(f\,x\,f) \leq \varphi(f\,x\,f)$ for all nonnegative x in M. Since both the set of all nonnegative normal functionals and the set of all nonnegative singular functionals are cones in the state space that is invariant under transformations of functionals, we see that the functional ψ defined by $\psi(x) = \varrho(f\,x\,f)$ $(x \in M)$ is zero. It implies immediately that $\varrho(f) = 0$. □

By the previous theorem, the singular functionals admit an abundance of zero projections. Indeed, using Zorn's Lemma we obtain:

6.2.4. Corollary. *A positive functional ϱ on a von Neumann algebra M is singular if, and only if, there is a system $(e_\alpha)_{\alpha \in I}$ of orthogonal projections in M such that*

$$\sum_{\alpha \in I} e_\alpha = 1 \qquad \textit{and} \quad \varrho(e_\alpha) = 0 \qquad \textit{for each } \alpha \in I\,. \tag{6.15}$$

Corollary 6.2.4 and Theorem 6.2.3 give a transparent description of the facial complement of the normal state space. A σ-additive version of this result is not as easy and it seems to be connected with some problems in set theory. However, natural candidate for a state "singular" in the σ-additive theory is the state ϱ for which (6.15) holds with countable index set I.

6.2.5. Definition. A positive functional ϱ on a von Neumann algebra M is said to be *countably singular* if there exists a sequence (e_n) of projections in M such that

$$\sum_{n=1}^{\infty} e_n = 1 \qquad \text{and} \quad \varrho(e_n) = 0 \text{ for each } n\,. \tag{6.16}$$

The next proposition tells us that the facial complement of the σ-additive state space is generated by the set of all countably singular states. In the sequel, the symbol $S_{cs}(M)$ will be reserved for the set of all countably singular states on the von Neumann algebra M.

6.2.6. Proposition. *Let M be a von Neumann algebra, then the split face $S_\sigma(M)^{\#}$ is the smallest split face containing all countably singular states.*

Proof: Since $S_\sigma(M)$ is a split face of $S(M)$ and no countably singular state can majorize a positive multiple of a σ-additive state, we have that $S_{cs}(M) \subset S_\sigma^{\#}(M)$. Hence $F \subset S_\sigma^{\#}(M)$, where F is the split face generated by $S_{cs}(M)$. We want to prove that $F \supset S_\sigma^{\#}(M)$. We will verify the inclusion $F^{\#} \subset S_\sigma(M)$. For this, take $\varrho \in F^{\#}$. Suppose that ϱ is not σ-additive and try to reach a contradiction. There is a sequence of orthogonal projections (e_n) such that $\sum_n e_n = 1$ and $\sum_n \varrho(e_n) < 1$. Let $(\pi_\varrho, H_\varrho, x_\varrho)$ be the G.N.S. data corresponding to the state ϱ. Then $f = \sum_n \pi_\varrho(e_n) \neq 1$ in $B(H_\varrho)$. Let us choose a unit vector $h \in (1-f)(H_\varrho)$. Then $(\omega_h \circ \pi_\varrho)(e_n) = 0$ for all n and hence $\omega_h \circ \pi_\varrho \in S_{cs}(M) \subset F$. Using the standard notation of the G.N.S. construction (see Chapter 2, The Gelfand-Naimark-Segal construction), the sets $\omega_h \circ \pi_\varrho$, where h is a unit vector in H_ϱ are in the norm closure E_ϱ of the set of all transformed states of ϱ. As a consequence,

$$\omega_h \circ \pi_\varrho \in E_\varrho \subset F^{\#}$$

by Proposition 6.2.1. So $\omega_h \circ \pi_\varrho \in F \cap F^{\#} = \emptyset$ - a contradiction. □

It is clear that if M is σ-finite, then $S_\sigma^{\#}(M)$ coincides with the set of all countably singular states. In the non-σ-finite case we only know that the face $S_\sigma^{\#}(M)$ is generated by countably singular states. However, if we assume the continuum hypothesis, we can show that any state in $S_\sigma^{\#}(M)$ is countably singular, which simplifies Proposition 6.2.6. By the continuum hypothesis we mean that each subset of $\mathbb{R}$ is either countable or has the cardinality continuum. (We believe the continuum hypothesis to be purely technical mathematical condition acceptable from the physical point of view because its adopting cannot affect any real physical measurement.)

We shall approach to the advertised result by a few preliminary steps. The following proposition is due to Akemann and Anderson [15]. We omit the proof because it is purely set theoretic.

6.2.7. Proposition. *Assume that the continuum hypothesis is true. Let μ be finitely additive probability measure on the power set of $\mathbb{R}$ vanishing on all singletons. Then there exists a sequence (A_n) of disjoint subsets of $\mathbb{R}$ with $\mathbb{R} = \bigcup A_n$ and $\mu(A_n) = 0$ for all n.*

The continuum hypothesis allows us to extend the equality between σ-additive and normal states to algebras of larger sizes. By a *countably generated von Neumann algebra* we mean a von Neumann algebra M such that $M = W^*(a_n)$ for some sequence $(a_n) \subset M$.

6.2.8. Lemma. *Assume that the continuum hypothesis is true. Let M be countably generated von Neumann algebra. Then all σ-additive states of M are normal and all singular states of M are countably singular.*

Proof: Let $(e_\lambda)_{\lambda \in I}$ be a family of orthogonal projections in M. We claim that the countable generation of M implies $\operatorname{card}(I) \leq \operatorname{card}(\mathbb{R})$. Indeed, choose, for each λ, a normal state, ϱ_λ, with a support projection majorized by e_λ to obtain an orthogonal family $(\varrho_\lambda)_{\lambda \in I}$ in M_*. But M contains a separable weak*-dense C^*-subalgebra, A. Since any normal functional is uniquely determined by its values on some fixed countable dense subset of A, we see that $\operatorname{card}(M_*) \leq \operatorname{card}(\mathbb{R})$ and the claim follows. Let ϱ be a singular state of M and let $(e_\lambda)_{\lambda \in \mathbb{R}}$ be an ($\mathbb{R}$-indexed) orthogonal family of projections in M such that $\sum e_\lambda = 1$ and $\varrho(e_\lambda) = 0$ for all $\lambda \in \mathbb{R}$. Then, as easily seen, with $e_A = \sum_{\lambda \in A} e_\lambda$ for $A \subset \mathbb{R}$, the assignment $\mu(A) = \varrho(e_A)$ defines a finitely additive probability measure on the power set of $\mathbb{R}$ vanishing on all singletons. Therefore, by Proposition 6.2.7, ϱ is countably singular. Hence, the faces of singular and σ-singular states coincide, and thus $S_\sigma(M) = S_n(M)$. $\square$

6.2.9. Theorem. *Let M be a von Neumann algebra. If M is σ-finite or if the continuum hypothesis is true, then $S^{\#}_{\sigma}(M) = S_{cs}(M)$.*

Proof: Assume that the continuum hypothesis is true. Let ϱ_1 and ϱ_2 be countably singular states on M and let (e_n) and (f_n) be sequences of projections in M with $\sum e_n = \sum f_n = 1$ and $\varrho(e_n) = \varrho(f_n) = 0$ for all n. Let N be the von Neumann subalgebra of M generated by all e_n and f_n. As ϱ_1 and ϱ_2 are singular on N, so is any convex combination, φ, of ϱ_1 and ϱ_2. Indeed, since N is countably generated, φ restricts itself to a countably singular state of N, by Lemma 6.2.8. Therefore, φ is a countably singular state of M, proving that $S_{cs}(M)$ is convex. It immediately follows that $S_{cs}(M)$ is a face of $S(M)$. It remains to show that $S_{cs}(M)$ is norm closed. Let ϱ be a uniform limit of a sequence (ϱ_n) of countably singular states of M. For each n we have a sequence $(e_{m,n})_n$ of orthogonal projections in M which sums up to 1 and which satisfies $\varrho(e_{m,n}) = 0$ for all m. Passing to the countably generated von Neumann subalgebra generated by the set $\{e_{m,n} : m, n \in \mathbb{N}\}$ and using Lemma 6.2.8 again, we see that ϱ must be countably singular. In other words, since $S_{cs}(M)$ is a unitarily invariant norm closed face, it is a split face by Proposition 6.2.1, and hence $S_{cs}(M) = S^{\#}_{\sigma}(M)$ by Proposition 6.2.6. □

6.3. Convergence theorems

There are important theorems in the classical measure and probability theory on pointwise convergence of sequences of measures. A prominent one is the Vitali-Hahn-Saks Theorem. It can be formulated as follows (see e.g. [105, Theorem 2, p.160]). Let μ be a σ-additive measure on the σ-field $\mathcal{A}$ of subsets of the underlying nonempty set S. Denote by

$$var(\mu, E) = \sup\left\{\sum_{i=1}^{n} |\mu(E_i)| \mid E_1, \ldots, E_n \text{ is a partition of } E\right\}$$

the variation of μ on $E \in \mathcal{A}$. Let (μ_n) be a sequence of μ-continuous finitely additive measures on $\mathcal{A}$ such that the sequence $(\mu_n(E))$ is convergent for each $E \in \mathcal{A}$. Then, (μ_n) are uniformly μ-continuous in the following sense: for each $\varepsilon > 0$ there is a $\delta > 0$, such that, for each $E \in \mathcal{A}$, $\sup_n |\mu_n(E)| < \varepsilon$ whenever $|\mu(E)| < \delta$. If all μ_n's are σ-additive, then the set function $\nu(E) = \lim_{n\to\infty} \mu_n(E)$, $(E \in \mathcal{A})$, forms a σ-additive measure. Another corollary, called the *Nikodým Convergence Theorem* (see e.g. [105, 3 Corollary p. 159]) says that any pointwise convergent sequence of σ-additive measures on the σ-field $\mathcal{A}$ converges to a σ-additive measure. Since the Nikodým Convergence Theorem is valid on algebras generated by orthogonal projections, its generalization to von Neumann projection lattices is quite straightforward. On the other hand, the noncommutative generalization of the Vitali-Hahn-Saks Theorem seems to be more difficult and requires the Generalized Gleason Theorem. This is the main goal of this section. First, we shall define a natural analog to the concept of absolute continuity of measures.

6.3.1. Definition. Let M be a von Neumann algebra. Let μ_1 be a positive finitely additive measure on $P(M)$ and $\mu_2\colon P(M) \to X$ be a finitely additive measure on $P(M)$ with values in a Banach space X. We say that μ_2 if μ_1-*continuous*, (in symbols $\mu_2 << \mu_1$) if, for each $\varepsilon > 0$, there is a $\delta > 0$ such that, for each $p \in P(M)$,

$$\|\mu_2(p)\| \leq \varepsilon \qquad \text{whenever} \quad \mu_1(p) < \delta\,.$$

It is clear from the definition that if $\mu_2 << \mu_1$, then $\mu_1(p) = 0$ implies $\mu_2(p) = 0$. It is well known that the converse is true provided that μ_1 and μ_2 are σ-additive scalar measures on the σ-field of sets (see e.g. [105, Lemma 13, p. 131]). We shall now establish the noncommutative version of this fact that will be based on the Generalized Gleason Theorem.

6.3.2. Proposition. *Let M be a von Neumann algebra without Type I_2 direct summand. Suppose that $\mu_2\colon P(M) \to X$ is a bounded completely additive measure with values in a finite-dimensional Banach space X and that μ_1 is a positive completely additive measure on $P(M)$. Then $\mu_2 << \mu_1$ if, and only if, for each $p \in P(M)$, $\mu_1(p) = 0$ implies $\mu_2(p) = 0$.*

Proof: Suppose that $\mu_2(p) = 0$ whenever p is a projection such that $\mu_1(p) = 0$. By the Generalized Gleason Theorem, μ_2 extends to a bounded linear map $T\colon M \to X$. T is weak*-weak continuous and therefore weak*-norm continuous by finite dimension of X. On the other hand, μ_1 extends to a positive normal functional $\hat{\mu}_1$ on M. First we prove that, for each nonnegative element x in M, $\hat{\mu}_1(x) = 0$ implies that $T(x) = 0$. For this it is sufficient to assume that $0 \le x \le 1$. Then $x = \sum \frac{1}{2^n} e_n$, where e_n are projections in M. Since μ_1 is positive, we obtain $\mu_1(e_n) = 0$ for each n and hence, by the hypothesis, $T(e_n) = 0$ for each n. By continuity of T we infer that $T(x) = 0$. Looking for a contradiction, let us suppose that there is a sequence of projections (p_n) in M and $\varepsilon > 0$ such that

$$\lim_n \mu_1(p_n) = 0 \qquad \text{and} \qquad \|\mu_2(p_n)\| \ge \varepsilon \quad \text{for all } n \in \mathbb{N}.$$

Since the unit ball of a von Neumann algebra M is weak*-compact, we can assume, by passing to a subsequence, that there is a positive operator x such that $p_n \to x$ in the weak*-topology. Using the continuity of T and $\hat{\mu}_1$ we deduce that $\hat{\mu}_1(x) = 0$ while $\|T(x)\| \ge \varepsilon > 0$. This is a contradiction. □

6.3.3. Remark. Let us remark that in view of Theorem 6.1.18 the assumption on the boundedness of μ_2 in the previous proposition is redundant if the von Neumann algebra M does not contain, in addition to the assumptions of the previous proposition, any finite-dimensional direct summand.

Besides, under the continuum hypothesis, Proposition 6.3.2 holds for σ-additive measures μ_1 and μ_2. Indeed, if the conclusion of Proposition 6.3.2 were false, then using the notation of the previous proof, we can find a sequence of projections (p_n) in M such that $\mu_1(p_n) \to 0$ while $\|\mu_2(p_n)\| \ge \varepsilon > 0$ for all n. Therefore, it is enough to consider the countably generated subalgebra $W^*(p_n)$. By Lemma 6.2.8, both μ_1 and μ_2 are completely additive on $W^*(p_n)$ and so Proposition 6.3.2 applies.

We are now ready to prove a partial extension of the Vitali-Hahn-Saks Theorem to projection lattices of von Neumann algebras that is based on Proposition 6.3.2.

6.3.4. Theorem. *Let M be a von Neumann algebra without Type I_2 direct summand. Let (μ_n) be a pointwise convergent sequence of bounded completely additive vector measures on $P(M)$ with values in a finite-dimensional space X. Let all μ_n be μ-continuous, where μ is a positive completely additive measure on $P(M)$. Then the following statement holds: The function $\nu\colon P(M) \to X$, $\nu(p) = \lim_{n\to\infty} \mu_n(p)$ $(p \in P(M))$ is a μ-continuous completely additive measure on $P(M)$.*

Proof: The complete additivity of μ is valid on orthogonal systems of projections that generate always a commutative algebra. Hence, the complete additivity of ν follows from classical Nikodým Convergence Theorem [105]. It is also bounded by Theorem 6.1.19. If $\mu(p) = 0$ for a projection $p \in P(M)$, then $\mu_n(p) = 0$ for all n and hence $\nu(p) = 0$. Thus, ν is μ-continuous by Proposition 6.3.2. □

6.3.5. Remark. Like in the Remark 6.3.3, we can conclude that Theorem 6.3.4 holds if we only assume that μ_n are σ-additive and the continuum hypothesis is true. Also, the assumption of boundedness can be relaxed provided that M has no finite-dimensional direct summand.

6.4. Historical remarks and comments

The boundedness principle for completely additive measures was established first by Dorofeev and Sherstnev for Hilbert space logics $L(H)$, where $\dim L(H) \geq 3$ (see Chapter 3). Then it was proved by Dorofeev [101, 102] that any completely additive measure on the projection lattice of a von Neumann algebra M having no Type I_n, $n < \infty$, direct summand is bounded. Dorofeev's nontrivial ideas are basically followed in our exposition. In the work [66], Bunce, Saito and Wright have shown that any completely additive measure on a von Neumann algebra with atomless centre is bounded. This result has considerably simplified arguments of Dorofeev Theorem for Type I algebras and established the best possible result: Any completely additive measure μ on $P(M)$ is bounded if, and only if, the von Neumann algebra M does not contain any finitely-dimensional direct summand of dimension at least two. This is the main result of Section 6.1 that is obtained by combination of methods of the papers [101] and [66]. The Nikodým Boundedness Theorem, presented here as a direct consequence of Dorofeev Theorem, was obtained by Hamhalter in [174] as a generalization of the former result concerning the Hilbert space logic [120]. For the Nikodým Boundedness Theorem for orthomodular structures we refer to [121]. It is worth noting that the Boundedness Principle for measures has an interesting parallel concerning continuity of linear forms. Indeed, the Boundedness Principle can be seen as follows: Let μ be a completely additive measure on $P(M)$, where M is a von Neumann algebra with no finite-dimensional direct summand. Then μ is locally bounded by the classical Nikodým Boundedness Theorem and this forces in a nontrivial way that μ is bounded on the whole of $P(M)$. The following remarkable result has been obtained by Cuntz and Ringrose [87, 276]. Let f be a linear form (possibly unbounded) on a C^*-algebra A. Then f is bounded if, and only if, it is bounded on every abelian C^*-subalgebra on A. Thus, again, the boundedness is determined by commutative substructure.

Various characterizations of the facial structure in duals of operator algebras may be found in [19, 129] and a recent monograph [20]. The result on decomposition of a functional into its normal and singular part (Theorem 6.2.3) is a folklore result of the operator theory (see e.g. [307, Theorem 3.8, p. 134]). Characterization of the face $S_\sigma^\#(M)$ given in Proposition 6.2.6 as well as Proposition 6.2.1 are due to L.J.Bunce and J.Hamhalter [54]. A stronger version of this result had been previously known in the classical finitely additive measure theory. In [31] it was shown that if B is a Boolean algebra, then any state in the facial complement of the σ-additive state space is a σ-convex combination of countably singular states. Proposition 6.2.7 was proved by Akemann and Anderson in [15], where other results on connections of properties of singular states and axioms of set theory are explained. Theorem 6.2.9 showing that the continuum hypothesis is enough to force that $S_\sigma(M)^\# = S_{cs}(M)$ for any von Neumann algebra M was obtained by Bunce and Hamhalter [54]. The following problem remains open.

6.4.1. Problem. *Is* $S_\sigma^\#(M) = S_{cs}(M)$ *for any von Neumann algebra* M*?*

From the proof of Theorem 6.2.9 it is apparent that for establishing the positive answer it is enough to prove Proposition 6.2.7 without the help of continuum hypothesis. So the previous problem can be in fact reduced to a problem of the classical measure theory. Let us remark in this connection that the position of σ-additive and completely additive states is linked with the following set-theoretic concept. A cardinal I is said to be *real measurable* if the power set of I admits a probability measure vanishing on all one-dimensional subsets. If such a cardinal exists it has to be very large. The question of whether real measurable cardinals exist at all is one of the most important open questions in set theory. It can be shown that all σ-additive probability measures are completely additive if, and only if, the conjecture on the nonexistence of real measurable cardinal holds true (see [15, 116, 131]).

The decomposition of measures have also been studied in the context of general orthomodular posets (see e.g. [93, 94, 95, 285, 286]). In this setup the Yosida-Hewitt decomposition of a state is defined as the decomposition into completely additive positive measure and so-called weakly purely finitely additive measure. A positive measure on na orthomodular poset P is called *weakly purely finitely additive* if it does not majorize any nonzero completely additive positive measure. It has turned out that this decomposition always exists but it need not be unique (unlike the situation in the projection lattices). With the aim to study the uniqueness of Yosida-Hewitt decomposition, Rüttimann [286] introduced the concept of filtering measure. A positive measure μ on an orthomodular poset P is called *filtering* if for any nonzero $p \in P$ there is a nonzero $q \in P$ with $q \leq p$ and $\mu(q) = 0$. (Using this terminology, Theorem 6.2.3 says that singular states on von Neumann algebras are filtering.) Any filtering state is of course weakly purely finitely additive. The decomposition of a state into completely additive and filtering part, called today Rüttimann's decomposition, is always a Yosida-Hewitt decomposition. Rüttimann's decomposition is always unique but it need not exist. It was shown by De Simone and Navara that even the uniqueness of the Yosida-Hewitt decomposition does not guarantee that it coincides with the Rüttimann decomposition. For this and further results along the line let us refer to [94, 95].

The partial noncommutative generalization of Vitali-Hahn-Saks Theorem (Theorem 6.3.4) has been proved for Hilbert space projection lattice and for a special case of complex measures in [116, Theorem 3.10.4, p. 190]. The present proof is based on Proposition 6.3.2 that has been established by Dorofeev in [103]. We coveres a slightly more general case of finite-dimensional vector measures. It is not known so far whether the Vitali-Hahn-Saks theorem can be extended to noncommutative projection lattices in its full strength. A considerable effort has been exercised in investigating Vitali-Hahn-Saks Theorem in the quantum logics setup (see e.g. [25, 85, 96, 106]). The results in this arena show that Vitali-Hahn-Saks Theorem does not hold for general orthomodular posets and thereby the results in the present section cannot be deduced directly from the classical principles without using deeper results of operator theory.

The theory of von Neumann algebras can be viewed as the noncommutative measure theory. That is why there are many results in the theory of von Neumann algebras inspired by basic principles of measure and probability theory. They have been obtained originally for linear functionals on von Neumann algebras and their measure-theoretic forms follow from the Generalized Gleason Theorem. Let us mention some of them.

Rather deep results have been proved for convergence of elements in von Neumann algebras that can be used for deriving strong limit theorems [198]. The following classical result [288, 310] generalizes Egoroff Theorem (see also [198, 307]).

6.4.2. Theorem (Noncommutative Egoroff Theorem). *Let M be a von Neumann algebra without Type I_2 direct summand. Suppose that μ is a faithful completely additive measure on $P(M)$. Let (x_n) be a sequence in M convergent to x in the strong operator topology. Then, for every projection $p \in M$ and any $\varepsilon > 0$, there exist a projection $q \leq p$ in M and a subsequence (x_{n_k}) of (x_n) such that*

$$\mu(p-q) < \varepsilon \qquad \text{and} \qquad \|(x_{n_k} - x)\, q\| \to 0 \text{ as } k \to \infty .$$

Egoroff's Theorem can be formulated as a result on the strong operator closure of bounded sets and it is connected with the noncommutative Lusin's Theorem. Elements of a C^*-algebra A acting on a Hilbert space can be viewed as "continuous functions" while the elements of its strong operator closure can be though of as analogy to bounded measurable functions. Lusin's Theorem on approximating measurable functions by continuous ones generalizes in the noncommutative setup as follows [289, 307].

6.4.3. Theorem (Noncommutative Lusin Theorem). *Let A be a C^*-algebra of operators acting on a Hilbert space H and let M be its closure in the weak operator topology. Suppose that M has no Type I_2 direct summand. Take an arbitrary nonzero completely additive measure μ on $P(M)$, nonzero projection $e \in P(M)$, $\varepsilon > 0$ and $\delta > 0$. Then for every $a \in M$, there exists a projection $e_0 \leq e$ and $a_0 \in A$ such that*

$$a\, e_0 = e_0\, a\,, \qquad \mu(e - e_0) < \varepsilon\,, \text{ and } \qquad \|a_0\| \leq (1+\delta)\, \|a\, e_0\| .$$

Many results in the theory of operator algebras are devoted to various Radon-Nikodym type theorems for linear states and traces (see e.g. [206, 259, 307]). The linear version of the Radon-Nikodym theorem due to Sakai [290] can be reformulated in the following way.

6.4.4. Theorem (Linear Radon-Nikodým Theorem). *Let M be a von Neumann algebra without Type I_2 direct summand. Let μ and ν be positive completely additive measures on $P(M)$ such that $\mu(p) \leq \nu(p)$ for all $p \in P(M)$. Then there is a positive element $h \in M_+$, $\|h\| \leq 1$ such that*

$$\mu(p) = \frac{1}{2}\,\hat{\nu}(h\,p) + \frac{1}{2}\,\hat{\nu}(p\,h)\,, \qquad p \in P(M)\,,$$

where $\hat{\nu}$ denotes the linear extension of ν. Moreover, if ν is faithful, then h is unique.

The standard Radon-Nikodým Theorem holds under assumption that μ is absolutely continuous with respect to ν. It was shown by Navara [244] that this form of Radon-Nikodým theorem holds for a given orthomodular lattice L if, and only if, the states of L are uniquely determined by their values on the centre of L. It was then observed by Hamhalter that this is not the case in noncommutative von Neumann algebras [168]. Therefore, stronger conditions posed on the measures are necessary for noncommutative versions of Radon-Nikodym Theorem.

Another basic principle of standard measure theory and its applications is the Lyapunov theorem on the convexity of the range of nonatomic measure. Various interesting results concerning Lyapunov Theorem for von Neumann algebras have been obtained by Akemann and Anderson [15] (see also [185]). One of them is the following theorem proved for a very special case of tracial state in Chapter 5.

6.4.5. Theorem (Noncommutative Lyapunov Theorem). *Let M be a von Neumann algebra without Type I_2 direct summand. Let M contain no atomic projection. Let $\mu\colon P(M) \to X$ be a completely additive measure with values in a finite-dimensional space X. Then $\mu(P(M))$ is a compact convex subset of X.*

7. APPLICATIONS OF GLEASON THEOREM

Chapter 6 treated the consequences of the Generalized Gleason Theorem that are parallel to fundamental principles of standard measure theory and, simultaneously, they form the infrastructure of mathematical foundations of quantum theory. In this chapter we would primarily like to discuss those applications of Gleason Theorem that have direct physical meaning and provide solutions to problems originally posed in physics.

Three more or less independent topics will be considered. The point one, Section 7.1, is devoted to the Gleason Theorem for multimeasures. This makes it possible to describe decoherence functionals in new quantum histories approach to quantum physics and to resolve open problems of local measurement in quantum information theory. This part shows the power of Gleason Theorem in clarifying basic features of not only quantum mechanics but also of contemporary quantum disciplines. Secondly, in Section 7.2, dynamical aspects of quantum formalism are studied. It is shown that derivations on von Neumann algebras are in one-to-one correspondence with velocity maps on projection lattices. This contributes to axiomatics of quantum theory by showing that the dynamical structure including the Hamiltonian of the system can be recovered entirely from the order structure of projections. Finally, the last Section 7.3 deals with the famous problem of hidden variables in quantum theory. By studying dispersions of states we derive general no-go-results on hidden variables in operator-algebraic approach to quantum theory based on characterization of traces as states with the smallest dispersions. As a consequence we exclude the existence of approximate hidden variables (states with arbitrarily small dispersions) on von Neumann algebras. Recent results on multiplicative quasi-states on C^*-algebras relevant to problem of hidden variables will also be presented.

7.1. Multiform Gleason Theorem and decoherence

Generalized Gleason Theorem establishes the one-to-one correspondence between bounded measures on projection lattices of von Neumann algebras and linear functionals. A natural question arises of whether or not there exists a similar correspondence between functions on the products of projection lattices that are separately finitely additive and multilinear maps on the products of the corresponding algebras. For example, if $F: M \times M \to \mathbb{C}$ is a bounded bilinear form on a von Neumann algebra M, then its restriction, m, to the product $P(M) \times P(M)$ of projection lattices is a map, called bimeasure, satisfying the following conditions:

(i) $m(p_1+p_2,q) = m(p_1,q)+m(p_2,q)$ whenever p_1 and p_2 are orthogonal projections in M and q is an arbitrary projection in M.

(ii) $m(p,q_1+q_2) = m(p,q_1)+m(p,q_2)$ whenever q_1 and q_2 are orthogonal projections in M and p is an arbitrary projection in M.

The problem of whether every bimeasure arises from bilinear form in the above described way is a variant of the Mackey-Gleason problem in quadratic setup. In the present part we would like to address this question and its possible physical consequences. In the physical interpretation, the value, $F(a,b)$, of a bilinear form F for operators a and b usually represents a "degree of interference" between the observables a and b. From the statistical point of view, $F(a,b)$ embodies the correlation between random variables a and b. Much of recent interest in multimeasures on projections is motivated by the following example taken from the standard quantum theory. Suppose that the quantum system is represented by the algebra $B(H)$, where H is a Hilbert space, and by a normal state ϱ of $B(H)$ with the associated tracial operator t_ϱ. Let $p_{t_1}, p_{t_2}, \ldots, p_{t_n}$ be a finite sequence of projections, labeled by a discrete set of time parameters $\{t_1, t_2, \ldots, t_n\}$ that represents a series of possible measurement outcomes. Such a sequence is called the *homogeneous history of order n*. In the Heisenberg picture it can be seen as a development of the initial two-valued observable p_{t_1}. By the standard rules of quantum mechanics, the probability $d_\varrho(p_{t_1}, p_{t_2}, \ldots, p_{t_n})$ to obtain the history $p_{t_1}, \ldots, p_{t_n}$ can be computed by the following formula, first given by Wigner [320],

$$\bar{d}_\varrho(p_{t_1}, p_{t_2}, \ldots, p_{t_n}) = \varrho(p_{t_1}\, p_{t_2} \cdots p_{t_n}\, p_{t_{n-1}}\, p_{t_{n-2}} \cdots p_{t_1}), \qquad (7.1)$$

or, in a physically more popular form,

$$\bar{d}_\varrho(p_{t_1}, p_{t_2}, \ldots, p_{t_n}) = \mathrm{tr}(p_{t_n}\, p_{t_{n-1}} \cdots p_{t_1}\, t_\varrho\, p_{t_1}\, p_{t_2} \cdots p_{t_n})\,. \tag{7.2}$$

The probability formula (7.2) can be naturally associated with a multimeasure, d_ϱ, with $2n$ arguments

$$d_\varrho(h_{t_1}, h_{t_2}, \ldots, h_{t_n}, k_{t_1}, k_{t_2}, \ldots, k_{t_n}) = \mathrm{tr}(h_{t_n}\, h_{t_{n-1}} \cdots h_{t_1}\, t_\varrho\, k_{t_1}\, k_{t_2} \cdots k_{t_n})\,. \tag{7.3}$$

It represents the "influence" between histories $h = \{h_1, \ldots, h_n\}$ and $k = \{k_1, \ldots, k_n\}$ that is equal to probability of obtaining the history h in the case of $h = k$. The above defined map, d_ϱ, called the *standard decoherence functional*, constitutes one of the most important examples of multimeasure and has inspired many results presented here.

In the present section we first extend the Generalized Gleason Theorem to multimeasures. Based on this general result, we will further discuss the representations of the decoherence functionals in the scheme of quantum histories approach to quantum theory and their consequences. Now we want to stabilize the terminology.

7.1.1. Definition. Let $A_1, A_2, \ldots, A_n$ be C^*-algebras with the projection structures $P(A_1), \ldots, P(A_n)$, respectively. Let X be a Banach space. The map $m\colon P(A_1) \times P(A_2) \times \cdots \times P(A_n) \to X$ is called an *X-valued multimeasure*(in short a *multimeasure*) if m is separately finitely additive, meaning that, for each j,

$$\begin{aligned} &m(p_1, \ldots, p_{j-1}, q_1 + q_2, p_{j+1}, \ldots, p_n) \\ &= m(p_1, \ldots, p_{j-1}, q_1, p_{j+1}, \ldots, p_n) + m(p_1, \ldots, p_{j-1}, q_2, p_{j+1}, \ldots, p_n)\,, \end{aligned}$$

where $p_i \in P(A_i)$, q_1 and q_2 are orthogonal projections in $P(A_j)$. If $X = \mathbb{C}$, we call m the *complex multimeasure*. *Completely additive* and *σ-additive* multimeasures are defined in the standard way as measures separately completely additive and σ-additive, respectively. In case when $n = 2$ we call the multimeasure the *bimeasure*. Moreover, if $A = A_1 = A_2$ we say that m is a bimeasure on A. A bimeasure m on A is said to be *hermitean* if

$$m(p, q) = \overline{m(q, p)} \qquad \text{for all projections } p, q \in P(A)\,.$$

Now let us define the linear counterparts of bimeasures.

7.1.2. Definition. Let $X, X_1, X_2, \dots, X_n$ be Banach spaces. Let $F\colon X_1 \times X_2 \times \cdots \times X_n \to X$ be an n-linear map. F is said to be *bounded* if there is a constant C such that, for each $x_i \in X_i$,

$$\|F(x_1, x_2, \dots, x_n)\| \le C \|x_1\| \cdot \|x_2\| \cdots \|x_n\| .$$

A smallest constant C for which this inequality holds is said to be the *norm* of F (in symbols $\|F\|$).

The next theorem can be viewed as another generalization of Gleason Theorem, this time in terms of multimeasures.

7.1.3. Theorem. *Let $M_1, \dots, M_n$ be von Neumann algebras, each having no direct summand of Type I_2. Suppose that $m\colon P(M_1) \times \cdots \times P(M_n) \to X$ is a bounded X-valued multimeasure. There is a unique bounded n-linear map $F\colon M_1 \times M_2 \times \cdots \times M_n \to X$ which extends m.*

Proof: First observe that m has at most one n-linear bounded extension. This is a consequence of the fact that cartesian product of linear spans of projections is dense in $M_1 \times M_2 \times \cdots \times M_n$.

For $n = 1$ Theorem 7.1.3 reduces to the Generalized Gleason Theorem. Further, we shall proceed by induction on n. Suppose that the statement of the theorem is true for $n-1$. For each fixed projection $p \in P(M_1)$, the map $m_p\colon P(M_2) \times \cdots \times P(M_n) \to X$, defined by

$$m_p(p_2, \dots, p_n) = m(p, p_2, \dots, p_n) ,$$

is a bounded multimeasure. By the induction hypothesis, m_p extends to a bounded $(n-1)$-linear map, F_p, on $M_2 \times \cdots \times M_n$. Denote by Y the Banach space of all bounded $(n-1)$-linear forms on $M_2 \times \cdots \times M_n$ with values in X. We shall consider the map $T\colon P(M_1) \to Y\colon p \to F_p$. Suppose that p_1, p_2 are orthogonal projections in M_1. Then, for projections $(q_2, \dots, q_n) \in P(M_2) \times \cdots \times P(M_n)$ the following holds true.

$$\begin{aligned} F_{p_1+p_2}(q_2, \dots, q_n) &= m(p_1 + p_2, q_2, \dots, q_n) \\ &= m(p_1, q_2, \dots, q_n) + m(p_2, q_2, \dots, q_n) \\ &= F_{p_1}(q_2, \dots, q_n) + F_{p_2}(q_2, \dots, q_n). \end{aligned}$$

Since multilinear continuous functionals are uniquely determined by their values on projections, the previous equality means that

$$F_{p_1+p_2} = F_{p_1} + F_{p_2}\,.$$

In other words, T is a finitely additive Y-valued measure on $P(M_1)$. We show that T is bounded. Let

$$C = \sup\{\|m(p_1, \ldots, p_n)\| \mid p_1 \in P(M_1), \ldots, p_n \in P(M_n)\} < \infty.$$

Then, for $p \in P(M_1)$, $\|F_p\| \le 4^{n-1}\, C$. Indeed, for $0 \le x_i \le 1$, $x_i \in M_i$, we can write $x_i = \sum_n \frac{1}{2^n} p_{n,i}$, where $(p_{n,i})_n$ are commuting projections from M_i. It implies, by multilinearity,

$$\|F_p(x_2, x_3, \ldots, x_n)\| \le C\,. \tag{7.4}$$

Since every norm-one element in a C^*-algebra is a combination of four positive elements with norm less than one, we see from (7.4) that

$$\|F_p(y_2, y_3, \ldots, y_n)\| \le 4^{n-1}\, C,$$

whenever $y_2, \ldots, y_n$ are norm-one elements of the corresponding algebras. Now we can appeal to the Generalized Gleason Theorem and find a bounded linear map $H\colon\ P(M_1) \to Y$ which extends T. Set finally

$$F(z_1, z_2, \ldots, z_n) = H(z_1)(z_2, \ldots, z_n)\,.$$

It is clear that F is the desired multilinear bounded map extending m. □

By specializing to hermitean bimeasures, we obtain the following corollary. It basically says that there is a canonical correspondence between hermitean bimeasures and hermitean bilinear forms.

7.1.4. Corollary. *Let M be a von Neumann algebra without Type I_2 direct summand. Let $m\colon\ P(M) \times P(M) \to \mathbb{C}$ be a hermitean bounded bimeasure. Then there is a bounded hermitean bilinear form F which extends m.*

Proof: Theorem 7.1.3 guarantees the existence of the bounded bilinear extension F of m. It remains to verify that F is hermitean. Suppose first that $0 \leq x, y \leq 1$. Then $x = \sum_n \frac{1}{2^n} p_n$, $y = \sum_n \frac{1}{2^n} q_n$, where p_n and q_n are projections. This gives us

$$F(x,y) = \sum_n \sum_m \frac{1}{2^{m+n}} m(p_n, q_m) = \sum_n \sum_m \frac{1}{2^{n+m}} \overline{m(q_m, p_n)} = \overline{F(y,x)}\,.$$

By the standard decomposition of an arbitrary element into a linear combination of four positive ones, we obtain

$$F(x,y) = \overline{F(y,x)}$$

for all $x, y \in M$. □

As we have seen in Chapter 3, the assumption of boundedness in the Generalized Gleason Theorem is necessary. However, if we assume that the measure is completely additive, the boundedness can be relaxed. In the next theorem we prove that the boundedness principle holds also for completely additive multimeasures. Although, as it is apparent from the proof, the following theorem holds for all multimeasures, it will be formulated for bimeasures only.

7.1.5. Theorem. *Let M_1 and M_2 be von Neumann algebras without any Type I_2 direct summand and any finite-dimensional direct summand. Suppose that $m\colon P(M_1) \times P(M_2) \to \mathbb{C}$ is a completely additive bimeasure. Then m is bounded. Furthermore, there exists a bounded bilinear form, $F\colon M_1 \times M_2 \to \mathbb{C}$, which extends m and, moreover, both $F(x,\cdot)$ and $F(\cdot,y)$ are normal functionals for all $x \in M_1$ and $y \in M_2$.*

Proof: For a fixed $p \in P(M_1)$, the map $\varphi_p\colon q \to m(p,q)$ is a completely additive measure on $P(M_2)$. By Theorem 6.1.18 it extends to a normal functional on M_2 denoted again by φ_p. The resulting map $T\colon P(M_1) \to (M_2)_*\colon p \to \varphi_p$ is certainly a finitely additive vector measure on the projection lattice $P(M_1)$ with values in the predual of M_2. To prove its boundedness it is sufficient, according to the Banach-Steinhaus Uniform Boundedness Principle, to verify that the set

$$\{\varphi_p(x) \mid p \in P(M_1)\}$$

is bounded, for an arbitrary $0 \leq x \leq 1$, $x \in M_2$. Let us pick such an x and represent it, as usual, in the form $x = \sum_{m=1}^{\infty} \frac{1}{2^m} q_m$, where q_m are projections in M_2. We are going to prove first that the map $T_x : p \to \varphi_p(x)$ is a (complex) completely additive measure on $P(M_1)$. For this, let us consider a collection $(p_i)_{i \in I}$ of orthogonal projections in M_1. Then, for $e = \sum_{i \in I} p_i$,

$$\begin{aligned}\varphi_e(x) &= \varphi_e\Big(\sum_{m=1}^{\infty} \frac{1}{2^m} q_m\Big) = \sum_{m=1}^{\infty} \frac{1}{2^m} \varphi_e(q_m) = \sum_{m=1}^{\infty} \frac{1}{2^m} m\Big(\sum_{i \in I} p_i, q_m\Big) \\ &= \sum_{m=1}^{\infty} \sum_{i \in I} \frac{1}{2^m} m(p_i, q_m) .\end{aligned}$$

Since the series on the right-hand side converges absolutely, by a rearrangement we obtain,

$$\varphi_e(x) = \sum_{i \in I} \sum_{m=1}^{\infty} \frac{1}{2^m} \varphi_{p_i}(q_m) = \sum_{i \in I} \varphi_{p_i}(x) .$$

Whence, the measure T_x is completely additive and therefore bounded by Theorem 6.1.18. Hence, the vector measure $T\colon P(M_1) \to (M_2)_*$ is bounded, and so, by the Generalized Gleason Theorem, it has a bounded linear extension, again denoted by T. Letting

$$F(x, y) = [T(x)](y) ,$$

we obtain a bilinear extension of m with the properties of the statement of the theorem. □

After obtaining the general results on the structure of multimeasures, we shall return to the Wigner's formula (7.3) which constitutes an important example in the so-called histories approach to quantum mechanics. The general quantum histories theory can be characterized by the space of histories and the space of decoherence functionals. The histories generalize the time ordered strings of projections, while decoherence functionals describe the probabilistic correlations between various histories. The axioms for decoherence functional, stated in the following definition, are abstracted from the properties of the standard homogeneous decoherence functional d_ϱ appearing in the Wigner's formula (7.3).

7.1.6. Definition. Let $M_1, M_2, \dots, M_n$ be von Neumann algebras. A *decoherence functional (of order n)* is a complex multimeasure

$$d\colon P(M_1) \times P(M_2) \times \cdots \times P(M_n) \times P(M_1) \times P(M_2) \times \cdots \times P(M_n) \to \mathbb{C}$$

such that

(i) $d(p_1, p_2, \dots, p_n, q_1, q_2, \dots, q_n) = \overline{d(q_1, q_2, \dots, q_n, p_1, p_2, \dots, p_n)}$ for all p_i and q_i in $P(M_i)$ (hermiteanetness).

(ii) $d(p_1, \dots, p_n, p_1, \dots, p_n) \geq 0$ for all $p_i \in P(M_i)$ (positivity).

(iii) $d(1, \dots, 1, 1 \dots, 1) = 1$ (normalization).

The decoherence functional is called *completely additive* if it is a completely additive multimeasure. If all algebras $M_1, M_2, \dots, M_n$ equals to the algebra M, we say that d is the *decoherence functional (of order n)* on M.

It can be easily seen that the standard decoherence functional d_ϱ (7.3) obeys all axioms of the previous definition and it is thus an example of a completely additive decoherence functional on $B(H)$ of order n. If the system evolves in the initial state ϱ, the projections $p_{t_1}, p_{t_2}, \dots, p_{t_n}$ depend on the Hamiltonian of the system if one works in the Heisenberg picture. More explicitly, $p_{t_i} = e^{i\,(t_i - t_1)\,H}\, p_1\, e^{-i\,(t_i - t_1)\,H}$, where H is the Hamiltonian operator. Therefore, the formula (7.3) encodes also the dynamics of the system. The decoherence functional has the advantage over the state on algebras in standard quantum mechanics that it represents both kinematical and dynamical aspects of the system.

7.1.7. Corollary. *Let*
$d\colon P(M_1) \times P(M_2) \times \cdots \times P(M_n) \times P(M_1) \times P(M_2) \times \cdots P(M_n) \to \mathbb{C}$
be a bounded decoherence functional such that all von Neumann algebras $M_1, M_2, \dots, M_n$ have no Type I_2 direct summand. Then there is a unique bounded multilinear form
$F\colon M_1 \times M_2 \times \cdots \times M_n \times M_1 \times M_2 \times \cdots \times M_n \to \mathbb{C}$
which extends d.

Proof: The result follows directly from Theorem 7.1.3. □

The set of all homogeneous histories of order n itself has a priori no structure of propositional logic. Such a structure is useful for working with "disjoint histories". For example using notation of (7.3), it is desirable to consider the "disjoint sum" of two homogeneous histories, h and k, corresponding to the sum of the operators $h_{t_1} h_{t_2} \cdots h_{t_n} + k_{t_1} k_{t_2} \cdots k_{t_n}$, transforming the initial state ϱ in the Schrödinger picture. Nevertheless, we can identify each homogeneous history $h = (h_1, h_2, \ldots, h_n)$ with the simple tensor $h_1 \otimes \cdots \otimes h_n$ in the n-fold tensor product $B(H_1)\overline{\otimes} \cdots \overline{\otimes} B(H)$. So, the structure of homogeneous histories can be embedded into the projection lattice of the algebra $B(H)\overline{\otimes} \cdots \overline{\otimes} B(H)$ which endows homogeneous histories with the structure of propositions. It is therefore natural to consider *general history* as the (general) projection in the lattice $P(B(H)\overline{\otimes} B(H)\overline{\otimes} \cdots \overline{\otimes} B(H))$. In this connection, a natural question arises whether also the decoherence functional could be expressed in terms of the tensor product and viewed as extended from homogeneous histories to all general histories. If the decoherence functional of order n on a von Neumann algebra M extends to a multilinear form F on the $2n$-fold cartesian product $M \times \cdots \times M$, then there is a unique linear form f on the algebraic tensor product $M \otimes_{alg} M \otimes_{alg} \cdots \otimes_{alg} M$ such that

$$F(x_1, x_2, \ldots, x_{2n}) = f(x_1 \otimes \cdots \otimes x_{2n}), \tag{7.5}$$

for all $x_1, \ldots, x_{2n} \in M$. This establishes a natural correspondence between decoherence functionals and (possibly unbounded) linear forms on algebraic tensor products. We shall call the linear form f uniquely determined by (7.5) the *linear form corresponding* to d. The following theorem by Isham, Linden and Schreckenberg describes decoherence functionals for finite-dimensional Hilbert spaces. It can be considered as the analog of Gleason Theorem to quantum histories.

7.1.8. Theorem (Isham-Linden-Schreckenberg). *Let H be a finite-dimensional Hilbert space of dimension at least three. Let d be a bounded decoherence functional of order n on $B(H)$. Then there is a unique operator X_d on the tensor product $K \otimes K$, where $K = H \otimes H \otimes \cdots \otimes H$ is the n-fold tensor product such that*

$$d(p_1, p_2, \ldots, p_n, q_1, q_2, \ldots, q_n) = \operatorname{tr}((p \otimes q)\, X_d), \tag{7.6}$$

where $p = p_1 \otimes p_2 \otimes \cdots \otimes p_n$ and $q = q_1 \otimes q_2 \otimes \cdots \otimes q_n$.

Proof: By Theorem 7.1.3 and the reasoning preceeding the present theorem the mapping d can be represented by a linear form f on $K \otimes_{alg} K$ such that

$$d(p_1, p_2, \ldots, p_n, q_1, q_2, \ldots, q_n) = f(p_1 \otimes p_2 \otimes \cdots \otimes p_n \otimes q_1 \otimes q_2 \otimes \cdots \otimes q_n)$$

for all $p_i, q_i \in P(H)$. Since any linear form on finite-dimensional Hilbert space is normal, we see that (7.6) holds for the trace class operator X_d representing f. □

The formula (7.6) is referred to as the *Isham-Linden-Schreckenberg representation*. It says that the decoherence functional can be extended to the space of all histories by taking

$$d(p, q) = \mathrm{tr}((p \otimes q)\, X_d) \qquad \text{for all } p, q \in P(\otimes_{i=1}^{n} H)\,.$$

Now we shall deal with the problem of extending the Isham-Linden-Schreckenberg representation to infinite dimensions. First observe that any bounded decoherence functional is uniquely represented by a linear form on the corresponding algebraic tensor product. However, this form does not have to be continuous, even if the decoherence functional is completely additive and bounded. This is exhibited in the subsequent example. We shall use the following conventions. Let $K(H)$ be the ideal of compact operators on the Hilbert space H. The algebraic tensor product $K(H) \otimes_{alg} K(H)$ will be viewed as subalgebra of $B(H \otimes H)$. It can be proved that $K(H) \otimes K(H) = K(H \otimes H)$.

7.1.9. Example. Let H be a separable Hilbert space of infinite dimension. Let ψ be a unit vector in H. Then

$$d_\psi(p, q) = (p\,\psi, q\,\psi), \qquad p, q \in P(H)\,,$$

is a bounded, completely additive decoherence functional on $B(H)$ of order one such that the corresponding linear form f_ψ is unbounded on the algebraic tensor product of $K(H) \otimes_{alg} K(H)$.

Proof: It is easy to verify that d_ψ is a completely additive decoherence functional that extends to the bilinear form

$$B_\psi(x, y) = (x\,\psi, y^*\,\psi) \qquad x, y \in B(H)\,.$$

Let (ψ_n) be an orthonormal basis of H such that $\psi_1 = \psi$. For each i we shall define one-dimensional operators a_i, b_i on H by equalities

$$\begin{aligned} a_i(\xi) &= (\xi, \psi)\, \psi_i \,, \\ b_i(\xi) &= (\xi, \psi_i)\, \psi \,. \end{aligned}$$

Let us now put $S_n = \sum_{i=1}^n a_i \otimes b_i$. Since

$$(a_i \otimes b_i)\,(\psi_k \otimes \psi_l) = (\psi_k, \psi)\, \psi_i \otimes (\psi_l, \psi_i)\, \psi = \delta_{k1} \delta_{li}\, (\psi_i \otimes \psi) \,,$$

we see that

$$S_n(\psi \otimes \psi_r) = \psi_r \otimes \psi \,,$$

whenever $r \leq n$, and that S_n vanishes on every other vector in the orthonormal basis $(\psi_k \otimes \psi_l)_{k,l}$ of $H \otimes H$. Therefore $\|S_n\| \leq 1$ for all n. However, for each i,

$$f_\psi(a_i \otimes b_i) = (a_i\, \psi, b_i^*\, \psi) = (b_i\, a_i\, \psi, \psi) = (\psi, \psi) = 1 \,,$$

and so

$$f_\psi(S_n) = n \,.$$

We conclude that f_ψ is unbounded on the algebraic tensor product $K(H) \otimes_{alg} K(H)$. □

The decoherence functional d_ψ from the foregoing example cannot be implemented by any trace class operator in the sense of the Isham-Linden-Schreckenberg representation (7.6). Indeed, if X_{d_ψ} were the corresponding trace class operator acting on $H \otimes H$, then

$$f_\psi(x \otimes y) = \mathrm{tr}((x \otimes y) X_{d_\psi}) \,, \qquad \text{for all } x, y \in B(H) \,,$$

and so

$$f_\psi(z) = \mathrm{tr}(z\, X_{d_\psi}) \qquad \text{for all } z \in K(H) \otimes_{alg} K(H) \,.$$

It means that f_ψ would be bounded on $K(H) \otimes_{alg} K(H)$ – a contradiction.

If μ is a completely additive probability measure on the projection lattice of an infinite-dimensional Hilbert space H, then μ extends to a normal functional which is described by the trace class operator on H. This representation corresponds to a decomposition of μ into convex mixture of pure states. The previous example shows that the relation between completely additive measures and trace-class operators breaks down for completely additive bimeasures. It indicates the different geometric structure of decoherence functionals.

It turns out that the boundedness of the corresponding linear form on compact operators is necessary and sufficient for the existence of the Isham-Linden-Schreckenberg representation. We say that the decoherence functional d of order one on $B(H)$ is *tensor bounded* if the corresponding linear form f_d on $B(H) \otimes_{alg} B(H)$ is bounded when restricted to $K(H) \otimes_{alg} K(H)$.

7.1.10. Proposition. *Let H be a Hilbert space,* $\dim H \geq 3$. *Let d be a bounded decoherence functional of order one on $B(H)$. Then d is tensor bounded if, and only if, there exists a trace class operator X_d on $H \otimes H$ such that*

$$d(p, q) = \operatorname{tr}((p \otimes q)\, X_d) \tag{7.7}$$

for all finite-dimensional projections $p, q \in B(H)$.

Proof: If (7.7) holds, then it is clear that $f_d(z) = \operatorname{tr}(z\, X_d)$, and hence d is tensor bounded. Suppose now that d is tensor bounded. Then, by definition, the corresponding linear form f_d is bounded on $K(H) \otimes_{alg} K(H)$ and it has a unique extension to a bounded linear functional on $K(H \otimes H) = K(H) \otimes K(H)$. However, any bounded functional on the ideal of compact operators uniquely extends to the normal functional $\tilde{f}_d$ on the algebra of all bounded operators (see Chapter 2, Compact and trace class operators). Therefore, the trace class operator X_d acting on $H \otimes H$ for which

$$\tilde{f}_d(a) = \operatorname{tr}(a\, X_d), \qquad \text{for all } a \in B(H \otimes H)\,,$$

satisfies also (7.7). □

7.1.11. Corollary. *Let H be a Hilbert space with* $\dim H \geq 3$. *Let d be a completely additive decoherence functional on $B(H)$ of order one. There exists a trace class operator X_d on $H \otimes H$ such that*

$$d(p, q) = \operatorname{tr}((p \otimes q)\, X_d)$$

for all projections $p, q \in B(H)$ if, and only if, d is tensor bounded.

Proof: Since the case of $\dim H < \infty$ is covered by the Isham-Linden-Schreckenberg formula, we may assume that $\dim H = \infty$. If d is completely additive then it is bounded by Theorem 7.1.5. Moreover, if d is tensor bounded, then there is a trace class operator X_d on $B(H \otimes H)$ such that

$$d(p,q) = \mathrm{tr}((p \otimes q)\, X_d)$$

for all finite-dimensional projections $p, q \in B(H)$. By complete additivity

$$d(p,q) = \mathrm{tr}((p \otimes q)\, X_d)$$

for all projections $p, q \in B(H)$. □

Although completely additive decoherence functionals cannot not in general be described by the trace class operators, it can be in many important cases computed from the formula appearing in the Isham-Linden-Schreckenberg representation. In order to formulate this fact we shall need the following concept. Let d be a bounded decoherence functional on $B(H)$ of order one with the corresponding linear form f_d on $B(H) \otimes_{alg} B(H)$. Let us say that d is *tracially bounded* if

$$\sup\{|f_d(p_\xi)| \mid \xi \in H \otimes_{alg} H\,, \|\xi\| = 1\} < \infty\,,$$

where p_ξ denotes the projection of $H \otimes H$ onto the one-dimensional space $\mathrm{sp}\{\xi\}$. (Observe that $p_\xi \in K(H) \otimes_{alg} K(H)$, whenever $\xi \in H \otimes_{alg} H$.)

7.1.12. Example. The decoherence functional d_ψ defined in the Example 7.1.9 is tracially bounded.

Proof: Consider again the orthonormal basis (ψ_n) of H with $\psi_1 = \psi$. Let $x, y \in B(H)$. Expanding inner product $(x\,\psi, y^*\,\psi)$ with respect to orthonormal basis (ψ_i) we have that

$$\begin{aligned}(x\,\psi, y^*\,\psi) &= \sum_{i=1}^{\infty} (x\,\psi, \psi_i)\,\overline{(y^*\,\psi, \psi_i)} = \sum_{i=1}^{\infty} (x\,\psi, \psi_i)\,(y\,\psi_i, \psi) \\ &= \sum_{i=1}^{\infty} ((x \otimes y)\,(\psi \otimes \psi_i),\, \psi_i \otimes \psi)\,.\end{aligned}$$

It means that

$$f_\psi(a) = \sum_{i=1}^{\infty} (a\,(\psi \otimes \psi_i), \psi_i \otimes \psi) \tag{7.8}$$

for any $a \in K(H) \otimes_{alg} K(H)$ (f_ψ is uniquely determined by its values on simple tensors). Let now u be a unitary map on $H \otimes H$ which sends $\psi_i \otimes \psi_j$ to $\psi_j \otimes \psi_i$. Let p be the projection of $H \otimes H$ onto the closed linear span of $\{\psi \otimes \psi_i \mid i = 1, \ldots\}$. If a is a one-dimensional projection, then $a \in L^1(H \otimes H)$ and we can rewrite (7.8) in the form

$$f_\psi(a) = \operatorname{tr} a\,p\,u\,.$$

Using the inequality (2.9) we see that

$$|f_\psi(a)| \le \|p\,u\| \cdot \operatorname{tr}(|a|) = \|p\,u\| \le 1\,,$$

implying immediately that d_ψ is tracially bounded. □

7.1.13. Theorem. *Let the decoherence functional d be of order one on the algebra $B(H)$, $\dim H \ge 3$. Let d be tracially bounded. Then there is a bounded operator T on $H \otimes H$ such that*

$$d(p,q) = \operatorname{tr}((p \otimes q)\,T)\,,$$

whenever p and q are finite-dimensional projections on H.

Furthermore, let d be completely additive. Then for all projections $p, q \in B(H)$ we have

$$d(p,q) = \sum_i \sum_j \operatorname{tr}((p_i \otimes q_j)\,T)\,, \tag{7.9}$$

whenever $p = \sum_i p_i$ and $q = \sum_j q_j$, where p_i, q_i are one-dimensional projections in H.

Proof: The linear form f_d on $K(H) \otimes_{alg} K(H)$ corresponding to d induces a weak frame function h on the inner product space $H \otimes_{alg} H$ by setting

$$h(\xi) = f_d(p_\xi)\,,$$

where p_ξ is a one-dimensional projection onto the span of the unit vector $\xi \in H \otimes_{alg} H$. Indeed, let $(e_i)_{i=1}^k$ be an orthonormal basis of the subspace

K of $H \otimes_{alg} H$. Denote by p_K the projection of $H \otimes H$ onto K. Then $p_K = p_{e_1} + \cdots + p_{e_k} \in K(H) \otimes_{alg} K(H)$ and

$$\sum_{i=1}^{k} h(e_i) = \sum_{i=1}^{k} f_d(p_{e_i}) = f_d(p_K)\,.$$

Whence, $f_d|K$ is a frame function with the weight $f_d(p_K)$. Since h is bounded by the tracial boundedness, there is a bounded operator T on $H \otimes H$ (see Theorem 5.2.4) such that

$$h(\xi) = (T\xi, \xi) = \operatorname{tr} p_\xi T\,.$$

As $p_\xi \otimes p_\mu$ is a projection onto $\operatorname{sp}\{\xi \otimes \mu\}$, we infer that

$$d(p,q) = \operatorname{tr}(p \otimes q)\, T)\,,$$

whenever p and q are finite-dimensional projections. The formula (7.9) is now a direct consequence of the complete additivity of d. □

In the concluding part of this section we shall show another application of the multi-form Gleason Theorem. We shall provide a solution to one open problem of quantum information theory concerning local operations and measurements on multipartite systems. Let $H_1 \otimes H_2 \otimes \cdots \otimes H_n$ be the tensor product of Hilbert spaces. Let us denote by $\sum(H_1, H_2, \ldots, H_n)$ the set of all unit vectors in this product space that are of the form $x_1 \otimes x_2 \otimes \cdots \otimes x_n$, where $x_i \in H_i$ are unit vectors. In quantum information theory the vector states corresponding to elements of $\sum(H_1, H_2, \ldots, H_n)$ are called *unentangled.* They describe the local measurements of n-particle system modelled by the Hilbert spaces $H_1, H_2, \ldots, H_n$. A natural question in measurement quantum theory is whether the knowledge of structure of probabilities associated to local measurements allows one to determine uniquely the state of the whole system. Viewed mathematically, let us represent probabilities on local measurements by the function

$$f : \sum(H_1, H_2, \ldots, H_n) \to \mathbb{R}^+$$

with the property

$$\sum_{i \in I} f(e_i) = w,$$

whenever $(e_i)_{i\in I}$ is an orthonormal basis of $H_1 \otimes H_2 \otimes \cdots \otimes H_n$ contained in $\sum(H_1, H_2, \ldots, H_n)$. The function f is called *unentangled frame function* with the *weight* w. Mathematical part of the problem of local measurements is the question of whether unentangled frame function can be extended to the frame function on the whole of $H_1 \otimes H_2 \otimes \cdots \otimes H_n$. First we show that we can always find a multimeasure representing an unentangled frame function.

7.1.14. Proposition. *Let* $f\colon \sum(H_1, H_2, \ldots, H_n) \to \mathbb{R}^+$ *be an unentangled frame function. Then there is a completely additive bounded multimeasure* m *on* $P(H_1) \times \cdots \times P(H_n)$ *such that*

$$f(x_1 \otimes \cdots \otimes x_n) = m(p_1, p_2, \ldots, p_n)\,,$$

whenever $x_1 \otimes \cdots \otimes x_n \in \sum(H_1, H_2, \ldots, H_n)$ *and* p_i $(i = 1, \ldots, n)$ *is the projection with the range* $\mathrm{sp}\{x_i\}$.

Proof: For simplicity, let us prove the proposition for $n = 2$. (This restriction is not essential.) Let p_1 and p_2 be projections in $P(H_1)$ and $P(H_2)$, respectively. Assume that $(x_i)_{i\in I}$ and $(y_j)_{j\in J}$ are orthonormal bases of the ranges of p_1 and p_2, respectively. The only candidate for the value $m(p_1, p_2)$ is the sum $\sum_{(i,j)\in I\times J} f(x_i \otimes y_j)$. We have to show that this value does not depend on the choice of the corresponding orthonormal bases. For this, let w be the weight of f. Let $(x'_i)_{i\in I}$, $(y'_i)_{j\in J}$ be other orthonormal bases of $p_1(H_1)$ and $p_2(H_2)$, respectively. Both the bases $(x_i \otimes y_j)_{(i,j)\in I\times J}$ and $(x'_i \otimes y'_j)_{(i,j)\in I\times J}$ of $p_1(H_1) \otimes p_2(H_2)$ can be completed to orthonormal bases of $H_1 \otimes H_2$ by adding the same orthonormal system composed of vectors in $\sum(H_1, H_2, \ldots, H_n)$. By the standard argument we obtain

$$\sum_{(i,j)\in I\times J} f(x_i \otimes y_j) = \sum_{(i,j)\in I\times J} f(x'_i \otimes y'_j)\,.$$

Now, setting

$$m(p_1, p_2) = \sum_{(i,j)\in I\times J} f(x_i \otimes y_j)\,,$$

we obtain the desired completely additive bimeasure. □

Having established a representation of unentangled frame function by a multimeasure, we can now employ the previous results to obtain new Gleason type theorems in this direction. Let us remark that tensor boundedness can be defined for multimeasures as well as for decoherence functionals. Then all results formulated above for decoherence functionals hold true for all multimeasures.

7.1.15. Theorem. *Let $H_1, H_2, \ldots, H_n$ be Hilbert spaces of dimension at least three. Let $f : \sum(H_1, \ldots, H_n) \to \mathbb{R}^+$ be an unentangled frame function such that the corresponding multimeasure is tensor bounded. Then there is a trace class operator X on $B(H_1 \otimes \cdots \otimes H_n)$ such that*

$$f(x_1 \otimes \cdots \otimes x_n) = \operatorname{tr}((p_1 \otimes \cdots \otimes p_n)\, X)\,,$$

for all unit vectors $x_i \in H_i$ and corresponding one-dimensional projections p_i $(i = 1, \ldots, n)$.

Proof: The proof directly follows from Proposition 7.1.14 and Corollary 7.1.11. □

7.1.16. Corollary. (Wallach Theorem) *Let $H_1, \ldots, H_n$ be finite-dimensional Hilbert spaces, each of dimension at least three. Let $f : \sum(H_1, H_2, \ldots, H_n) \to \mathbb{R}^+$ be an unentangled frame function. Then there exists a self-adjoint operator $T \in B(H_1 \otimes \cdots \otimes H_n)$ such that whenever $x_1 \otimes \cdots \otimes x_n$ is in $\sum(H_1, H_2, \ldots, H_n)$ and p_j is the projection of H_j onto the one-dimensional subspace generated by x_j, the following equality is in force:*

$$f(x_1 \otimes x_2 \otimes \cdots \otimes x_n) = \operatorname{tr}((p_1 \otimes p_2 \otimes \cdots \otimes p_n)\, T)\,.$$

7.2. Velocity maps and derivations

In this section an interesting application of the Generalized Gleason Theorem to axiomatics of quantum dynamics will be shown. In the Heisenberg picture of quantum evolution, the rate of change, $\delta(X)$, of an observable X (viewed as a self-adjoint operator on a Hilbert space) is determined by the following formula

$$\delta(X) = i\,[H, X]\,, \tag{7.10}$$

where H is the Hamiltonian operator. (We use the standard notation $[a, b] = a\,b - b\,a$ for the commutant of elements a, b.) The properties of the map (7.10) have been abstracted to define the concept of derivation which plays an important role in the theory of dynamical systems. Let A be a C^*-algebra. The *derivation* on A is a linear map $d\colon\ A \to A$ such that

$$d(a\,b) = d(a)\,b + a\,d(b) \qquad \text{for all } a, b \in A\,. \tag{7.11}$$

Moreover, d is called the $*$-*derivation* if $d(x^*) = d(x)^*$ for all $x \in A$. Given a derivation d let us denote by d^* the derivation $d^*(x) = d(x^*)^*$. Then

$$d = \frac{d + d^*}{2} + i\,\frac{i\,d^* - i\,d}{2} \tag{7.12}$$

is the canonical decomposition of d into a sum of two $*$-derivations. The formula (7.11) is the Leibnitz formula for derivation and it can be verified easily that it is satisfied for the map δ defined by the commutant formula (7.10). On the other hand, it has been proved by Kadison and Sakai that any derivation on a von Neumann algebra is *inner*, i.e. it can be written in the form (7.10). This result is far from being true for general C^*-algebras. In case of general algebras we know by the result of Sakai that any derivation on a C^*-algebra is continuous.

From physical point of view, the conditions in the definition of derivation are too strong and cannot be fully justified. Indeed, the rate of change of an observable is meaningful only for a self-adjoint operator. Since the product $a\,b$ of self-adjoint operators a and b is self-adjoint if, and only if, a and b commute, the Leibnitz formula (7.11) has sense only for commuting self-adjoint operators a and b. Also, the sum $a + b$ of the observables a and b has physical meaning only if a and b are simultaneously measurable (i.e. commuting). Therefore, it is natural to postulate the rate of change to be only a quasi-linear map. This motivated the following definition.

7.2.1. Definition. Let A be a C^*-algebra. The map $d\colon A_{sa} \to A_{sa}$ defined on the self-adjoint part A_{sa} of A is called the *velocity map* if the following conditions are satisfied for all commuting $a, b \in A_{sa}$ and $\lambda \in \mathbb{R}$:

(i) $d(\lambda\, a) = \lambda\, d(a)$

(ii) $d(a) + d(b) = d(a + b)$,

(iii) $d(a\, b) = d(a)\, b + a\, d(b)$.

We shall deal with the problem of whether or not all velocity maps are restrictions of derivations on the whole algebra. It turns out that the answer is in the positive for all von Neumann algebras for which the Gleason Theorem holds true. Moreover, we show that derivations are characterized only by their actions on the projection lattice. This provides an explanation of the formula (7.10) purely in terms of simultaneously measurable yes-no observables.

7.2.2. Proposition. *Let M be a von Neumann algebra without Type I_2 part and let $d\colon P(M) \to M$ be a bounded map such that*

$$d(e\, f) = d(e)\, f + e\, d(f)\,, \qquad \textit{whenever } e\, f = f\, e\,.$$

Then d extends to a linear map $\tilde{d}\colon M \to M$ such that

$$\tilde{d}(a\, b) = \tilde{d}(a)\, b + a\, \tilde{d}(b)$$

for all commuting self-adjoint elements in $a, b \in A$.

Proof: In the first step we show that d is a finitely additive measure with values in M. Let e and f be orthogonal projections in M. Thus, as $e\, f = f\, e = 0$, we see that

$$d(e)\, f + e\, d(f) = d(f)\, e + f\, d(e) = 0\,.$$

In addition, since $e = e\,(e + f) = (e + f)\, e$, we have

$$d(e) = d(e)\,(e + f) + e\, d(e + f) = d(e + f)\, e + (e + f)\, d(e)\,,$$

which results in

$$2\, d(e) = e\, d(e + f) + d(e + f)\, e + d(e)\,(e + f) + (e + f)\, d(e)\,.$$

Since the last two terms sum up to $d(e)$, we obtain

$$d(e) = e\,d(e+f) + d(e+f)\,e\,.$$

Similarly,

$$d(f) = f\,d(e+f) + d(e+f)\,f\,.$$

It implies that

$$d(e) + d(f) = (e+f)\,d(e+f) + d(e+f)\,(e+f) = d(e+f)\,.$$

By the Generalized Gleason Theorem, d extends to a continuous linear map, denoted again by d. It implies that the assignment

$$h(x,y) = d(x\,y) - d(x)\,y - x\,d(y)$$

is continuous and bilinear on $M \times M$. Let A be a maximal abelian subalgebra of M. By our assumption, $h(e,f) = 0$ for all projections $e, f \in A$. Hence, by continuity of d and the spectral theorem,

$$d(x\,y) = d(x)\,y + x\,d(y)\,,$$

for all commuting self-adjoint elements $x, y \in M$. □

Although the following lemma is more or less a special case of the well known Derivation Theorem (see e.g [290]) saying that any derivation on von Neumann algebra is inner, let us provide a complete proof for the sake of completeness and reader's convenience.

7.2.3. Lemma. *Let $\delta\colon M \to M$ be a continuous linear map that acts on a von Neumann algebra M. Let $\delta(a\,b) = \delta(a)\,b + a\,\delta(b)$ for all $a, b \in A$, where A is an abelian von Neumann subalgebra of M. Then there is an element $x_0 \in M$ such that*

$$\delta(a) = [x_0, a]\,, \qquad \textit{for all } a \in A\,,$$

and $\|x_0\| \le \|\delta\|$. Moreover, if $\delta(x^*) = \delta(x)^*$ for all $x \in A$, then x_0 can be chosen as $i\,k$, where $k \in M$ is a self-adjoint element.

Proof: Denote by $U(A)$ the unitary group of A. For a unitary element $u \in U(A)$ we define an affine map $T_u : M \to M$ by

$$T_u(x) = (u\,x + \delta(u))\,u^{-1}\,.$$

Then, if $u, v \in U(A)$, we have

$$\begin{aligned} T_u T_v(x) &= \{u\,(v\,x + \delta(v))\,v^{-1} + \delta(u)\}\,u^{-1} \\ &= u\,v\,x\,v^{-1}\,u^{-1} + u\,\delta(v)\,v^{-1}\,u^{-1} + \delta(u)\,u^{-1} \\ &= \{u\,v\,x + u\,\delta(v) + \delta(u)\,v\}(u\,v)^{-1} \\ &= \{u\,v\,x + \delta(u\,v)\}(u\,v)^{-1} = T_{u\,v}(x)\,. \end{aligned}$$

In other words, $T_u\,T_v = T_{u\,v}$. Let K be the weak*-closed convex hull of the set $\{T_u(0) \mid u \in U(A)\}$. Obviously, each T_u leaves K invariant. As $\|T_u(0)\| = \|\delta(u)\,u^{-1}\| \le \|\delta\|$, we see that $\sup_{x\in K}\|x\| \le \|\delta\|$. Therefore K is a weak*-compact convex set and the set $\{T_u \mid u \in U(A)\}$ is a family of commuting weak*-continuous affine maps taking K into K. By employing the Markov-Kakutani fixed point theorem, we can find an element $x_0 \in K$ such that $T_u(x_0) = x_0$ for all $u \in U(A)$. Hence, for each $u \in U(A)$,

$$(u\,x_0 + \delta(u))u^{-1} = x_0\,, \tag{7.13}$$

and therefore

$$\delta(u) = [x_0, u]\,.$$

Since A is linearly spanned by unitary elements, we infer that

$$\delta(a) = [x_0, a] \qquad \text{for all } a \in A\,.$$

Suppose now that δ is a $*$-derivation on A. Then, for each $u \in U(A)$,

$$(\delta(u)\,u^{-1})^* = u\,\delta(u^*) = \delta(u\,u^*) - \delta(u)\,u^* = -\delta(u)\,u^{-1}\,.$$

Because x_0 is a weak* limit of convex combinations of the elements of the form $\delta(u)\,u^{-1}$ for some $u \in U(A)$, we have that $x_0^* = -x_0$. Therefore, the fixed point x_0 can be chosen in the form $i\,k$, where $k \in M$ is self-adjoint and $\|k\| \le \|\delta\|$. □

Now we are in the position to prove the main result of the present section.

7.2.4. Theorem. *Let M be a von Neumann algebra without Type I_2 direct summand and let $d\colon P(M) \to M$ be a bounded function such that*

$$d(e\,f) = d(e)\,f + e\,d(f)\,,$$

for all commuting $e, f \in P(M)$. Then d extends uniquely to a derivation on M. In particular, any velocity map on M_{sa} bounded on the unit ball extends to a derivation on M.

Proof: By Proposition 7.2.2, d extends to a continuous linear map on M that satisfies the Leibnitz formula on every abelian von Neumann subalgebra. Let us denote this map again by d. Using identity (7.12) we can write d as a linear combination of two linear maps that, in addition to the Leibnitz formula, preserve $*$-operation on normal elements. Without loss of generality we can therefore assume that the same holds for d. Fix a normal state φ on M. We want to show that there is a self-adjoint element $h \in M$ such that

$$\varphi(d(x)) = \varphi([i\,h, x]) \qquad \text{for all } x \in M\,. \tag{7.14}$$

Suppose, on the contrary, that there exists no self-adjoint element h fulfilling (7.14). In other words, for the adjoint map $d^*: M_* \to M_*$ we have supposed that

$$d^*(\varphi) \notin \{\delta^*_{i\,h}(\varphi) \mid h \in M_{sa}, \|h\| \le \|d\|\} = K\,,$$

where $\delta_{i\,h}$ denotes the inner derivation

$$\delta_{i\,h}(x) = [i\,h, x] \qquad \text{for all } x \in M\,.$$

The map $h \to \delta^*_{i\,h}\varphi : M_{sa} \to (M_*)_{sa}$ that maps the self-adjoint part of M into the self-adjoint part of its predual is weak*-continuous on bounded parts of M_{sa}. The set K is weak* compact in the self-adjoint part of M_*. By appealing to the Bipolar theorem there is a self-adjoint element $a \in M$ such that

$$\sup\{|\delta^*_{i\,h}(\varphi)(a)| \mid h \in M_{sa}, \|h\| \le \|d\|\} < |d^*(\varphi)(a)|\,.$$

Thus,

$$d(a) \notin \{\delta_{i\,h}(a) \mid h \in M_{sa}, \|h\| \le \|d\|\}\,,$$

which is a contradiction by Lemma 7.2.3 applied to the algebra $W^*(a,1)$. The equality (7.14) implies that, given $x, y \in M$,

$$\varphi(d(x\,y)) = \varphi(d(x)\,y + x\,d(y)),$$

for each normal state φ on M. This immediately gives us that d is a derivation on M. □

7.3. Approximate hidden variables

In the present section we resume the discussion in Chapter 5 on the theory of hidden variables in operator algebraic approach to foundations of quantum theory. The aim is to extend no-go theorems on hidden variables on the Hilbert space logic to more general projection structures of von Neumann algebras and C^*-algebras, and to expose new results on dispersions of states and measures on operator algebras that enable to derive general results on approximative hidden variables.

First we investigate the existence of dispersion-free states. As any state on the projections of a von Neumann algebra without Type I_2 direct summand extends to a linear state, the existence of a state with values 0 and 1 on projections is in this way effectively excluded by the Generalized Gleason Theorem. Nevertheless, though the Generalized Gleason Theorem gave a considerable impetus to the theory of hidden variables, the nonexistence of 0-1 states on projection structures can be established by much simpler arguments based on Theorem 3.4.1.

7.3.1. Theorem. *The projection lattice $P(M)$ of a von Neumann algebra M that has neither nonzero abelian direct summand nor Type I_2 direct summand admits no 0-1 state.*

Proof: First we prove that there is a unital von Neumann subalgebra A of M such that, for $z \in P(Z(A))$,

$$A = z\,A \oplus (1 - z)\,A,$$

where $z\,A$ is either zero or a copy of $M_2(\mathbb{C})$ and $(1 - z)\,A$ is either zero or a copy of $M_3(\mathbb{C})$. For this, it suffices to find a unital von Neumann subalgebra of M that is of the form

$$\sum_{\alpha} \oplus M_\alpha \,, \tag{7.15}$$

where each M_α is isomorphic either to $M_2(\mathbb{C})$ or $M_3(\mathbb{C})$. Let p be a central projection in M such that $p\,M$ is of finite Type I and $(1-p)\,M$ has no nonzero Type I_n, $n < \infty$, direct summand. If $1-p \neq 0$, then $1-p$ can be halved (see Chapter 2, Classification of von Neumann algebras), and we can find a unital subalgebra of $(1-p)\,M$ isomorphic to $M_2(\mathbb{C})$. Let us now consider $p\,M$. By the structure theory and the hypothesis, $p\,M$ is either zero or a direct sum of Type I_n algebras, $3 \leq n < \infty$. Each such direct summand contains a unital subalgebra isomorphic to $M_n(\mathbb{C})$. As every integer $n \geq 2$ can be written in the form $n = 2\,k + 3\,l$, where k, l are nonnegative integers, we can find in each $M_n(\mathbb{C})$ a unital subalgebra of the form (7.15). Let μ be a 0-1 probability measure on $P(M)$. If $\mu(1-z) = 1$, then μ restricts to a 0-1 probability measure on $P((1-z)\,A) \simeq M_3(\mathbb{C})$ that is excluded by Theorem 3.4.1. It remains to consider the possibility $\mu(z) = 1$. In the course of the proof of Proposition 5.3.6 we have seen that any copy of $M_2(\mathbb{C})$ in M lives in some subalgebra isomorphic to $M_3(\mathbb{C})$, provided that M does not have any Type I_2 direct summand. So again, μ induces a 0-1 probability measure on $M_3(\mathbb{C})$, which is absurd. □

It is worth noting that the assumption on Type I_2 direct summand in the previous theorem is indispensable. To see this let us represent Type I_2 algebra as $C(X) \otimes M_2(\mathbb{C}) \simeq C(X, M_2(\mathbb{C}))$, where X is a hyperstonean space. Take any fixed point $x \in X$ and a 0-1 measure μ on $P(M_2(\mathbb{C}))$. The map

$$m \colon p \to \mu(p(x)) \,, \qquad p \in P(M) \,,$$

gives a 0-1 probability measure on $P(M)$. Hence, Type I_2 algebras admit plenty of 0-1 measures on their projection lattices.

The nonexistence of 0-1 states on projections means that we cannot think of excluding any possible embedding of the projection lattice into a Boolean structure that preserves logical operations on compatible propositions. Let us recall that a mapping $\varphi \colon L_1 \to L_2$ between orthomodular posets L_1 and L_2 is said to be a *unital orthomorphism* if $\varphi(1_{L_1}) = 1_{L_2}$ and if, for each orthogonal $a, b \in L_1$, the following conditions hold:

(i) $\varphi(a \vee b) = \varphi(a) \vee \varphi(b)$

(ii) $\varphi(a)$ and $\varphi(b)$ are orthogonal.

In other words, φ is a "local homomorphism" in the sense of being a homomorphism on any Boolean subalgebra of L_1. The following consequence of Theorem 7.3.1 says that von Neumann projection lattice cannot be mapped nontrivially into a Boolean lattice (classical model) such that the mapping respects every Boolean substructure.

7.3.2. Corollary. *Let M be a von Neumann algebra without Type I_2 direct summand and with zero abelian direct summand. There is no nonzero unital orthomorphism $\varphi\colon P(M) \to B$, where B is a Boolean algebra.*

Proof: Take a nonzero (finitely additive) 0-1 probability measure μ on B. If $\varphi\colon P(M) \to B$ is a unital orthomorphism that maps $P(M)$ into B, then

$$\mu \circ \varphi\colon P(M) \to [0,1]$$

is a finitely additive 0-1 probability measure on $P(M)$, which is excluded by Theorem 7.3.1. (Of course, the same argument allows for an obvious generalization to any L (instead of B) which possesses a 0-1 state.) □

Having excluded 0-1 states on von Neumann algebras, we want now to discuss the problem of existence of dispersion free states on C^*-structures. Suppose that A is a C^*-algebra such that the projection poset $P(A)$ generates linearly a dense subset of A. Let us suppose that μ is a 0-1 probability measure on $P(A)$ that extends to a linear functional φ. As $\varphi(p)$ is 0 or 1 for every projection $p \in P(A)$ we have $\varphi(p\,x) = \varphi(p)\,\varphi(x)$ for all $x \in A$ by the Cauchy-Schwarz inequality (see Chapter 2, The Gelfand-Naimark-Segal construction). Hence, $\varphi(y\,x) = \varphi(y)\,\varphi(x)$ whenever y is a linear combination of projections and hence the density of $\operatorname{sp} P(A)$ implies that φ is multiplicative on A. Therefore the problem of existence of 0-1 measures on projections is connected with a more general problem of the existence of multiplicative states or multiplicative quasi-states on C^*-algebras. The linear version of this question is more or less clear: C^*-algebra A admits a multiplicative state if, and only if, it contains a closed two-sided ideal of codimension one, or equivalently, if the enveloping von Neumann subalgebra A has one-dimensional direct summand. On the other hand, the description of C^*-algebras admitting no multiplicative quasi-state is still open.

Below we derive some particular results concerning infinite simple algebras and real-rank zero algebras. Another no-go theorem on C^*-algebras due to Misra stems from the problem of integral representation of measures in terms of dispersion free states. We shall now formalize this approach in the general form.

7.3.3. Definition. Let S be a nonempty set. Let Φ be a nonempty set of complex functions on S. Let $\Psi \subset \Phi$. We say that Φ admits a *hidden variable theory with respect to* Ψ if the following holds: There is a probability space $(X, \mathcal{A}, \mu)$ and a map $\alpha\colon\ \Phi \times X \to \Psi :\ (\varphi, \xi) \to \alpha(\varphi, \xi)$ such that the function $\xi \to \alpha(\varphi, \xi)(a)$ is μ-integrable for each $a \in S$, and

$$\varphi(a) = \int_X \alpha(\varphi, \xi)\,(a)\,d\,\mu(\xi) \qquad \text{for all } a \in S\,. \tag{7.16}$$

The equality (7.16) says that $\varphi \in \Phi$ is expressible as "integral average" of elements from a special subset $\Psi \subset \Phi$. From this point of view ξ can be thought of as a hidden parameter of the system. In particular, we will be interested in special cases when Φ is the set of probability measures on the projection poset $P(A)$ of a C^*-algebra A and Ψ is a subset of Φ consisting of dispersion-free probability measures or measures with small dispersions. The existence of a hidden variable theory in these cases means that any state can be obtained from dispersion-free states as the "classical" statistical average. It is also natural to consider this approach for special quasi-states on C^*-algebras. In this connection, it has been proved that the set of monotone quasi-states on a noncommutative C^*-algebra admits no hidden variable theory with respect to the set of multiplicative monotone quasi-states. This result is based on the following characterization of commutative C^*-algebra due to Ogasawara that we prove first.

7.3.4. Theorem (Ogasawara). *Let A be a C^*-algebra. If $0 \le x \le y$ implies $x^2 \le y^2$ for all $x, y \in A$, then A is commutative.*

Proof: Take arbitrary $x, y \in A_+$ and $\varepsilon > 0$. According to the hypothesis $x \le x + \varepsilon\, y$ implies that

$$x^2 \le x^2 + \varepsilon\,(x\,y + y\,x) + \varepsilon^2\,y^2\,,$$

and hence

$$0 \le x\,y + y\,x + \varepsilon\,y^2\,.$$

Since ε is arbitrary, we obtain that

$$x\,y + y\,x \geq 0 \qquad \text{for all } x, y \geq 0\,. \tag{7.17}$$

For $x, y \geq 0$ write $x\,y = a + i\,b$, where a, b are self-adjoint. By (7.17) $a = (x\,y + y\,x)/2$ is always nonnegative. As $x \geq 0$ and $y\,x\,y \geq 0$ we infer by applying (7.17) to the pair x and $y\,x\,y$ that

$$x\,y\,x\,y = a^2 - b^2 + i\,(a\,b + b\,a) \quad \text{and hence } a^2 - b^2 \geq 0\,. \tag{7.18}$$

It yields the following claim:

$$\begin{array}{c}\text{If } x\,y = a + i\,b \text{ for nonnegative } x \text{ and } y \\ \text{and self-adjoint } a \text{ and } b, \text{ then } a^2 \geq b^2\,.\end{array} \tag{7.19}$$

We shall consider the set

$$\begin{aligned} E \;=\; \{\lambda \geq 1 \mid \lambda\,b^2 \leq a^2 \text{ whenever } a, b \in A_{sa} \text{ with} \\ x\,y = a + i\,b \text{ for some } x, y \geq 0\}\,. \end{aligned}$$

The implication (7.19) says that $1 \in E$. Suppose that $\lambda \in E$. On applying (7.17) to $b^2 \geq 0$ and $a^2 - \lambda\,b^2 \geq 0$ we arrive to

$$\begin{aligned} 0 &\leq b^2\,(a^2 - \lambda\,b^2) + (a^2 - \lambda\,b^2)\,b^2 \\ &= b^2\,a^2 + a^2\,b^2 - 2\,\lambda\,b^4 \end{aligned} \tag{7.20}$$

On the other hand, the assumption that $\lambda \in E$ and the identity (7.18) imply that

$$\lambda\,(a\,b + b\,a)^2 \leq (a^2 - b^2)^2\,, \tag{7.21}$$

i.e.

$$\begin{aligned} &\lambda\,[a\,b^2\,a + b\,a^2\,b + a\,(b\,a\,b) + (b\,a\,b)\,a] \\ &\leq a^4 + b^4 - a^2\,b^2 - b^2\,a^2\,. \end{aligned} \tag{7.22}$$

Taking into account that

$$a\,b^2\,a \geq 0, \qquad b\,a^2\,b \geq \lambda\,b^4, \qquad a\,(b\,a\,b) + (b\,a\,b)\,a \geq 0,$$

$$\text{and } a^2\,b^2 + b^2\,a^2 - 2\lambda\,b^4 \geq 0$$

(last inequality is (7.20)), the inequality (7.22) yields

$$\lambda^2 b^4 \leq a^4 + (1 - 2\lambda) b^4$$

and hence

$$(\lambda^2 + 2\lambda - 1) b^4 \leq a^4 .$$

Using the fact that the square root is operator monotone, we derive from the previous inequality that

$$(\lambda^2 + 2\lambda - 1)^{1/2} b^2 \leq a^2 .$$

Therefore $(\lambda^2 + 2\lambda - 1)^{1/2} \in E$. Since $(\lambda^2 + 2\lambda - 1)^{1/2} > \lambda$, we see that the set E is not upper bounded. Hence, there is an increasing sequence $\lambda_n \to \infty$ such that

$$a^2 \geq \lambda_n b^2 \text{ for all } n .$$

It gives $\|a^2\| \geq |\lambda_n| \cdot \|b^2\|$ for all n and hence $\|b^2\| = 0$. We conclude that, for $x, y \geq 0$ with $x\,y = a + i\,b$, $a, b \in A_{sa}$, the "imaginary part" b is always zero. In other words, $x\,y$ is self-adjoint, meaning that $x\,y = y\,x$. Since the positive elements generate linearly the whole of A, A has to be commutative. □

The following no-go theorem for hidden variables says that nonabelian C^*-algebras do not have order determining set of multiplicative monotone quasi-functionals. We say that a quasi-functional ϱ on a C^*-algebra is *monotone* if $\varrho(a) \leq \varrho(b)$ whenever $a \leq b$ in A_{sa}.

7.3.5. Theorem (Misra). *Let A be a nonabelian C^*-algebra. Then the set Φ of all quasi-states on A admits no hidden variable theory with respect to the set Ψ of all multiplicative monotone quasi-states on A.*

Proof: Suppose on the contrary that Φ admits a hidden variable theory with respect to Ψ given by the probability space $(X, \mathcal{A}, \mu)$ and a map $\alpha : \Phi \times X \to \Psi$. Then, for each (linear) state φ on A,

$$\varphi(a) = \int_X \alpha(\varphi, \xi)(a)\, d\mu(\xi) ,$$

where each $\alpha(\varphi, \xi)$ $(\xi \in X)$ is a monotone multiplicative map on A. For $x \leq y$ we can write

$$\begin{aligned} \varphi(y^2 - x^2) &= \varphi(y^2) - \varphi(x^2) = \int_X (\alpha(\varphi,\xi)(y^2) - \alpha(\varphi,\xi)(x^2))\, d\mu(\xi) \\ &= \int_X (\alpha^2(\varphi,\xi)(y) - \alpha^2(\varphi,\xi)(x))\, d\mu(\xi) \geq 0 , \end{aligned}$$

as $\alpha(\varphi,\xi)(y) \geq \alpha(\varphi,\xi)(x)$ for all $\xi \in X$. Since the state space is order determining, it implies that $x^2 \geq y^2$ and hence A is abelian by Theorem 7.3.4. □

Hence, only commutative C^*-algebras can be represented by multiplicative functionals (Gelfand theory). It is worth observing that the previous proof implies in fact that nonabelian C^*-algebras have no order determining set of quasi-states that are monotone and multiplicative on all singly generated subalgebras.

Physically, the question of the existence of dispersion free state has to be decided by an experiment. But, in a real measurement situation every experiment shows some dispersion on propositions. Therefore it seems to be more realistic to interpret the existence of hidden variables as the existence of a state with arbitrarily small dispersion. For this reason we introduce a quantitative measure of the dispersion by defining an overall dispersion of states and measures.

7.3.6. Definition. Let φ be a state on the projection orthoposet $P(A)$ of a C^*-algebra A or a state on A. *Overall dispersion*, $\sigma(\varphi)$, of φ is defined by

$$\sigma(\varphi) = \sup\{\varphi(p) - \varphi(p)^2 \mid p \in P(A)\}.$$

It is noteworthy that overall dispersion $\sigma(\varphi)$ is zero exactly when φ is a dispersion-free state. For example, for a (linear) state φ on a von Neumann algebra we have that $\sigma(\varphi) = 0$ exactly when φ is multiplicative. We shall study first the question of establishing the lower bound for dispersions of linear states on matrix algebras.

7.3.7. Lemma. *Let A be unital C^*-algebra admitting a projection p with $p \sim 1-p$. Then, for any state φ on A, there is a projection q in A such that $\varphi(q) = 1/2$. Hence, $\sigma(\varphi) = 1/4$ for any state φ on A.*

Proof: If $p \sim 1-p$, then the projections p and $1-p$ induce a matrix units system and so a unital subalgebra isomorphic to $M_2(\mathbb{C})$. Therefore it suffices to prove the lemma in case of $A = M_2(\mathbb{C})$. Since there is no (linear) 0-1 state on $M_2(\mathbb{C})$, we can find, for a fixed state φ, a one-dimensional projection e with $0 < \varphi(e) < 1$. One can assume that $\varphi(e) \leq 1/2 \leq \varphi(1-e)$. We will show that there is a continuous path $\{\psi(t) \mid t \in [0,1]\}$ in $P(A)$

connecting e and $1-e$. Indeed, by a suitable unitary transform of $M_2(\mathbb{C})$, we can, without any loss of generality, assume that

$$e = \begin{pmatrix} 0 & 0 \\ 0 & 1 \end{pmatrix} \text{ and } 1-e = \begin{pmatrix} 1 & 0 \\ 0 & 0 \end{pmatrix}.$$

Then

$$\psi(t) = \begin{pmatrix} t & \sqrt{t-t^2} \\ \sqrt{t-t^2} & 1-t \end{pmatrix} \qquad t \in [0,1]$$

is a path in $P(M_2(\mathbb{C}))$ with $\psi(0) = e$ and $\psi(1) = 1-e$. Therefore $\varphi(\psi(t_0)) = 1/2$ for some $t_0 \in [0,1]$. □

7.3.8. Proposition. *Let τ be the normalized trace on $M_{2k+1}(\mathbb{C})$, $k > 0$. Then τ is the only state such that $\sigma(\tau) \leq \sigma(\varphi)$ for all states φ on $M_{2k+1}(\mathbb{C})$.*

Proof: The trace τ takes the following values on the projections:

$$\frac{n}{2k+1}, \quad n = 0, \ldots 2k+1.$$

Hence, the dispersion of τ attained on projections of rank k and $k+1$ is

$$\sigma(\tau) = \frac{k}{2k+1} - \frac{k^2}{(2k+1)^2} = \frac{k(k+1)}{(2k+1)^2}.$$

Let φ be a state on $M_{2k+1}(\mathbb{C})$ such that

$$\sigma(\varphi) \leq \sigma(\tau). \tag{7.23}$$

Take a minimal projection $p \in M_{2k+1}(\mathbb{C})$. Then $1-p$ projects onto subspace of even dimension and so, by Lemma 7.3.7, there is a projection $q \leq 1-p$ with

$$\varphi(q) = \frac{1}{2}\varphi(1-p).$$

Thus,

$$\varphi(p+q) = \frac{1}{2}(\varphi(p)+1).$$

By (7.23)

$$\frac{1}{2} + \frac{1}{2}\varphi(p) \geq \frac{k+1}{2k+1},$$

and, in turn,

$$\varphi(p) \geq \frac{1}{2k+1}.$$

Since the unit of $M_{2k+1}(\mathbb{C})$ is a sum of $2k+1$ (orthogonal) minimal projections including p, we derive that

$$\varphi(p) = \frac{1}{2k+1},$$

meaning that $\varphi = \tau$. □

The previous result says that for any state φ on an algebra $M_n(\mathbb{C})$, $n > 1$, the following estimation holds true for any state φ

$$\sigma(\varphi) \geq \frac{2}{9}.$$

So the dispersions of states on matrix algebras are uniformly bounded from below. Next theorem says that the same holds for states on all C^*-algebras that have rich projection structure.

7.3.9. Theorem. *Let A be a unital real rank zero C^*-algebra having no nonzero representation onto an abelian C^*-algebra. Then*

$$\sigma(\varphi) \geq \frac{2}{9}$$

for any state φ on A. In particular, if M is a von Neumann algebra with zero abelian part and no Type I_2 part, then

$$\sigma(\mu) \geq \frac{2}{9}$$

for any probability measure μ on $P(M)$.

Proof: First we prove the second statement. Like in the proof of Theorem 7.3.1 we can find a unital subalgebra A of M such that

$$A = zA \oplus (1-z)A, \tag{7.24}$$

where $z \in P(Z(A))$, zA is either zero or a copy of $M_2(\mathbb{C})$, and $(1-z)A$ is either zero or a copy of $M_3(\mathbb{C})$. Let φ be a state of M. By Proposition 7.3.8 there is a projection $e \in (1-z)A$ such that

$$\varphi(e) \in [1/3\, \varphi(1-z), 2/3\, \varphi(1-z)].$$

By taking a projection $f \in z\,A$ with $\varphi(f) = 1/2\ \varphi(z)$ (Lemma 7.3.8), we obtain

$$\frac{1}{3} \leq \varphi(e+f) \leq \frac{2}{3}.$$

Whence,

$$\sigma(\varphi) \geq \frac{2}{9}$$

and the proof of the von Neumann algebra case is complete using Gleason Theorem.

Now we will prove the result for a unital real rank zero algebra A. Let A^{**} denote the second dual of A (identified with the universal enveloping von Neumann algebra of A). By the hypothesis A^{**} has no abelian direct summand. By the previous part of the proof there is a projection $p \in A^{**}$ such that for the canonical normal extension of φ to A^{**}, denoted by the same letter, we have

$$\frac{1}{3} \leq \varphi(p) \leq \frac{1}{2}.$$

The fact that A has real rank zero implies that any projection in A^{**} is a limit in the strong operator topology of a net of projections in A [9]. Hence there is a sequence (p_n) of projections in A such that $\varphi(p_n) \to \varphi(p)$ and so

$$\sigma(\varphi) \geq \frac{2}{9}.$$

□

Theorem 7.3.9 implies that there are no dispersion free states on C^*-algebras with reasonably many projections. Moreover, by deriving the uniform lower bound for dispersions we provide the global explanation of the nonexistence of dispersion-free states that is quantified and experimentally easily verifiable. For a wide class of operator algebras it excludes the existence of approximate hidden variables. Hence it is not possible, in principle, to have a state in quantum measure theory with an arbitrarily small dispersion. While in statistical physics the dispersion is only a degree of our lack of information on the system, in quantum mechanics it is an intrinsic property of the system itself. By Proposition 7.3.8 the estimation $\sigma(\varphi) \geq 2/9$ cannot be improved in general. Nevertheless, by modifying slightly the arguments in the proof of the previous theorem, we can obtain that

$$\sigma(\varphi) \geq \frac{\frac{n-1}{2}\left(\frac{n-1}{2}+1\right)}{n^2}$$

for each state φ on a real rank zero C^*-algebra having no cyclic representation on a Hilbert space of dimension less or equal to $n-1$. Another aspect is this: Since the normalized trace τ has the smallest possible dispersion on all matrix algebras, it cannot be an integral mixture (7.16) of states $\alpha(\varphi, \xi)$ with dispersions strictly smaller then that of τ. So the uncertainty cannot be decreased by any integral representation.

Theorem 7.3.9 and Proposition 7.3.8 demonstrate the difference between quantum and classical measure theory. In the probability model given by the probability space $(X, \mathcal{A}, \mu)$, the states with the smallest dispersion are the Dirac measures on X. It also holds that any probability measure on $\mathcal{A}$ can be decomposed in the sense of (7.16) into an integral average over dispersion-free states. In contrast to this, having a finite-dimensional Hilbert space as a model for quantum observables attaining finitely many values (such as e.g. the spin of electron), the probability distributions with the smallest possible dispersion are not pure states but normalized traces, i.e. the uniform convex mixtures of orthogonal pure states. Roughly speaking, while in the classical theory the minimal dispersion is zero and it is realized by "sharp" Dirac measures, in quantum probability theory the smallest dispersion is attained by a kind of quantum uniform distribution in which all atomic propositions have the same probability. We want now to discuss this phenomenon from a more general point of view. It will be proved that traces can be characterized as states having the minimal dispersion on all matrix subalgebras. For this we shall need the following concept.

7.3.10. Definition. Let φ be a state on a C^*-algebra A. We say that φ has the *minimal dispersion property* if the following condition holds. Let B be a subalgebra of A (unital or not) that is isomorphic to $M_3(\mathbb{C})$. If φ is nonzero on B then, for $\omega = \varphi|B$, $\sigma(\frac{\omega}{\|\omega\|}) \leq \sigma(\psi)$ for all states ψ on B.

Let us remark that a state φ has the minimal dispersion property if, and only if, its restriction to any 3x3 matrix subalgebra is a tracial functional (Proposition 7.3.8).

7.3.11. Theorem. *Let M be a von Neumann algebra with zero abelian direct summand and no Type I_2 direct summand. A state φ on M is tracial if, and only if, φ has the minimal dispersion property.*

Proof: Let us suppose that φ has the minimal dispersion property. In order to establish the tracial property of φ it is sufficient to verify that φ is constant on classes of projections exchangeable by a symmetry. Indeed, suppose that $\varphi(s\,p\,s) = \varphi(p)$ whenever p is a projection in M and s a symmetry in M. By the spectral theorem, $\varphi(s\,x\,s) = \varphi(x)$ for all $x \in M$. Then, $\varphi(s\,x) = \varphi(s\,s\,x\,s) = \varphi(x\,s)$ for all $x \in M$ and a symmetry $s \in M$. Since $2\,p - 1$ where p is a projection is a symmetry, we derive that $\varphi(p\,x) = \varphi(x\,p)$ for all $x \in M$ and $p \in P(M)$. By the spectral theorem again, we see that $\varphi(x\,y) = \varphi(y\,x)$ for all $x, y \in M$, and hence φ is a trace.

Let p and q are projections such that $q = s\,p\,s$, where s is a symmetry. Then $p = s\,q\,s$ and therefore $s\,(p \wedge q)\,s = p \wedge q$. The projection $p - p \wedge q$ and $q - q \wedge p$ are exchangeable by symmetry s. It is therefore sufficient to assume that $p \wedge q = 0$. Moreover, we also may assume that p and q are orthogonal. Indeed, by the parallelogram law we have

$$p \vee q - p \sim q - q \wedge p = q \sim p.$$

In other words, $p \vee q - p$ and p are orthogonal equivalent projections. Therefore, we are done if we show that φ is constant on orthogonal equivalent projections. Indeed, if this is true, then

$$\varphi(p) = \frac{\varphi(p \vee q)}{2},$$

and, by symmetry, $\varphi(p) = \varphi(q)$.

Let us now fix orthogonal equivalent projections p and q. By the comparison theory of projections in von Neumann algebras there are projections $p_1 \sim p_2 \sim q_1 \sim q_2$ and abelian projections $p_3 \sim q_3$ such that

$$\begin{aligned} p &= p_1 + p_2 + p_3 \\ q &= q_1 + q_2 + q_3. \end{aligned}$$

Equivalent orthogonal projections p_1, p_2, q_1 are contained, in the only nontrivial case of $p_1 \neq 0$, in a subalgebra of M isomorphic to $M_3(\mathbb{C})$. On applying Proposition 7.3.8 we obtain

$$\varphi(p_1) = \varphi(p_2) = \varphi(q_1) \text{ and, similarly, } \varphi(p_1) = \varphi(p_2) = \varphi(q_2).$$

It remains to prove that $\varphi(p_3) = \varphi(q_3)$. Suppose that p_3 is nonzero. Then p_3 and q_3 are contained in some subalgebra of M isomorphic to $M_2(\mathbb{C})$.

But M has no Type I_2 part and it means that both p_3 and q_3 belong to some subalgebra isomorphic to $M_3(\mathbb{C})$ (Proposition 5.3.6). Therefore, $\varphi(p_3) = \varphi(q_3)$ by Proposition 7.3.8 and hence $\varphi(p) = \varphi(q)$. □

It is worth noting that the assumption on the absence of Type I_2 direct summand is needed for the existence of 3x3 substructures in M. The previous theorem is a stronger version of many no-go theorems in hidden variable theory built on projections in von Neumann algebras. If the von Neumann algebra M under consideration is infinite, then M admits no tracial state and so the dispersion of any state is strictly greater then $2/9$ on some 3x3 matrix subalgebra. This can be applied in a special case to quasi-states on infinite factors. We conclude this section by showing that the same holds for quasi-states on simple infinite unital C^*-algebras which can be viewed as the C^*-algebraic version of infinite von Neumann factors. (The dispersion of quasi-state is defined in the same way as for the state.)

7.3.12. Proposition. *There is no quasi-state on an infinite simple unital C^*-algebra A having the minimal dispersion property.*

Proof: Seeking a contradiction, suppose that φ is a quasi-state on A with the minimal dispersion property. We first show that there are six orthogonal infinite projections in A such that φ is nonzero on at least one of them. To this end, let us realize that A contains a sequence $g_1, g_2, \ldots, g_6$ of orthogonal infinite projections (see Chapter 2, Projection structure of C^*-algebras). If φ is nonzero on at least one of them, then we are done. If not, then $\varphi(\sum_{i=1}^{6} g_i) = 0$ and so $\varphi(1 - \sum_{i=1}^{6} g_i) \neq 0$. Set $s = 1 - \sum_{i=1}^{5} g_i$. Obviously, s is infinite and $\varphi(s) \neq 0$. Since any projection in a simple infinite algebra is subequivalent to any infinite projection (see Chapter 2, Projection structure of C^*-algebras) we can extend s, by taking subprojections of $g_1, g_2, \ldots, g_5$ equivalent to s, to a system of six orthogonal equivalent infinite projections.

Let us now fix orthogonal equivalent infinite projections $p_1, p_2, \ldots, p_6$ with $\varphi(p_1) \neq 0$. Then A contains a subalgebra B isomorphic to $M_2(p_1 A p_1)$ which is, simultaneously, embedded into a copy of $M_3(B)$. Let e and f be unitarily equivalent projections in B. Let us identify B with the left upper corner in $M_3(B)$. It is clear that the projections

$$g = \begin{pmatrix} 0 & 0 & 0 \\ 0 & e & 0 \\ 0 & 0 & 0 \end{pmatrix} \quad h = \begin{pmatrix} 0 & 0 & 0 \\ 0 & 0 & 0 \\ 0 & 0 & e \end{pmatrix}$$

are unitarily equivalent in $M_3(B)$. Therefore the system of equivalent orthogonal projections $e \sim g \sim h$ is contained in a C^*-subalgebra, C, of A isomorphic to $M_3(\mathbb{C})$. By Gleason Theorem φ is linear on C and hence $\varphi(e) = \varphi(g) = \varphi(h)$ (Proposition 7.3.8). By the same reasoning $\varphi(f) = \varphi(g) = \varphi(h)$. It means that φ is unitarily invariant on projections in B. The algebra $p_1 A p_1$ (viewed as a corner in B) is infinite and so it possesses two orthogonal projections q_1, q_2 both equivalent to p_1. Hence, the projections p_1, q_1, q_2 are unitarily equivalent in B (see Chapter 2, Projection structure of C^*-algebras), and hence

$$\varphi(p_1) = \varphi(q_1) = \varphi(q_2)$$

by the minimal dispersion property. But

$$2\,\varphi(p_1) = \varphi(q_1 + q_2) \leq \varphi(p_1),$$

which is a contradiction. □

7.3.13. Proposition. *There is no dispersion free quasi-state on any simple unital infinite C^*-algebra.*

Proof: A consequence of Proposition 7.3.12

7.4. Historical remarks and comments

Multimeasures and decoherence functional

The multilinear form of Gleason Theorem has been inspired, besides its purely mathematical motivations, by the recent effort to generalize quantum theory in order to address the problem of time in quantum gravity. The general idea has been to develop the abstract scheme such that time can be encoded into intrinsic properties of the system and, in turn, could play a subsidiary role. Consistent history approach was initiated by Griffiths [147] and has been further developed by Omnes [251, 252, 253, 254, 255]. In a series of articles Gell-Mann and Hartle [138, 139, 189] studied quantum cosmology and the path integral formulation of relativistic quantum theory in terms of the concept of the histories approach. They have proposed a system of axioms in which the history of the system need not be regarded as a time ordered string of events or propositions and it is considered to be a fundamental theoretical entity in its own right. In the Gell-Mann and Hartle formalism the space of histories is an ordered set P in which the order corresponds to "coarse graining". The second component of this scheme is the space of decoherence functionals that assign to each pair of histories their "degree of interference". This approach has been brought to the present form by Isham [194]. He has shown the connection of quantum histories approach with quantum logic approach and has embedded the structure of homogeneous histories into the projection lattice of the tensor product. Isham, Linden and Schreckenberg then proved the Isham-Linden-Schreckenberg Theorem (Theorem 7.1.8) describing the decoherence functionals in finite dimensions. Multiform Gleason Theorem was proved by Wright and Rudolph and Wright in [279, 324]. Once the multilinear extension of multimeasures is established, the deep results of functional analysis, including remarkable Grothendieck-Pisier-Haagerup inequality [156], can be employed to describe multimeasure in terms of collections of linear forms or inner products. For these results we refer to [324, 325, 326, 327, 328, 333, 334]. One of the most important result on the structure of multilinear forms on C^*-algebras is due to Goldstein [143]. We would like to slightly digress and to comment on this remarkable result and its consequences related to orthogonal vector measures.

7.4.1. Theorem (Goldstein). *Let $B\colon A \times A \to \mathbb{C}$ be a separately continuous conjugate-bilinear form on a C^*-algebra A such that*

$$B(x,y) = 0 \quad \text{whenever } x, y \in A_{sa} \text{ with } xy = 0\,. \tag{7.25}$$

Then there are $\psi_1, \psi_2 \in A^*$ such that

$$B(x,y) = \psi_1(y^* x) + \psi_2(x y^*)\,, \qquad \text{for all } x, y \in A\,.$$

Moreover if B is positive, then ψ_1 and ψ_2 can be chosen positive.

The sesquilinear form $B(x,y)$ satisfying (7.25) is called *orthogonal.* Goldstein Theorem has an important application to establishing a one-to-one correspondence between the orthogonally scattered mapping on the group C^*-algebra, $C^*(G)$, of a locally compact group G. Moreover, this abstract result enables one to describe the noncommutative Fourier transform of maps on group C^*-algebras [143, 332]. The proof of Goldstein theorem is long and complex and we refer the reader to the original paper [143]. In interplay with the Generalized Gleason theorem the Goldstein Theorem makes it possible to describe covariance functions of orthogonal vector measures and to show that such measures are associated with Jordan homomorphisms. The orthogonal vector measures defined below are generalization of stochastic processes with independent increments.

7.4.2. Definition. Let M be a von Neumann algebra and H a Hilbert space. A finitely additive measure $m\colon P(M) \to H$ is called *orthogonal vector measure* (or *orthogonally scattered vector measure*) if

$$(m(p), m(q)) = 0 \qquad \text{for all orthogonal projections } p, q \in M\,.$$

If the von Neumann algebra M has no Type I_2 direct summand, then an orthogonal vector measure $m\colon P(M) \to H$ extends to a bounded linear map $F\colon M \to H$ such that the conjugate-bilinear form

$$B(x, y) = (F(x), F(y))\,, \quad x, y \in M\,,$$

is orthogonal. Therefore, by evoking Theorem 7.4.1 we can find two positive functionals ψ_1, ψ_2 on M such that

$$(m(p), m(q)) = \psi_1(q\,p) + \psi_2(p\,q) \qquad \text{for all } p, q \in P(M)\,.$$

This enables one to describe the structure of covariance function of orthogonal vector measures and to give the following lucid description of orthogonal vector measures obtained in [143, 197, 331].

7.4.3. Theorem. *Suppose that M is a von Neumann algebra without Type I_2 direct summand. Let $m\colon P(M) \to H$ be an orthogonal vector measure. Then there is a Hilbert space K, with a vector $\xi \in K$, a linear isometry $V\colon H \to K$, and a $*$-Jordan homomorphism $\pi\colon M \to B(K)$ such that*

$$m(p) = V^*\,\pi(p)\,\xi \qquad \textit{for all } p \in P(M)\,.$$

Many other geometric characterizations of orthogonal vector measures can be found in [161, 162, 165, 164, 223]. Let us now return to the main topic of Section 7.1. The boundedness principle for completely additive multimeasures is due to Wright [325]. In the present book it is formulated in a stronger form (Theorem 7.1.5). The analysis of the possible extensions of Isham-Linden-Schreckenberg representation in terms of tensor boundedness and tracial boundedness has been carried on by Rudolph and Wright in [280, 281, 282], where further information on extensions of the standard decoherence functional and its representation by a unique unbounded linear form on the tensor product may be found. Unlike original paper [282] the present proof of Theorem 7.1.13 is elementary and avoids using Haagerup-Pisier-Grothendieck inequality. Theorem 7.1.16 was proved by Wallach in [316] and extended by Rudolph and Wright in [283]. The progress in quantum histories approach has generated many further results, for the recent development we refer e.g. to [196, 284].

Another interesting line of the research, related to bilinear form of Gleason Theorem, is the work of Edwards and Rütimann [128] who applied the multiform Gleason Theorem to finding a description of measures on the structure of centrally equivalent projections in von Neumann algebra. In case of the Hilbert space algebra, their results reduce to "classical" Gleason Theorem.

Velocity maps

The concept of the derivation is well established in the theory of operator algebras and C^*-dynamical systems. For the wealth of information we refer the reader to standard monographs [206, 259, 291, 290, 301, 307, 308]. The axioms for the velocity map were suggested by Paratharasy and investigated by Bhat. Bhat proved in [33] that if $\delta\colon A_{sa} \to A_{sa}$ is a continuous map where A is a von Neumann algebra such that (i) $\delta(a\,1) = 0$ for all $a \in \mathbb{R}$ (ii) $\delta(a\,b) = a\,\delta(b) + \delta(a)\,b$ for all $a, b \in A$, then δ is a velocity map. Moreover, he showed that if H is a separable Hilbert space and $\delta\colon B(H)_{sa} \to B(H)_{sa}$ is a continuous linear map satisfying

$$\delta(x\,y) = x\,\delta(y) + \delta(x)\,y \qquad \text{for all } x, y \in B(H)_{sa} \text{ with } x\,y = y\,x\,,$$

then δ extends to a derivation. Besides, he succeeded in finding inversion formula for the representing operator H in (7.10) in terms of d, provided that M is a Type I factor acting on a separable space. (Let us recall that all proofs of Derivation Theorem are far from being constructive.) Theorem 7.2.4 is due to Bunce and Wright [62] and extends the result of Bhat considerably. The proof of Theorem 7.2.4 presented here is purely analytical. The original proof in [62] is more algebraic as it uses Herstein theorem on Jordan derivations on rings. Since the connection with Jordan derivation is very interesting we would like to comment on it briefly. The *Jordan derivation* on a ring A is an additive map $\delta\colon A \to A$ such that

$$\delta(a \circ b) = \delta(a) \circ b + \delta(b) \circ a\,, \qquad \text{for all } a, b \in A\,,$$

where $a \circ b = \frac{1}{2}(a\,b + b\,a)$ is the Jordan product on A. The following remarkable result was obtained by Herstein [191]. (Recall that a ring A is called *prime* if $x\,A\,y = \{x\,a\,y \mid a \in A\} = \{0\}$ for all $x, y \in A$ implies that either $x = 0$ or $y = 0$.

7.4.4. Theorem (Herstein). *If A is a prime ring of characteristic different from 2, then any Jordan derivation of A is an ordinary derivation of A.*

Following Herstein theorem, Sinclair showed that continuous Jordan derivations on semi-simple Banach algebras are derivations [296]. His idea is to prove that any Jordan derivation leaves invariant the primitive ideals in the algebra. It allows one to reduce the problem to Herstein theorem. With the help of these results, the proof of Theorem 7.2.4 can be done by using the following proposition taken from [62].

7.4.5. Proposition. *Let A be a C^*-algebra and let $d\colon A_{sa} \to A_{sa}$ be a (real) linear map such that*

$$d(x\,y) = d(x)\,y + x\,d(y)\,,$$

whenever $x\,y = y\,x$. Then d extends to a derivation of A.

Proof: The hypothesis implies that $d(a^2) = 2\,a \circ d(a)$ for all $a \in A_{sa}$. Denote by $\circ$ the Jordan derivation on A_{sa}. Given $x, y \in A_{sa}$ we have

$$\begin{aligned} d(x \circ y) &= \frac{1}{2}\Big[d((x+y)^2) - d(x^2) - d(y^2)\Big] \\ &= (d(x+y)) \circ (x+y) - d(x) \circ x - d(y) \circ y \\ &= d(x) \circ y + d(y) \circ x. \end{aligned}$$

By a routine calculation we see that $d(x \circ y) = d(x) \circ y + x \circ d(y)$, for all $x, y \in A$. So d is a Jordan derivation. Now the result of Sinclair applies. □

Hidden variables

The material in Section 7.3 follows one stream in the theory of hidden variables which was initiated by von Neumann. After the von Neumann impossibility proof, various kinds of hidden variable theory have been obtained. The well known negative result was proved by Plymen in [266] to the effect that there is no dispersion free normal state on a von Neumann algebra without a central one-dimensional summand. It was then shown by Hamhalter that any von Neumann algebra without central abelian summand admits no dispersion free state [169]. The results concerning nonexistence of dispersion free completely additive states on abstract orthomodular lattices can be found in [201]. The question of the existence of approximate hidden variables is due to Mackey and has been discussed in [201, 274]. Theorem 7.3.9 as well as Theorem 7.3.11, which are strongest results in the theory of hidden variables based on the concept of dispersion, have been obtained by Hamhalter in [184], where characterization of unbounded traces in terms of dispersions and further results on the properties of quasi-traces on C^*-algebras may be found. Theorem 7.3.5 was proved in [233] with the help of the result of Ogasawara [250] (see also [259]). The application of Gleason Theorem to no-go theorem for infinite simple C^*-algebras (Proposition 7.3.12) has been proved by Hamhalter in [184].

The literature which deals with to the hidden variable problems is vast and steadily growing since this topic synthesizes mathematical, physical, and philosophical ideas. In our treatment we mention only one approach which has played an important role in the discussion. Many other important aspects, like contextual hidden variable theories, etc., are dealt with elsewhere (see e.g. [153, 274] and the references therein).

8. ORTHOMORPHISMS OF PROJECTIONS

The standard formalism of quantum mechanical system consists of three basic components: the algebra of observables, the set of states on this algebra (which describes the probability structure in question), and the group of automorphisms of the algebra (which expresses the time development of the system). It is the ambition of the logico-algebraic approach to quantum mechanics, as it was articulated by Mackey [224], to recover all these aspects from the structure of propositions, i.e. from the quantum logic of projections. The observables can be reconstructed from projections by making use of the spectral resolution. As seen before, linear states are in a one-to-one correspondence with finitely-additive measures on projections (the Generalized Gleason Theorem discussed in Chapter 5). The aim of the present chapter is to complete this program by showing that also the dynamics of the system is uniquely determined by the automorphisms and homomorphisms of the projection lattices. In the following Section 8.1 we shall apply the Generalized Gleason Theorem to prove that orthomorphisms between projection lattices of von Neumann algebras extend uniquely to Jordan homomorphisms. The Dye Theorem on automorphisms of projection lattices as well as the celebrated Wigner Unitary-Antiunitary Theorem will then be derived as special consequences of this result. In Section 8.2 we present the theorem on the σ-additivity of lattice homomorphisms between projection lattices of von Neumann algebras. This result makes it possible to explain the use of normal homomorphisms in quantum theory by referring only to the natural property of preserving suprema of finitely many projections. The theorem on the σ-additivity of lattice homomorphisms and its consequences complete the description of quantum mechanical formalism purely in terms of the projection lattice.

The topics of this chapter found themselves in the centre of attention of many researchers working in the mathematical foundations of quantum mechanics. Results of the current research and historical aspects are briefly surveyed in Section 8.3.

8.1. Orthomorphisms of projection lattices

Let L_1 and L_2 be two orthoposets. Let us recall that a mapping $\varphi\colon L_1 \to L_2$ is called *orthomorphism* if, and only if, for all orthogonal $a, b \in L_1$,

(i) $\varphi(a)$, $\varphi(b)$ are orthogonal,

(ii) $\varphi(a \vee b) = \varphi(a) + \varphi(b)$.

We say that an orthomorphism is *completely additive* if $\varphi(\bigvee_\alpha a_\alpha) = \bigvee_\alpha \varphi(a_\alpha)$ for all orthogonal families (a_α) in L_1, and the σ-additivity is defined analogously by considering the disjoint countable families only. If both L_1 and L_2 are lattices we say that φ is a *lattice homomorphism* if $\varphi(a \vee b) = \varphi(a) \vee \varphi(b)$ for all $a, b \in L_1$. Moreover, a lattice homomorphism φ is a *σ-lattice homomorphism* if $\varphi\left(\bigvee_{n=1}^\infty a_n\right) = \bigvee_{n=1}^\infty \varphi(a_n)$ for *all* sequences (a_n) of elements of L_1. By *automorphism of the orthoposet L* we mean an orthomorphism $\varphi\colon L \to L$ that is surjective, one-to-one, and such that φ^{-1} is also an orthomorphism. In that case φ is an order isomorphism that is completely additive. By employing the Generalized Gleason Theorem, the general form of orthomorphisms between von Neumann projection lattices can be obtained in the following manner.

8.1.1. Theorem. *Let M be a von Neumann algebra without Type I_2 direct summand and A be a C^*-algebra. For every orthomorphism $\varphi\colon P(M) \to P(A)$ there is a $*$-Jordan homomorphism $\Phi\colon M \to A$ such that, for all $p \in P(M)$,*

$$\Phi(p) = \varphi(p).$$

In other words, there is a $*$-homomorphism $\pi_1\colon M \to A$ and a $*$-antihomomorphism $\pi_2\colon M \to A$ such that

$$\varphi(p) = \pi_1(p) \oplus \pi_2(p) \qquad \text{for all } p \in P(M).$$

Proof: By employing the Generalized Gleason Theorem one can extend φ to a bounded linear map $\Phi: M \to A$. It remains to show that Φ is a $*$-Jordan homomorphism. Let us take orthogonal projections $p_1, \ldots, p_n \in M$ and form their real linear combination $x = \sum_{i=1}^n \lambda_i p_i \in M$, $\lambda_i \in \mathbb{R}$. Hence, x is a self-adjoint element with finite spectrum and, according to the assumption, the same applies to $\Phi(x) = \sum_{i=1}^n \lambda_i \Phi(p_i)$, because $\Phi(p_1), \ldots, \Phi(p_n)$ are orthogonal projections in A. Upon computing $\Phi(x^2)$ we see that

$$\begin{aligned}\Phi(x^2) &= \Phi\left(\sum_{i=1}^n \lambda_i^2 p_i\right) = \sum_{i=1}^n \lambda_i^2 \Phi(p_i) = \left(\sum_{i=1}^n \lambda_i \Phi(p_i)\right)^2 \\ &= \Phi(x)^2 .\end{aligned}$$

By the density of self-adjoint elements with finite spectrum in M_{sa}, we see that Φ is a Jordan homomorphism of M_{sa} into A_{sa}. It implies immediately that Φ is a $*$-Jordan homomorphism. The second statement of the theorem is a consequence of the representation of $*$-Jordan homomorphism from a von Neumann algebra by a sum of a $*$-homomorphism and a $*$-antihomomorphism (see Chapter 2, Morphisms of C^*-algebras). □

By specializing the previous theorem to von Neumann algebras, we obtain the result that any orthomorphism φ between projection lattices of von Neumann algebras without Type I_2 direct summand is a sum of a $*$-homomorphism π_1 and a $*$-antihomomorphism π_2. Moreover, if φ is completely additive then both π_1 and π_2 are normal. If M is an abelian von Neumann algebra, then the orthomorphism $\varphi: P(M) \to P(N)$ can be viewed as (finitely additive) spectral measure on M. The classical spectral theory says that φ extends to a $*$-homomorphism of M into N which, in disguise, is nothing but the function calculus. In view of that, Theorem 8.1.1 can be interpreted as the existence of "quantum function calculus" corresponding to noncommutative spectral measure. The power of this theorem can be demonstrated by showing that Dye Theorem on automorphism of quantum system effortlessly follows from it.

8.1.2. Theorem (Dye). *Let M be a von Neumann algebra without Type I_2 direct summand. Then for any automorphism φ of the projection lattice $P(M)$ there is a $*$-Jordan isomorphism Φ of the algebra M such that $\varphi = \Phi|P(M)$. That is, Φ is induced by the direct sum of $*$-isomorphism and $*$-antiisomorphism.*

Proof: Let Φ be a $*$-Jordan homomorphism mapping M onto M extending φ. Suppose now that $0 \leq x \leq 1$ in M and $\Phi(x) = 0$. Let us write, for a system of commuting projections (e_n), $x = \sum_n \frac{1}{2^n} e_n$. Then $0 = \Phi(x) \geq \frac{1}{2^n} \Phi(e_n) \geq 0$ and so $\Phi(e_n) = 0$ for all n. Therefore $e_n = 0$ for all n, and hence $x = 0$. It implies that Φ is injective and so Φ has to be a Jordan isomorphism. □

Suppose now that Φ is a $*$-Jordan homomorphism mapping a von Neumann algebra M onto a von Neumann factor N and that π_1 and π_2 is the $*$-homomorphic and $*$-antihomomorhic part of Φ, respectively. Then $N = \pi_1(M) \oplus \pi_2(M)$ and hence π_1 or π_2 has to be zero since N is a factor. In other words, Φ is either $*$-homomorphism or $*$-antihomomorphism. It makes it possible to simplify the representation of surjective factor orthomorphism. In particular, we obtain the classical Wigner Unitary-Antiunitary Theorem in its lattice theoretic form.

8.1.3. Theorem. *Let H be a Hilbert space with* $\dim H \geq 3$. *Suppose that φ is an automorphism of its projection lattice $P(H)$. Then*

$$\varphi(p) = u^* p\, u\,, \tag{8.1}$$

where u is either unitary or antiunitary operator acting on H.

Proof: By the previous discussion φ is implemented by a $*$-Jordan homomorphism that is either a $*$-isomorphism or a $*$-antiisomorphism. By Chapter 2 (Compact and trace class operators), every $*$-isomorphism of $B(H)$ is implemented by a unitary in $B(H)$. This competes the proof. □

The Wigner Unitary-Antiunitary Theorem has played very important role in foundations of quantum mechanics because it provides a clear explanation of the fact that the symmetry group of standard quantum model consists of unitaries or anti-unitaries. The Wigner Theorem has many forms, some of them will be discussed in Section 8.3.

Now we want to generalize the Wigner Theorem to orthomorphisms between the standard Hilbert-space quantum logics.

8.1.4. Theorem. *Let H_1 and H_2 be Hilbert spaces,* $\dim H_1 \geq 3$, *and let $\varphi\colon P(H_1) \to P(H_2)$ be a completely additive orthomorphism. Then there are operators $s_\alpha\colon H_1 \to H_2$, $\alpha \in I$, such that each s_α is either unitary or antiunitary map of H_1 onto a closed subspace of H_2, and such that*

$$\varphi(p) = \bigoplus_{\alpha \in I} s_\alpha\, p\, s_\alpha^*\,, \qquad \textit{for all } p \in P(H_1)\,.$$

Proof: By the previous discussion, φ extends to the direct sum of normal $*$-homomorphism and normal $*$-antihomomorphism. Since any normal $*$-representation on $B(H_1)$ is unitarily equivalent to the sum of identity representations (see Chapter 2, Normal states and normal homomorphisms) we see that φ is of the required form. □

In the next section, devoted to the automatic σ-additivity of $*$-homomorphisms, we show that in the important case of separable Hilbert spaces the assumption on completely additivity in Theorem 8.1.4 can be relaxed.

8.2. Countable additivity of $*$-homomorphisms

The main goal of the present section is to show that any $*$-homomorphism that respects finite lattice operations on the projections is automatically σ-additive. This property of operator algebras indicates a nontrivial interplay between algebraic structure of the projection lattice and linear topological structure of the algebra. Let us recall that a $*$-homomorphism $\pi\colon M \to N$ between von Neumann algebras is said to be σ-additive (or countably additive) if $\pi\left(\sum_{n=1}^{\infty} e_n\right) = \sum_{n=1}^{\infty} \pi(e_n)$, whenever (e_n) is an orthogonal sequence of projections. Recall also that a $*$-homomorphism $\pi\colon M \to N$ is countably additive if, and only if, π is weak*-continuous on any σ-finite subalgebra of M. If π is σ-additive, then it has to be a lattice homomorphism. Indeed, consider projections $e, f \in P(M)$. Then $\pi(e) \vee \pi(f) \leq \pi(e \vee f)$. To show the reverse inequality let us consider a normal state ϱ of N such that $\varrho(\pi(e) \vee \pi(f)) = 0$. By spectral theory, there is an increasing sequence (p_n) of projections in M converging strongly to $e \vee f$ where, for each n, $e + f \geq \alpha_n p_n$ with some real number $\alpha_n > 0$. Hence, $0 = \varrho(\pi(e+f)) \geq \alpha_n \varrho(\pi(p_n))$ and so $\varrho(\pi(p_n)) = 0$ for each n. Hence $\varrho(\pi(e \vee f)) = 0$. It gives us $\pi(e) \vee \pi(f) = \pi(e \vee f)$. A natural question arises whether the property of being a lattice homomorphism is in fact equivalent to σ-additivity. The following simple example shows that we cannot expect a positive answer for any algebra with an infinite-dimensional abelian direct summand.

8.2.1. Example. (1) If M is an abelian von Neumann algebra and N is an arbitrary von Neumann algebra, then any $*$-homomorphism $\pi\colon\ M \to N$ is a lattice homomorphism. Indeed, for arbitrary projections $e, f \in M$,

$$\begin{aligned} \pi(e \vee f) &= \pi(e + f - e\,f) = \pi(e) + \pi(f) - \pi(e)\,\pi(f) \\ &= \pi(e) \vee \pi(f)\,. \end{aligned}$$

(2) If M is an infinite dimensional von Neumann algebra, then there is always a $*$-representation $\pi\colon\ M \to B(H)$ such that π is not σ-additive. Indeed, let (e_n) be a sequence of nonzero orthogonal projections such that $\sum_n e_n = 1$. Then $W^*(e_n) \simeq l^\infty$ and we can take a state on $W^*(e_n)$ such that $\varphi|c_0 = 0$. Then φ is not σ-additive and hence its G.N.S. representation is not σ-additive either (see Corollary 8.2.5 and its proof).

However, it turns out that all lattice homomorphisms on the projections of essentially noncommutative von Neumann algebras are already σ-additive. The following proposition will be helpful not only for studying orthomorphisms but also for pursuing other questions in subsequent chapters.

8.2.2. Proposition. *Let $M = l^\infty \otimes M_n(\mathbb{C})$, $n > 1$. Suppose that ϱ_1 is a state on l^∞ such that $\varrho_1|c_0 = 0$ and that ϱ_2 is a pure state on $M_n(\mathbb{C})$. There are projections p and q in M such that*

$$p \vee q = 1 \qquad \textit{and} \qquad (\varrho_1 \otimes \varrho_2)(p) = (\varrho_1 \otimes \varrho_2)(q) = 0\,.$$

Proof: We shall identify $l^\infty \otimes M_n(\mathbb{C})$ with $\sum_{n=1}^\infty \oplus M_n(\mathbb{C})$ via the map

$$(a_n) \otimes x \simeq \sum_{n=1}^{\infty} \oplus\, a_n\, x\,, \qquad (a_n) \subset l^\infty\,,\ x \in M_n(\mathbb{C})\,.$$

Let ξ be a unit vector in the n-dimensional Hilbert space H_n corresponding to the state ϱ_2. Take a sequence of unit vectors (ξ_n) in H_n such that $\xi_n \to \xi$, $\xi \neq \xi_n$ for each n. By denoting p_ν the projection of $M_n(\mathbb{C})$ onto the one-dimensional subspace sp$\{\nu\}$, we set

$$e = \sum_{n=1}^{\infty} \oplus p_\xi\,, \qquad f = \sum_{n=1}^{\infty} \oplus p_{\xi_n}\,.$$

Then, for a product state $\varphi = \varrho_1 \otimes \varrho_2$ we have

$$\varphi(e) = \varphi(1 \otimes p_\xi) = \varrho_1(1)\, \varrho_2(p_\xi) = 1$$

and

$$\varphi(f) = \varphi(e\, f\, e) = \varphi\left(\sum_{n=1}^{\infty} \oplus p_\xi\, p_{\xi_n}\, p_\xi\right) = \varphi\left(\sum_{n=1}^{\infty} \oplus \lambda_n\, p_\xi\right),$$

where $\lambda_n \to 1$. As $(1-\lambda_n)_n \in c_0$ we see that $\varrho_1((\lambda_n)) = 1$ by the assumption. Therefore

$$\varphi(f) = \varphi((\lambda_n) \otimes p_\xi) = \varrho_1((\lambda_n))\, \varrho_2(p_\xi) = 1\,.$$

Let us observe that $e \wedge f = \sum_{n=1}^{\infty} \oplus (p_\xi \wedge p_{\xi_n}) = 0$, and hence by taking projections $p = 1 - e$, $q = 1 - f$, we have $p \vee q = 1$ and $(\varrho_1 \otimes \varrho_2)(p) = (\varrho_1 \otimes \varrho_2)(q) = 0$. □

8.2.3. Lemma. *Let M be a von Neumann algebra, and let (p_n) be a sequence of orthogonal projections in M.*

(i) *If M is of Type I_n, where $2 \le n < \infty$, then there exist orthogonal and equivalent projections $e_1, \ldots, e_n$ in M which sum up to 1 and which satisfy $e_i\, p_j = p_j\, e_i$ for all i and j.*

(ii) *If M has no finite Type I direct summand, then there exists in M a projection $e \sim 1 - e$ such that $e\, p_j = p_j\, e$ for all j.*

Proof: Define a self-adjoint element $x = \sum_n \frac{1}{2^n}\, p_n$. In order to prove the lemma we will need results on diagonalizing stated in Chapter 2 (Classification of von Neumann algebras). If M is of Type I_n, then there are equivalent (abelian) projections $e_1, \ldots, e_n$ with $\sum_{i=1}^{n} e_n = 1$ that commute with x. As $p_1 = 1_{[1/2,1]}(x)$, we see that each e_i commutes with p_1. By the same argument e_i commute with each p_j and we are done. Similarly, if M has no Type I finite direct summand, then there is a projection $e \sim 1 - e$ commuting with x and so with every p_j. □

8.2.4. Theorem. *Let $\pi\colon M \to N$ be a $*$-homomorphism between von Neumann algebras, where M has no nonzero abelian direct summand. Then the following statements are equivalent.*

(i) $\pi(e \vee f) = \pi(e) \vee \pi(f)$ *for all projections e and f.*

(ii) π *is σ-additive.*

Proof: The implication (ii)⇒(i) has already been proved. Assume (i) and, without any loss of generality, assume also that $\pi(1) = 1$. Suppose first that M contains orthogonal equivalent projections $e_1, \ldots, e_n$ such that $e_1 + e_2 + \cdots + e_n = 1$, for some n where $2 \leq n < \infty$. Put $e = e_1$. Subject to a suitable choice of $n \times n$ matrix units (u_{ij}), there is a surjective $*$-isomorphism

$$\varphi\colon M \to (e\,M\,e) \otimes M_n(\mathbb{C})\,.$$

This implies that $(\pi(u_{ij}))$ is also a system of matrix units for N, inducing surjective $*$-isomorphism

$$\psi\colon N \to (f\,N\,f) \otimes M_n(\mathbb{C})\,,$$

where $f = \pi(e)$, and resulting in the commutative diagram

$$\begin{array}{ccc} M & \overset{\pi}{\to} & N \\ \downarrow \varphi & & \downarrow \psi \\ (e\,M\,e) \otimes M_n(\mathbb{C}) & \overset{\pi_1 \otimes \pi_2}{\to} & (f\,N\,f) \otimes M_n(\mathbb{C}) \end{array}$$

where $\pi_1 : e\,M\,e \to f\,N\,f$ satisfies $\pi_1(a) \otimes 1 = \psi\,\pi(a)$, $(a \in e\,M\,e)$, and π_2 is an automorphism of $M_n(\mathbb{C})$. As ψ is normal, we see that for establishing the σ-additivity of π on $e\,M\,e$ it is enough to prove the σ-additivity of π_1. Let us prove the latter property of π_1.

We will proceed by a contradiction. Suppose that π_1 is not σ-additive. Then there is a sequence (p_n) of orthogonal projections in $e\,M\,e$ such that

$$\sum_{n=1}^{\infty} p_n = e \qquad \text{and} \qquad \sum_{n=1}^{\infty} \pi_1(p_n) \neq \pi_1(e) = f\,.$$

Hence we can choose a normal state ϱ_1 on $f\,N\,f$ such that

$$\varrho_1\left(\sum_{n=1}^{\infty} \pi_1(p_n)\right) = 0\,.$$

Suppose that ϱ_2 is a pure state on $M_n(\mathbb{C})$. Let us observe that $\varrho_1\,\pi_1$ restricts to a state on $W^*(p_n) \simeq l^\infty$ vanishing on c_0. By applying Proposition 8.2.2 to the state

$$(\varrho_1 \otimes \varrho_2)(\pi_1 \otimes \pi_2) = (\varrho_1\,\pi_1) \otimes (\varrho_2\,\pi_2)\,,$$

we can find projections p and q in $(e\,M\,e) \otimes M_n(\mathbb{C})$ such that

$$p \vee q = 1 \qquad \text{and} \qquad (\varrho_1 \otimes \varrho_2)(\pi_1 \otimes \pi_2)(p) = (\varrho_1 \otimes \varrho_2)(\pi_1 \otimes \pi_2)(q) = 0\,.$$

Letting $g = \varphi^{-1}(p)$ and $h = \varphi^{-1}(q)$ we have $g \vee h = 1$ so that $\pi(g) \vee \pi(h) = 1$, by the hypothesis. On the other hand, $\tau = (\varrho_1 \otimes \varrho_2)\,\psi$ is a normal state of N with $\tau\,\pi(g) = \tau\,\pi(h) = 0$. But the kernel of a normal state induces a sublattices of the projection lattice. Indeed, as $p_n \nearrow \pi(g) \vee \pi(h)$, where p_n is an increasing sequence of projections with $\pi(g) + \pi(h) \geq (1/n)\,p_n$, we infer that $\tau(\pi(g) \vee \pi(h)) = 0$ – a contradiction.

So far we have proved that π is countably additive on each hereditary subalgebra $e_i\,M\,e_i$. We shall deduce the countable additivity from this by discussing each Type separately. Fix a sequence (f_n) of orthogonal projections in M.

(a) Assume that M is of Type I_n, $2 \leq n < \infty$. By (i) of Lemma 8.2.3 there are equivalent projections $e_1, \ldots, e_n$ with their sum 1 such that each e_i commutes with each f_j. Since π is countably additive on $e_i\,M\,e_i$ for each $i = 1, \ldots, n$, and $(e_i\,f_j)_j$ is an orthogonal sequence of projections in this algebra, we have

$$\pi(e_i)\,\pi\left(\sum_{n=1}^{\infty} f_n\right) = \pi\left(\sum_{n=1}^{\infty} e_i\,f_n\right) = \pi(e_i) \sum_{n=1}^{\infty} \pi(f_n)\,.$$

By summing up the previous equations for $i = 1, 2, \ldots, n$, we obtain that $\pi(\sum_{n=1}^{\infty} f_n) = \sum_{n=1}^{\infty} \pi(f_n)$.

(b) Suppose that M is of finite Type I. Then M is a direct sum of countably many algebras of Type I_n, where $2 \le n < \infty$. Hence, according to (a), it is clearly sufficient to establish the countable additivity of π on the centre, $Z(M)$, of M. We can always find projections e_1, e_2, e_3 and f in M such that $e_1 + e_2 + e_3 = 1$ with

$$e_1 \sim e_2 \qquad \text{and } e_3 \sim f \le e_1 + e_2 \,.$$

By the first part of the proof we see that π is σ-additive when restricted to any hereditary subalgebra $e_i\, M\, e_i$ and so, obviously, π is σ-additive on $Z(M)\, e_i$. This implies immediately that π is σ-additive on $Z(M)$, as required.

(c) Finally, it remains to deal with the case of M containing no finite Type I direct summand. Then Lemma 8.2.3 guarantees the existence of a projection $e \sim 1 - e$ commuting with all f_n. The calculations analogous to that applied in the case (a) will then establish the countable additivity of π. This completes the proof. □

As the above proof shows, for establishing the σ-additivity of a *unital* $*$-homomorphism π it is enough to assume that $\pi(e) \vee \pi(f) = 1$ whenever e and f are projections with $e \vee f = 1$. This has the following completely natural physical interpretation. The truth of proposition "e or f" implies the truth of the proposition "$\pi(e)$ or $\pi(f)$". Only the $*$-homomorphisms with a higher degree of continuity can qualify for this natural condition. Thus, apart from any independent mathematical interest, Theorem 8.2.4 explains the use of continuous homomorphisms and states in quantum mechanics. In order to relate the σ-additivity of $*$-homomorphism to that of states, which was investigated in Section 6.2, let us prove the following consequence of Theorem 8.2.4.

8.2.5. Corollary. *Let M be a von Neumann algebra without nonzero abelian direct summand. A state ϱ on M is σ-additive if, and only if, its G.N.S. representation π_ϱ is a lattice homomorphism.*

Proof: In view of Theorem 8.2.4 it suffices to show that ϱ is σ-additive if, and only if, π_ϱ is σ-additive. Let us recall that the G.N.S. representation π_ϱ acts on Hilbert space H_ϱ that is the completion of M/N_ϱ, where N_ϱ is the left ideal of ϱ. Let $\xi = a + N_\varrho$ be a unit vector in H_ϱ. The state ϱ_ξ given by the formula

$$\varrho_\xi(x) = (\pi_\varrho(x)\,\xi\,,\xi)\,, \qquad x \in M\,,$$

is the transformed state ϱ_a. Since the face of all σ-additive states is splitting (Section 6.2), we see that ϱ_a is σ-additive. Hence, for any sequence (p_n) of orthogonal projections in M we obtain

$$\omega_\xi\left(\pi_\varrho\left(\sum_n p_n\right)\right) = \omega_\xi\left(\sum_n \pi_\varrho(p_n)\right)$$

for a dense set of vectors ξ in H_ϱ. It implies immediately that $\pi_\varrho\left(\sum_n p_n\right) = \sum_n \pi_\varrho(p_n)$. □

Let us observe that using the result on decomposing Jordan $*$-homomorphisms into a direct sum of $*$-homomorphisms and $*$-antihomomorphisms, we can immediately extend our main result to $*$-Jordan homomorphisms.

8.2.6. Corollary. *Let $\pi\colon M \to N$ be a $*$-Jordan homomorphism between von Neumann algebras M and N where M has no nonzero abelian direct summand. The following statements are equivalent*

(i) $\pi(e \vee f) = \pi(e) \vee \pi(f)$ *for all projections* $e, f \in M$,

(ii) π *is countably additive.*

Finally, using Theorem 8.1.1 we can establish that almost all lattice orthomorphisms between projection lattices are in fact σ-orthomorphisms.

8.2.7. Corollary. *Let M and N be von Neumann algebras, where M has neither nonzero abelian nor Type I_2 direct summand. Then every lattice homomorphism $\varphi\colon P(M) \to P(N)$ is a lattice σ-homomorphism.*

Proof: By Theorem 8.1.1, φ extends to a $*$-Jordan homomorphism between M and N. Now Theorem 8.2.4 says that φ is σ-additive. It implies immediately that φ is a lattice σ-homomorphism. □

The results on automatic continuity of $*$-homomorphisms due to Feldmann and Fell, Takesaki and others (see Chapter 2, Classification of von Neumann algebras) pose restrictions on the structure of the domain algebra and on the size of the range of the homomorphisms. Unlike this, Theorem 8.2.4 derives σ-additivity only under a simple algebraic condition. At the end of this section we state some criteria for σ-additivity of important types of vector measures and homomorphisms, which enable further generalizations of Wigner Theorem. For this we shall need the following well known fact.

8.2.8. Proposition. *Let H be a separable Hilbert space. Any $*$-homomorphism $\pi\colon B(H) \to B(H)$ is σ-additive.*

Proof: Any normal $*$-representation π of $B(H)$ is a direct sum, $\pi = \pi_1 \oplus \pi_2$ of a normal $*$-homomorphism π_1 and a $*$-homomorphism π_2 that vanishes on the ideal of compact operators. Let (e_n) be an orthonormal basis of H. For a subset $A \subset \mathbb{N}$, let us denote by p_A the projection onto $\overline{\mathrm{sp}}(e_n)_{n\in A}$. By the classical argument of the set theory we can find a system $(A_\alpha)_{\alpha\in I}$ of uncountably many infinite subsets of $\mathbb{N}$ such that $A_\alpha \cap A_\beta$ is finite whenever $\alpha \neq \beta$. Suppose that $\pi_2 \neq 0$. Then $\pi_2(p_{A_\alpha}) \neq 0$ for each $\alpha \in I$ because the projection p_{A_α} is equivalent to the unit projection. On the other hand, for all $\alpha \neq \beta$ we obtain that $\pi(p_{A_\alpha})\,\pi(p_{A_\beta}) = \pi(p_{A_\alpha \cap A_\beta}) = 0$ because the projection $p_{A_\alpha \cap A_\beta}$ has finite rank. In summary, $(\pi(p_{A_\alpha}))_{\alpha\in I}$, $\alpha \in I$, forms an uncountable system of nonzero orthogonal projections in $B(H)$ – a contradiction. So $\pi = \pi_1$ and the proof is complete. □

8.2.9. Theorem. *Let $\varphi\colon P(H) \to P(H)$ be an orthomorphism where H is a separable Hilbert space, $\dim H \geq 3$. Then φ is a direct sum of at most countably many orthomorphisms φ_n's, each of the form*

$$\varphi_n(p) = s_n\, p\, s_n^*\,, \qquad p \in P(H)\,,$$

where each s_n is either unitary or antiunitary operator of H onto its closed subspace.

Let us recall that a map $m\colon P(M) \to H$ of the projection lattice $P(M)$ of a von Neumann algebra M into a Hilbert space H is called the *orthogonal measure* if it is finitely additive and $(m(p), m(q)) = 0$ whenever p and q are orthogonal (see Section 7.4). We are going to show a continuity of orthogonal vector measures between separable Hilbert spaces.

8.2.10. Lemma. *Let $m\colon P(H) \to K$ be an orthogonal vector measure, where H and K are Hilbert spaces, $\dim H \geq 3$, and K is separable. Then there is a state ϱ on $B(H)$ uniquely determined by the condition*

$$\varrho(p) = \|m(p)\|^2 , \qquad p \in P(H) .$$

Let us denote by $(\pi_\varrho, H_\varrho, \xi_\varrho)$ the G.N.S. data of ϱ. Then there is a linear isometry V of the real Hilbert space generated by $m(P(H))$ onto the real Hilbert space generated by $\{\pi_\varrho(a)\,\xi_\varrho \mid a \in B(H)_{sa}\}$ such that $V\,m(p) = \pi_\varrho(p)\,\xi_\varrho$ for all $p \in P(H)$. If, in addition, $\overline{\mathrm{sp}}\{m(p) \mid p \in P(H)\}$ is dense in K, then H_ϱ is separable.

Proof: The existence of ϱ is guaranteed by Generalized Gleason Theorem. By the same principle m extends to a bounded linear map $\hat{m}\colon B(H) \to K$. Let $x = \sum_{i=1}^n \lambda_i\, p_i$, where $\lambda_i \in \mathbb{R}$ and $p_1, \ldots, p_n$ are orthogonal projections in $P(H)$. Then

$$\begin{aligned} \|\hat{m}(x)\|^2 &= \left(\sum_{i=1}^n \lambda_i\, \hat{m}(p_i), \sum_{j=1}^n \lambda_j\, \hat{m}(p_j)\right) = \sum_{i,j=1}^n \lambda_i\, \overline{\lambda_j}\, (\hat{m}(p_i), \hat{m}(p_j)) \\ &= \sum_{i=1}^n |\lambda_i|^2\, \|m(p_i)\|^2 . \end{aligned}$$

For the map $\hat{\psi}\colon x \to \pi_\varrho(x)\,\xi_\varrho$ of $B(H)$ into K we obtain by similar calculations

$$\begin{aligned} \|\hat{\psi}(x)\|^2 &= \sum_{i=1}^n |\lambda_i|^2\, \|\pi_\varrho(p_i)\xi_\varrho\|^2 = \sum_{i=1}^n |\lambda_i|^2\, \varrho(p_i) \\ &= \sum_{i=1}^n |\lambda_i|^2\, \|m(p_i)\|^2 . \end{aligned}$$

By continuity of $\hat{\psi}$ and $\hat{m}$, $\|\hat{\psi}(a)\| = \|\hat{m}(a)\|$ for each self-adjoint element $a \in B(H)$. Define now the map $V\colon \hat{m}(B(H)_{sa}) \to \hat{\psi}(B(H)_{sa})$ by

$$V\,\hat{m}(a) = \hat{\psi}(a) \qquad \text{for all } a \in B(H)_{sa} .$$

Then V is a well defined real linear isometric map. Indeed, if $\hat{m}(a) = \hat{m}(b)$ for self-adjoint a and b, then $\hat{m}(a-b) = 0$, implying $\hat{\psi}(a) = \hat{\psi}(b)$.

If $\overline{\mathrm{sp}}\, m(P(H)) = K$, we can find a sequence (p_n) in $P(H)$ such that $\overline{\mathrm{sp}}\{m(p_n) \mid n \in \mathbb{N}\} = K$. As $\overline{\mathrm{sp}}\{\hat{\psi}(a) \mid a \in B(H)_{sa}\}$ is H_ϱ by the cyclicity of G.N.S. representation, we see that $\overline{\mathrm{sp}}\{\hat{\psi}(p_n) \mid n \in \mathbb{N}\} = H_\varrho$, and hence H_ϱ is separable. □

8.2.11. Theorem. *Let $m\colon P(H) \to H$ be an orthogonal vector measure where H is a separable Hilbert space. Then m is σ-additive.*

Proof: By the previous lemma there is no loss of generality in assuming that

$$m(p) = \pi(p)\,\xi\,,$$

where $\pi\colon B(H) \to B(H)$ is a $*$-homomorphism and ξ is a unit vector in H. By virtue of Proposition 8.2.8, π is σ-additive and the same holds for m. □

8.3. Historical remarks and comments

The study of orthomorphisms of projections lattices has been motivated by Wigner Theorem that is considered to be one of the most important results on symmetries of quantum system and it is well known to quantum physicists. Wigner Theorem was first published in Wigner's book [319], where it was formulated in terms of transformations of pure states which preserve the transition probabilities. However, the original proof had some gaps. The problem of characterizing transformations of quantum systems have been attacked by many authors. This has resulted in a number of descriptions of automorphisms of various structures associated with the Hilbert space logic such as, for instance, the state space, the normal state space with orthogonality relations, the pure state space with transition probabilities, the effect algebras, and the projection lattices [29, 220, 292, 313, 314]. A few interesting recent attempts have also been made to obtain simpler proof of Wigner Theorem [71, 234, 318]. In connection with it we will now present the proof of Wigner Theorem for maps preserving transition probabilities that is based on ideas of Molnar [234].

8.3.1. Theorem. *Let H be a complex Hilbert space and let $T\colon H \to H$ be a surjective map (linearity is not assumed) with the property that*

$$|(T(x),T(y))| = |(x,y)| \qquad \text{for all } x,y \in H\,.$$

Then T is of the form

$$T(x) = \varphi(x)\,U\,x\,, \qquad x \in H\,,$$

where $U\colon H \to H$ is either unitary or antiunitary operator and $\varphi\colon H \to \mathbb{C}$ is a function (so-called phase-function) with values in the unit circle of the complex plane.

Proof: For vectors $f,g \in H$ let us denote by $f \odot g$ the finite rank operator on H such that

$$(f \odot g)(h) = (h,g)\,f \qquad \text{for all } h \in H.$$

All finite rank operators are linear combinations of the operators of this form. Moreover, all self-adjoint finite-dimensional operators are real linear combinations of one-dimensional projections. We shall now lift the mapping T to the space of all self-adjoint finite-rank operators. Let $S = \sum_{k=1}^{n} \lambda_k\,(f_k \odot f_k)$, $\lambda_k \in \mathbb{R}$, where $f_1,\ldots,f_n$ is a finite subset of unit vectors in H. (Note that $f_k \odot f_k$ is in fact the projection onto $\mathrm{sp}\{f_k\}$.) Let us define

$$\hat{T}(S) = \sum_{k=1}^{n} \lambda_k\,(\,T(f_k) \odot T(f_k)\,)\,.$$

We prove that $\hat{T}$ is well defined on the (real) space of self-adjoint finite-rank operators. For any $h \in H$ we have

$$\begin{aligned}
&\left(\sum_{k=1}^{n} \lambda_k\,(T(f_k) \odot T(f_k)\,T(h)T(h)\right) = \\
&= \sum_{k=1}^{n} \lambda_k\,(T(h),T(f_k)\,(T(f_k),T(h)) = \sum_{k=1}^{n} \lambda_k\,|(T(h),T(f_k))|^2 \\
&= \sum_{k=1}^{n} \lambda_k\,|(h,f_k)|^2 = \sum_{k=1}^{n} \lambda_k\,(h,f_k)\,(f_k,h) \\
&= \left(\sum_{k=1}^{n} \lambda_k\,(f_k \odot f_k)\,h,h\right) = (S\,h,h)\,.
\end{aligned}$$

Let

$$S = \sum_{l=1}^{m} \mu_l \, (g_l \odot g_l) \, ,$$

be another expression of S, where $g_1, \ldots, g_m$ is a finite sequence of unit vectors. By the previous calculation

$$\left(\sum_{k=1}^{n} \lambda_k \, (T(f_k) \odot T(f_k)) \, T(h), T(h) \right) = \left(\sum_{l=1}^{m} \mu_l \, (T(g_l) \odot T(g_l)) \, T(h), T(h) \right)$$

for all $h \in H$. Since T is surjective, we obtain

$$\sum_{k=1}^{n} \lambda_k \, (T(f_k) \odot T(f_k)) = \sum_{l=1}^{m} \mu_l \, (T(g_l) \odot T(g_l)) \, .$$

This proves that $\hat{T}$ is well defined and, simultaneously, (real) linear. Moreover, $\hat{T}$ is a Jordan homomorphism because for any

$$S = \sum_{k=1}^{n} \lambda_k \, (f_k \odot f_k)$$

with orthonormal sequence $f_1, \ldots, f_n$ we have

$$\hat{T}(S^2) = \hat{T}\left(\sum_{k=1}^{n} \lambda_k^2 \, (f_k \odot f_k) \right) = \sum_{k=1}^{n} \lambda_k^2 \, (T(f_k) \odot T(f_k)) = \hat{T}(S)^2 \, .$$

(We have used the fact that $(T f_k \odot T f_k)_k$ are pairwise orthogonal projections.) Since $\hat{T}$ is bounded it can be extended to the Jordan $*$-homomorphism on the algebra $K(H)$ of compact operators in $B(H)$. Moreover, $\hat{T}$ is injective. For this, it remains to take into account that $\hat{T}$ is nonzero on every one-dimensional projection. By Chapter 2 every such map is implemented by either unitary or antiunitary map. So there is an operator U on H that is either unitary or antiunitary and such that

$$T(f) \odot T(f) = U \, (f \odot f) U^{-1} \text{ for all unit vectors } f \, .$$

In other words $T f$ is a unit vector in $\mathrm{sp}\{U f\}$, meaning that $T f = c f$, where c is a complex unit. Setting $\varphi(f) = c$, we complete the proof. □

The methods of the previous proof were used in [235] for generalizing the Wigner Theorem to C^*-modulus. As we have seen therein, Wigner Theorem does not require the Gleason Theorem. A generalization of Wigner Theorem to automorphisms of the projection lattices of von Neumann algebras, known as Dye's Theorem, was proved in [127] (again without the aid of Gleason Theorem). The idea to apply Gleason Theorem to generalizing Dye's Theorem to orthomorphisms is due to Bunce and Wright. They proved Theorem 8.1.1 in the context of Jordan algebras by utilizing only to Gleason Theorem for nonnegative measures.

The description of completely additive orthomorphisms on separable Hilbert space logic (Theorem 8.1.4) was originally proved by Paszkiewicz [257]. The proof presented here is simpler and it is based on the insight of [332]. The main result of Section 7.2 on equivalence of lattice homomorphisms and σ-additivity (Theorem 8.2.4) was proved by Bunce and Hamhalter in [52]. The result on the automatic σ-additivity of orthogonal vector measures between separable Hilbert spaces was proved by Hamhalter in [166] where further results on the interplay between the size of ranges of vector measures and degree of their additivity is shown using some combinatorial results of the set theory. The results of Section 7.2 will be applied in Chapter 10 to the study of Jauch-Piron states.

In conclusion, let us remark that parallel results to "linearization" of orthomorphisms of projection lattices are, e.g., results on representing affine transformations of the state spaces of C^*-algebras by $*$-Jordan homomorphisms. For this interesting topic we refer the reader to [20, 204, 206, 207].

9. RESTRICTIONS AND EXTENSIONS OF STATES

The restrictions and extensions of states on operator algebras and ordered structures are mutually inverse but closely related processes. The relationship between them can be viewed as a part of the general question of the interplay between global and local systems that emerges in axiomatic foundations of quantum theory and elsewhere. In the present chapter we focus on some aspects of this problem concerning the duality between restrictions and extensions of pure states in the setting of operator algebras and orthomodular structures.

In the first section we analyze the restriction properties of pure states on C^*-algebras with respect to abelian subalgebras, which may have the interpretation of relating the C^*-algebraic quantum mechanic to classical one. It will be shown that, under circumstances usually met in physics, the system of finite pure states corresponding to mutually exclusive states of the physical system is uniquely given by the biorthogonal system of orthogonal elements in the algebra, i.e by a system of "mutually contradictory" observables. Moreover, we show that orthogonal pure states with separable G.N.S. representations are simultaneously multiplicative (i.e. pure) on some large abelian part of the given algebra. From the mathematical point of view, the basic line is the effort to generalize classical results on separating points and sets in the theory of locally compact Hausdorff spaces to their generalizations given by C^*-algebras.

In the second section we continue the study of consequences of the Generalized Gleason Theorem. The main result therein says that any complex measure on the projection lattice of a von Neumann algebra without Type I_2 direct summand extends to a measure on an arbitrary larger orthomodular structure. It says that the von Neumann projection structures can be embedded into larger order structures without any restriction on the state space properties. From the "metaphysical" point of view it asserts that the validity of the model of quantum physics based on the state spaces of von Neumann algebras cannot be refuted by any theoretical construction enlarging it. It is also shown that the C^*-algebras for which the Gleason Theorem holds have projection structures compatible with any larger generalized orthomodular poset. The results of this part are based on the analysis of linear structures associated with quantum logic which is interesting in its own right.

The perspectives of the current research in this area are commented on in the concluding section.

9.1. Restriction properties of pure states

In this section we shall deal with the restriction properties of orthogonal pure states on C^*-algebras. The principal motivation for us is to what extend basic features of pure states on function algebras can be extended to algebras of operators. In the physical interpretation it amounts to the following problem. What are the differences between topological properties of compact physical phase spaces, such as subsets of $\mathbb{R}^n$ used in the classical mechanics, and "quantum phase spaces" given by noncommutative algebras of operators? In order to specify the most striking differences between those structures we will review first basic facts on pure states on abelian C^*-algebras (see Chapter 2). Let A be an abelian C^*-algebra. Then there is a locally compact Hausdorff space X such that A is isomorphic to the algebra $C_0(X)$ of all continuous functions on X vanishing at infinity. The following conditions imposed on a state ϱ on $C_0(X)$ are equivalent:

(i) ϱ is a pure state,

(ii) ϱ is multiplicative (i.e. ϱ is a $*$-homomorphism of A into $\mathbb{C}$),

(iii) there is a unique point $x \in X$ such that

$$\varrho(f) = f(x) \qquad \text{for all } f \in C_0(X)\,.$$

The one-to-one correspondence between points of X and pure states on $C_0(X)$ specified above suggests that "basic objects" of "noncommutative topology" should be structures built on the set of pure states, or more generally, on the set of irreducible representations of the given algebra (on the so-called spectrum of the algebra). The equivalence of conditions (i) and (ii) embodies an interplay between geometric structure of states and their algebraic properties. As it turns out, this equivalence is far from being true for nonabelian algebras. It is well known that any nonabelian C^*-algebra has a pure state which is not multiplicative (for otherwise the atomic representation would be a faithful representation into abelian algebra of functions on the set of complex homomorphisms). At that, a simple C^*-algebra has no multiplicative states although it has a plenty of pure states. This raises a question of whether a pure state ϱ on a C^*-algebra A is multiplicative (i.e. pure) on at least some nontrivial abelian subalgebra of A. First we observe that one can always find a singly generated abelian C^*-subalgebra B (not necessarily unital) of A such that the prescribed pure state ϱ on A is multiplicative on B. Indeed, let $(\pi_\varrho, H_\varrho, \xi_\varrho)$ be the G.N.S. representation associated with ϱ. By the Kadison Transitivity Theorem there is a self-adjoint $x \in A$ such that $\pi_\varrho(x)\xi_\varrho = \xi_\varrho$, and hence $\varrho(x) = (\pi_\varrho(x)\,\xi_\varrho, \xi_\varrho) = 1$. It means that the spectrum Spec(x) of x contains 1. For a function $0 \le f \le 1$ continuous on $\mathbb{R}$, with $f(0) = 0$ and $f(1) = 1$, the operator $a = f(x)$ is a norm-one positive element. Therefore, since $\varrho(a) = (\pi_\varrho(f(x))\,\xi_\varrho, \xi_\varrho) = (f(1)\,\xi_\varrho, \xi_\varrho) = 1$, then we have, by the Cauchy-Schwarz inequality, that $\varrho(a\,b) = \varrho(a)\,\varrho(b)$ for all $b \in A$. Especially, ϱ is multiplicative on the subalgebra $C^*(a)$. (Let us remark that if A is unital, then we can trivially take $a = 1$.) As a result, for an individual pure state ϱ one can always find a nonzero abelian singly generated subalgebra $C^*(a)$ such that $\varrho | C^*(a)$ is a pure state. The main question studied in the present section is the problem of whether $C^*(a)$ can be replaced by a maximal abelian subalgebra and whether systems of orthogonal pure states enjoy similar multiplicativity properties. The following simple example shows that we cannot hope for obtaining a positive answer for pure states in a general configuration. Let H be a Hilbert space of dimension at least 2 and $A = K(H)$ be the algebra of all compact operators acting on H. Let us consider vector states ω_ξ and ω_μ on $K(H)$ corresponding to distinct nonorthogonal unit vectors ξ and μ in H, respectively. Suppose that B is a maximal abelian subalgebra of A such that both ω_ξ and ω_μ are multiplicative on B. As B is generated by one-dimensional projections, and both states are 0-1 states on $P(B)$, there are one-dimensional projections p and q in B such that $\omega_\xi(p) = \omega_\mu(q) = 1$. But then $p(H) = \mathrm{sp}\{\xi\}$, $q(H) = \mathrm{sp}\{\mu\}$. However, since ξ and μ are distinct and nonorthogonal the projections p

and q cannot commute, which is a contradiction. Whence, to find a maximal abelian subalgebra such that given pure states are multiplicative on it, we have to pose some restrictions on the position of the states considered that correspond to orthogonality in case of vector states. This leads us to the following definition. Pure states ϱ_1 and ϱ_2 on a C^*-algebra A are called *orthogonal* if $\|\varrho_1 - \varrho_2\| = 2$ (i.e. if they have the maximal possible distance). The following examples might be helpful for the intuition.

9.1.1. Example. (i) Let ϱ_1 and ϱ_2 be pure states on the abelian C^*-algebra A. Then ϱ_1 and ϱ_2 are orthogonal if, and only if, $\varrho_1 \neq \varrho_2$.
(ii) Vector states ω_ξ and ω_μ on the algebra $B(H)$, $\dim H \geq 2$, are orthogonal if, and only if, ξ and μ are orthogonal vectors.

Proof: (i) Let us represent A as $C_0(X)$, where X is a locally compact Hausdorff space and let $x_1, x_2 \in X$ be points corresponding to pure states ϱ_1 and ϱ_2, respectively. By the standard Tietze-Uryshon-like argument, if $x_1 \neq x_2$, then there is a continuous function $f \in C_0(X)$ with $\|f\| \leq 1$, $f(x_1) = 1$ and $f(x_2) = -1$. Hence $(\varrho_1 - \varrho_2)(f) = 2$, giving $\|\varrho_1 - \varrho_2\| = 2$.

(ii) If ξ and μ are orthogonal, then for $x = p_\xi - p_\mu$, where p_ξ and p_μ is the projection onto $\mathrm{sp}\{\xi\}$ and $\mathrm{sp}\{\mu\}$, respectively, we have $(\varrho_1 - \varrho_2)(x) = 2$, and thereby $\|\varrho_1 - \varrho_2\| = 2$. On the other hand, suppose that $\|\omega_\xi - \omega_\mu\| = 2$. By multiplying the vector μ by a suitable complex unit we may assume that the product (ξ, μ) is real. Then we can write

$$\begin{aligned} \|\xi + \mu\|^2 &= 2\,(1 + (\xi, \mu)) \\ \|\xi - \mu\|^2 &= 2\,(1 - (\xi, \mu))\,. \end{aligned}$$

For each $a \in B(H)$ we have

$$\begin{aligned} |\omega_\xi(a) - \omega_\mu(a)| &= |(a\,\xi, \xi) - (a\,\mu, \mu)| \\ &= \frac{1}{2}|(a\,(\xi - \mu), \xi + \mu) + (a\,(\xi + \mu), \xi - \mu)| \\ &\leq \|a\|\,\|\xi + \mu\|\,\|\xi - \mu\|\,. \end{aligned}$$

Hence,

$$2 = \|\omega_\xi - \omega_\mu\| \leq \|\xi + \mu\|\,\|\xi - \mu\| = 2\,(1 - (\xi, \mu)^2)\,)\,.$$

It implies immediately that $(\xi, \mu) = 0$. □

Example 9.1.1 (i) describes the classical situation where all distinct states are automatically orthogonal. The existence of distinct and simultaneously nonorthogonal pure states that follows from Example 9.1.1 (ii) is a key property of quantum system that captures the Bohr's concept of complementarity. In the standard Hilbert space quantum mechanics the sequence (finite or infinite) of mutually orthogonal vector states arises as a sequence corresponding to eigenvectors of some observable O. As the first step towards solving the problem of multiplicativity of orthogonal pure states, we shall focus on the problem whether, vice versa, a sequence $\varrho_1, \ldots, \varrho_n$ of orthogonal pure states comes into existence from the states corresponding to different levels of some observable O. Moreover, we shall be concerned with the question whether O can be chosen so that it is determining for $\varrho_1, \ldots, \varrho_n$. Our analysis of the latter questions will be based on the technique of the universal representation and its connection with the double dual (see Chapter 2, The enveloping von Neumann algebra). In order to facilitate the subsequent discussion, let us first recall a few facts on pure states and fix the notation. In the sequel, we shall always consider the C^*-algebra A in question to be embedded into its second dual, A^{**}, that is into its enveloping von Neumann algebra. Any bounded functional φ on A canonically extends to normal functional on A^{**}, by a harmless abuse of notation denoted by the same letter. This extended normal functional has the same norm (by the Kaplansky Density Theorem). Let us recall that a projection $p \in A^{**}$ is called *open* if there is an increasing net (a_γ) of elements in A such that $a_\gamma \nearrow p$. Equivalently, the projection $p \in A^{**}$ is open, if and only if, p is in the weak* closure of the algebra $p\,A^{**}\,p \cap A$. Let us also recall that the product of two commuting open projections is again an open projection. The projection $p \in A^{**}$ is called *closed* if $1 - p$ is open.

Let φ be a pure state on A. Its support projection, $s(\varphi)$, is a one-dimensional projection in A^{**} that is uniquely determined by φ and such that $s(\varphi)\, x\, s(\varphi) = \varphi(x)\, s(\varphi)$ for all $x \in A^{**}$. The map $\varphi \to s(\varphi)$ is a one-to-one map of the pure state space, $P(A)$, onto the set of all one-dimensional projections in A^{**}. If $\varrho_1, \ldots, \varrho_n$ are orthogonal pure states then $\sum_{i=1}^n s(\varrho_i)$ is a closed projection. Let $c(\varphi)$ denote the smallest central projection in A^{**} such that $s(\varphi) \le c(\varphi)$. Then $c(\varphi)\, A^{**}$ is a Type I factor isomorphic to $B(H_\varphi)$, where H_φ is the Hilbert space resulting in the G.N.S. construction corresponding to φ. We shall always identify the algebras $B(H_\varphi)$ and

$c(\varphi)\,A^{**}$. The G.N.S. representation $\pi_\varphi\colon\, A \to B(H_\varrho)$ is equivalent to the representation, denoted by the same letter, $\pi_\varphi\colon\, a \to c(\varphi)\,a : A \to c(\varphi)\,A^{**}$. Pure states φ and ϱ on A are equivalent if, and only if, $c(\varphi) = c(\varrho)$, and they are disjoint, if, and only if, $c(\varphi)$ and $c(\varrho)$ are orthogonal central projections in A^{**}. We shall now prove that pure states ϱ and φ on A are orthogonal if, and only if, $s(\varrho)\,s(\varphi) = 0$. First let us suppose that ϱ and φ are disjoint. In this case $s(\varrho)\,s(\varphi) = 0$ and, simultaneously, ϱ and φ are orthogonal because

$$2 \geq \|\varrho - \varphi\| \geq (\varrho - \varphi)(c(\varrho) - c(\varphi)) = 2\,.$$

Suppose now that ϱ and φ are equivalent. Then ϱ and φ can be viewed as vector states on $B(H_\varrho)$ and we can apply Example 9.1.1 (ii).

The dual concept to orthogonality of states is that of orthogonality of elements of the algebra. Let us say that self-adjoint elements, a, b, of the C^*-algebra A are *orthogonal* if $a\,b = 0$. This is equivalent to saying that the range projections $r(a)$ and $r(b)$ are orthogonal projections in A^{**}.

Let (ϱ_n) be a sequence (finite or infinite) of orthogonal pure states. Let us say that a sequence (a_n) of positive, norm-one, pairwise orthogonal elements is the *supporting sequence* for the sequence (ϱ_n), if $\varrho_i(a_j) = \delta_{ij}$ for all i, j. In case of finitely many states the existence of a supporting sequence is guaranteed by the following proposition.

9.1.2. Proposition. *Any finite sequence $\varrho_1, \ldots, \varrho_n$ of orthogonal pure states on a C^*-algebra A admits a supporting sequence.*

Proof: Let us take unit vectors $\xi_i \in s(\varrho_i)\,(H_i)$, $i = 1, \ldots, n$, where H_i is the G.N.S. Hilbert space of ϱ_i. Using the n-fold Transitivity Theorem we can find a self-adjoint element $y \in A$ such that

$$\pi_{\varrho_i}(y)\,\xi_i = \frac{1}{2^i}\,\xi_i \qquad i = 1, 2, \ldots, n\,,$$

where π_{ϱ_i} denotes the G.N.S. representation corresponding to ϱ_i. Then, for each $i = 1, \ldots, n$, and for each integer k,

$$\varrho_i(y^k) = (\pi_{\varrho_i}(y)^k\,\xi_i, \xi_i) = \left(\frac{1}{2^{i\,k}}\,\xi_i, \xi_i\right) = \frac{1}{2^{i\,k}} = \varrho_i(y)^k\,.$$

In other words, the states ϱ_i are multiplicative on the C^*-algebra generated by y. Let us now choose continuous functions $0 \le f_i \le 1$ $(i = 1, \dots, n)$ on the real line vanishing at zero with mutually disjoint supports and satisfying

$$f_i\left(\frac{1}{2^i}\right) = 1 \qquad \text{for all } i = 1, 2, \dots, n\,.$$

By putting $a_i = f_i(a)$ we obtain norm-one, pairwise orthogonal elements. Moreover,

$$\varrho_i(a_i) = \left(\pi_{\varrho_i}(f_i(y))\,\xi_i, \xi_i\right) = \left(f_i\left(\frac{1}{2^i}\right)\xi_i, \xi_i\right) = 1\,.$$

□

As the following simple example shows Proposition 9.1.2 does not hold for infinitely many orthogonal pure states. Consider the C^*-algebra $C(X)$ of all continuous complex functions on an infinite compact Hausdorff space X. Then X admits a sequence of distinct points (x_n) converging to the point x such that $x \neq x_n$ for all n. But x cannot be separated from (x_n) by any continuous function.

As a noncommutative example of a similar type one can consider the C^*-algebra A acting irreducibly on a separable Hilbert space H of infinite dimension such that A has zero intersection with the ideal of compact operators. Let (ω_{x_n}) be a sequence of vector orthogonal pure states corresponding to the orthonormal basis (x_n) of H. This sequence has no supporting sequence in A for otherwise it would consists of one-dimensional projections. In the physical interpretation, Proposition 9.1.2 says that finitely many orthogonal pure states can always be realized as states corresponding to distinct values of a single observable. In subsequent analysis we show that this observable can even be chosen in such a way that it determines completely the given system of orthogonal pure states, provided that the C^*-algebra in question is separable. For this reason we introduce the following concept. Let ϱ be a pure state of a C^*-algebra A. An element $a \in A_{sa}$ is called the *determining element* for ϱ if $\|a\| = 1$ and ϱ is the only pure state on A such that $\varrho(a) = 1$. Alternatively speaking, the element a uniquely determines the pure state ϱ by the property that the expectation value of the observable a is one exactly when the system is prepared in the state ϱ. This is a typical situation of standard quantum mechanic working with a vector state ω_x on a Hilbert space algebra $B(H)$ for which the projection onto $\mathrm{sp}\{x\}$ is a determining element. Let us observe that any determining element a for a pure state ϱ can be modified to be positive. To see this, let us decompose $a = a^+ - a^-$ into its positive and negative part a^+ and a^-, respectively. Then $\varrho(a) = \varrho(a^+) - \varrho(a^-) = 1$ and, in turn, $\varrho(a^+) = 1$. On the other hand, if φ is any pure state such that $\varphi(a^+) = 1$, then it is definite at a^+ and so

$$\varphi(a^-) = \varphi(a^+)\,\varphi(a^-) = \varphi(a^+\,a^-) = 0\,.$$

Hence, $\varphi(a) = 1$ and so $\varphi = \varrho$. A supporting sequence (a_n) for orthogonal pure states (ϱ_n) is called *determining supporting sequence* if each a_n is a determining element for ϱ_n. The following theorem establishes the existence of a supporting determining sequence for finitely many states on separable algebras.

9.1.3. Theorem. *Any finite sequence $\varrho_1, \ldots, \varrho_n$ of orthogonal pure states on a separable C^*-algebra A admits a determining supporting sequence.*

Proof: By Proposition 9.1.2 there is a sequence $u_1, \ldots, u_n$ of positive norm-one elements in A such that

$$\varrho_i(u_j) = \delta_{ij} \qquad \text{for all } i, j \leq n\,.$$

Denote by $r(u_i)$ the range projection of u_i in A^{**}. Then $r(u_i)$ is an open projection and, moreover, $u_i \leq r(u_i)$ implies $\varrho_i(r(u_i)) = 1$. It means that $s(\varrho_i) \leq r(u_i)$, whence $1 - s(\varrho_i)$ and $r(\varrho_i)$ are commuting open projections and so is their product $(1 - s(\varrho_i))\, r(\varrho_i)$. Let X_i be a hereditary subalgebra of A given by

$$X_i = r(u_i)\,(1 - s(\varrho_i))\,A^{**}\,r(u_i)\,(1 - s(\varrho_i)) \cap A\,.$$

By separability of A, X_i contains a strictly positive element, x_i, with norm 1. We want to prove that the sequence $a_i = u_i - x_i\,u_i\,x_i$, $i = 1\ldots, n$, has all properties of the supporting determining sequence except for the positivity. The orthogonality of $(a_i)_{i=1}^n$ follows from the fact that $a_i \in r(u_i)\,A^{**}\,r(u_i)$ and from the orthogonality of the projections $r(u_1), r(u_2), \ldots, r(u_n)$. Further, $a_i \leq u_i \leq 1$ and $a_i \geq -x_i\,u_i\,x_i \geq -x_i^2 \geq -1$ gives $\|a_i\| \leq 1$ $(i = 1, \ldots, n)$. Since ϱ_i vanishes on the hereditary subalgebra

$$(1 - s(\varrho_i))\,A^{**}\,(1 - s(\varrho_i))\,,$$

we have $\varrho_i(x_i\, u_i\, x_i) = 0$ because $x_i\, u_i\, x_i \in (1 - s(\varrho_i))\, A^{**}\, (1 - s(\varrho_i))$. Consequently, $\varrho_i(a_i) = \varrho_i(u_i) = 1$. It remains to prove that a_i is a determining element for ϱ_i. First we show that $r(x_i) = r(u_i)\,(1 - s(\varrho_i))$. Suppose, for a contradiction, that $r(x_i) < r(u_i)\,(1 - s(\varrho_i))$. Let us take a normal state ψ on A^{**} with $\psi(r(u_i)\,(1 - s(\varrho_i))) = 1$ and $\psi(r(x_i)) = 0$. Then $0 \leq \psi(x_i) \leq \psi(r(x_i)) = 0$, and hence ψ vanishes on the algebra X_i by strict positivity of x_i. However, X_i is weak* dense in the algebra $r(x_i)\,(1 - s(\varrho_i))\, A^{**}\, r(x_i)\,(1 - s(\varrho_i))$, and therefore $\psi(r(u_i)\,(1 - s(\varrho_i))) = 0$, which is a contradiction. Suppose now that ϱ is a pure state on A such that $\varrho(a_i) = 1$. Taking into account that $0 \leq \varrho(u_i), \varrho(x_i\, u_i\, x_i) \leq 1$, we conclude that $\varrho(u_i) = 1$ and $\varrho(x_i\, u_i\, x_i) = 0$. From the latter equality and normality of ϱ on A^{**} we infer that $0 = \varrho(x_i\, r(u_i)\, x_i) = \varrho(x_i^2)$. Hence, by the Cauchy-Schwarz inequality, $\varrho(x_i) = 0$ and so ϱ vanishes on $r(x_i) = r(u_i)\,(1 - s(\varrho_i))$. (Let us remark that $\varrho(r(x)) = 0$ whenever ϱ is a normal state vanishing at a positive element x. To this end it is enough to assume that $0 \leq x \leq 1$. Then $r(x)$ is a strong operator limit of a sequence $(x^{1/n})_{n=1}^{\infty}$. As ϱ is definite at x we have $\varrho(x^{1/n}) = 0$ for all n. By the normality of ϱ we see that indeed $\varrho(r(x)) = 0$.) However, as $\varrho(r(u_i)) = 1$ we have that $\varrho(1 - s(\varrho_i)) = \varrho(r(u_i)\,(1 - s(\varrho_i)) = 0$. In other words, $\varrho(s(\varrho_i)) = 1$ and therefore $\varrho = \varrho_i$. Finally, the sequence of positive parts $a_1^+, \ldots, a_n^+$ is the desired supporting determining sequence. □

9.1.4. Corollary. *Any pure state on a separable C^*-algebra admits a determining element.*

We can conclude that, under conditions often occurring in physics, a pure state of the C^*-algebraic quantum system is uniquely determined by a single observable, and thereby by the smallest possible abelian subalgebra. As a consequence we also obtain that orthogonal pure states can be reconstructed from their pure restrictions to some maximal abelian subalgebra.

9.1.5. Corollary. *Let $\varrho_1, \ldots, \varrho_n$ be orthogonal pure states on a separable C^*-algebra A. Then there is a maximal abelian subalgebra B of A such that, for each i,*

(i) *$\varrho_i | B$ is a pure state,*

(ii) *ϱ_i is the only pure state extension of $\varrho_i | B$ over A.*

Proof: Let $a_1, \ldots, a_n$ be a determining sequence for $\varrho_1, \ldots, \varrho_n$. Let B be any maximal abelian subalgebra containing $C^*(a_1, \ldots, a_n)$. Then, of course, each ϱ_i is uniquely determined by its pure restriction to $C^*(a_1, \ldots, a_n)$. By the Hahn-Banach and the Krein-Milman Theorem each $\varrho_i | C^*(a_1, \ldots, a_n)$ extends to a pure state φ_i on B. However, by the uniqueness of the extension, $\varphi_i = \varrho_i | B$. □

As we have seen a pure state on a noncommutative algebra is seldom multiplicative. On the other hand, it can be uniquely determined by multiplicative state on some maximal abelian subalgebra, or even by a singly generated subalgebra, on condition that the given algebra is separable. These results do not hold without assuming separability. Indeed, suppose that B is a C^*-algebra not containing any strictly positive element. Let $A = B + \mathbb{C}1$ be the C^*-algebra obtained by unitalization of B. Let ϱ be a pure state on A uniquely determined by the condition $\varrho | B = 0$. Suppose that $0 \leq x \leq 1$ is a determining element for ϱ. We show that $1 - x$ is a strictly positive element for B, which leads to a contradiction. For this, observe first that $\varrho(1-x) = 0$ and hence $1-x \in B$. Further, let φ be a pure state on B such that $\varphi(1-x) = 0$. Then φ extends to a pure state $\hat{\varphi}$ on A with $\hat{\varphi}(x) = 1$, and hence $\hat{\varphi} = \varrho$. It means that $\varphi = 0$ which is absurd. Thus, $1 - x$ is a strictly positive element of B. This reasoning also shows that determinacy of pure states by elements of the algebra is not possible in the nonseparable case even if one deals with abelian algebras. Indeed, put $B = C_0(X)$ in the above example, where X is a locally compact Hausdorff space that is not σ-compact. Let us also remark in this regard that a pure state on a von Neumann algebra has a determining element, if and only if, it is normal (see [180]). It means, for example, that no pure state on a Type *II* or *III* von Neumann algebra has a determining element. It can also be shown that Corollary 9.1.5 does not hold for nonseparable algebras. The construction is more complicated and we only outline its basic idea (for the details, the reader may consult [11]). The first step is to construct a C^*-algebra A_0 such that no maximal abelian subalgebra of A_0 contains an approximate unit for the whole of A_0. (Such an algebra has to be very nonseparable.) Let now $A = A_0 + \mathbb{C}1$ result by unitalization of A_0. Suppose that ϱ is a pure state on A that vanishes on A_0. Let B be a maximal abelian subalgebra of A. It has to be of the form $B = B_0 + \mathbb{C}1$, where B_0 is a maximal abelian subalgebra in A_0. Consider now an approximate unit

(a_α) for B_0. An approximate unit of the C^*-algebra is characterized by the property that any pure state of the algebra has to be nonzero at at least one element of it (see [8, 11, 180]). As (a_α) is not an approximate unit for A we see that there is a pure state φ on A_0 such that $\varphi(a_\alpha) = 0$ for all α. In other words, ϱ and the pure extension of φ over A are distinct pure states that coincide on B. Therefore, the state ϱ is not uniquely determined by any maximal abelian subalgebra in the sense of Corollary 9.1.5.

However, we can still ask whether there exists some maximal abelian subalgebra such that a given pure state is multiplicative on it. The following problem seems to be a natural operator-theoretic extension of the equivalence between pureness and multiplicativity for states on function algebras.

9.1.6. Problem. *Is a given sequence (finite or infinite) of orthogonal pure states on a C^*-algebra multiplicative on some maximal abelian subalgebra?*

There are both positive and negative partial results concerning the foregoing problem. First, let us prove that the answer is in the positive for any finite sequence of orthogonal pure states with the separable G.N.S. representations. For establishing this result we will need the following auxiliary lemma.

9.1.7. Lemma. *Let $\varrho_1, \dots, \varrho_n$ be orthogonal pure states such that $c(\varrho_i)\, A^{**}$ is σ-finite for each $i = 1, 2, \dots, n$. Suppose that $x \in A$ commutes with all support projections $s(\varrho_1), s(\varrho_2), \dots, s(\varrho_n)$. Then there is an element $a \in A$ with*

$$0 \le a \le r(x)\left(1 - \sum_{i=1}^{n} s(\varrho_i)\right)$$

satisfying

$$e\, r(a) = \left(e - \sum_{i=1}^{n} s(\varrho_i)\right) r(x), \qquad \textit{where} \quad e = \bigvee_{i=1}^{n} c(\varrho_i).$$

Proof: As all minimal projections $s(\varrho_1), s(\varrho_2), \dots, s(\varrho_n)$ commute with $r(x)$, an element $p = \left(e - \sum_{i=1}^{n} s(\varrho_i)\right) r(x)$ is a projection. If $p = 0$ we can set $a = 0$. Suppose that $p \neq 0$. By the assumption, p is σ–finite. Any projection in $e\,A^{**}$ is a union of atomic projections. Therefore p can be written as

$$p = \sum_{\alpha \in I} e_\alpha,$$

where each e_α is a minimal projection in A^{**} and I is a subset of the set of positive integers. Let us choose states ω_α's on A with the supports $s(\omega_\alpha) = e_\alpha$ $(\alpha \in I)$ and set

$$\omega = \sum_{\alpha \in I} \lambda_\alpha\, \omega_\alpha, \qquad \text{where } 0 < \lambda_\alpha \le 1, \ \sum_{\alpha \in I} \lambda_\alpha = 1.$$

It can be easily verified that the support of ω is exactly p. Since the projection

$$r(x)\left(1 - \sum_{i=1}^{n} s(\varrho_i)\right)$$

is open, we can select a net $(a_\gamma) \subset A$ of positive elements such that $a_\gamma \nearrow r(x)\left(1 - \sum_{i=1}^n s(\varrho_i)\right)$. As $\omega(p) = 1$ and $p \le r(x)\left(1 - \sum_{i=1}^n s(\varrho_i)\right)$, we have $\omega\left(r(x)\left(1 - \sum_{i=1}^n s(\varrho_i)\right)\right) = 1$ and so $\omega(a_\gamma) \nearrow 1$. Hence, there is a subsequence (a_j) (finite or infinite) of (a_γ) such that $\omega(a_j) \nearrow 1$. The sequence $(r(a_j))$ being increasing, there exists a projection $q \in A^{**}$ such that $r(a_j) \nearrow q$. Now $q \le r(x)\left(1 - \sum_{i=1}^n s(\varrho_i)\right)$ and $\omega(q) = 1$, which implies

$$r(x)\left(e - \sum_{i=1}^{n} s(\varrho_i)\right) \le q \le r(x)\left(1 - \sum_{i=1}^{n} s(\varrho_i)\right)$$

and, in turn,

$$e\,q = \left(e - \sum_{i=1}^{n} s(\varrho_i)\right) r(x)\,.$$

Putting

$$a = \sum_j \frac{1}{2^j}\, a_j$$

we obtain an element in A satisfying $r(a) = q$ and the proof can be completed by the equality above. □

9.1.8. Theorem. *Let $\varrho_1, \varrho_2, \ldots, \varrho_n$ be orthogonal pure states on a C^*-algebra A. Let $c(\varrho_i)\, A^{**}$ be σ-finite von Neumann algebra for all $i = 1, \ldots, n$. Then there is a maximal abelian subalgebra B of A such that every state ϱ_i restricts to a pure state on B.*

Proof: Let us take positive, norm-one, mutually orthogonal elements $x_1, x_2, \ldots, x_n \in A$ with the property

$$s(\varrho_j)\, x_i\, s(\varrho_j) = \delta_{ij}\, s(\varrho_j), \qquad i, j = 1, \ldots, n,$$

the existence of which is guaranteed by Proposition 9.1.2. If $i \neq j$, then

$$s(\varrho_j)\, x_i\, s(\varrho_j) = (x_i^{1/2}\, s(\varrho_j))^*\, (x_i^{1/2}\, s(\varrho_j)) = 0$$

and hence $x_i^{1/2}\, s(\varrho_j) = 0$. This results in $x_i\, s(\varrho_j) = 0$. For $i = j$ we have

$$s(\varrho_i)\, x_i\, s(\varrho_i) = s(\varrho_i)$$

and, in turn, $s(\varrho_i)\,(1 - x_i)\, s(\varrho_i) = 0$. This implies, as above, that $s(\varrho_i)\, x_i = s(\varrho_i)$. So we know that

$$x_i\, s(\varrho_j) = \delta_{ij}\, s(\varrho_j) \qquad i, j = 1, \ldots n.$$

According to Lemma 9.1.7 we can find elements $a_i \in A$ satisfying, for all $i = 1, \ldots, n$,

$$0 \leq a_i \leq r(x_i) \left(1 - \sum_{i=1}^{n} s(\varrho_i)\right),$$

and

$$e\, r(a_i) = \left(e - \sum_{i=1}^{n} s(\varrho_i)\right) r(x_i), \tag{9.1}$$

where $e = \bigvee_{i=1}^{n} c(\varrho_i)$. Let us put $x = x_1 + x_2 + \cdots + x_n$. Since $x_1, \ldots, x_n$ are mutually orthogonal we see that $r(x) = r(x_1) + \cdots + r(x_n)$, and by (9.1)

$$e \sum_{i=1}^{n} r(a_i) = \left(e - \sum_{i=1}^{n} s(\varrho_i)\right) r(x). \tag{9.2}$$

Letting $b_i = x_i - a_i$ $(i = 1, \ldots, n)$ we obtain a system of orthogonal (and thereby commuting) elements in A. Moreover, b_i's are norm one since

$1 \geq x_i \geq x_i - a_i \geq -1$ and $\varrho_i(b_i) = 1$. Let us choose a maximal abelian subalgebra $\mathcal{C}$ in the algebra

$$X = r(x)\left(1 - \sum_{i=1}^{n} s(\varrho_i)\right) A^{**}\, r(x)\left(1 - \sum_{i=1}^{n} s(\varrho_i)\right) \bigcap A \bigcap \{b_1, b_2, \ldots, b_n\}',$$

where $\{b_1, b_2, \ldots, b_n\}'$ denotes the commutant of the set $\{b_1, \ldots, b_n\}$. If X is zero, we set $\mathcal{C} = \{0\}$. Let $\mathcal{B}$ be the C^*-algebra generated by $b_1, b_2, \ldots, b_n$ and the set $\mathcal{C}$. We shall show that $\mathcal{B}$ is a maximal abelian C^*-subalgebra of $r(x)\, A^{**}\, r(x) \cap A$. For this, let us take an element $u \in r(x)\, A^{**}\, r(x) \cap A$ commuting with all elements in $\mathcal{B}$. In order to prove that $u \in \mathcal{B}$ we shall consider the following auxiliary element h:

$$\begin{aligned} h &= e\, r(x)\, (1 - b_1 - b_2 - \cdots - b_n) = \\ &= e\,(r(x) - x_1 - x_2 - \cdots - x_n) + e\,(a_1 + a_2 + \cdots + a_n) \\ &\geq e\,(a_1 + a_2 + \cdots + a_n). \end{aligned}$$

Hence,

$$e\, r(x) \geq r(h) \geq \left(e - \sum_{i=1}^{n} s(\varrho_i)\right) r(x) \tag{9.3}$$

by (9.2). As $a_i\, s(\varrho_j) = 0$, for all $i, j = 1, \ldots, n$, we conclude that

$$b_i\, s(\varrho_j) = (x_i - a_i)\, s(\varrho_j) = \delta_{ij}\, s(\varrho_j)\,,$$

and hence

$$h\, s(\varrho_i) = r(x)\, s(\varrho_i) - b_i\, s(\varrho_i) = s(\varrho_i) - s(\varrho_i) = 0 \tag{9.4}$$

for all $i = 1, \ldots, n$. In the last equality we have used the fact that $\varrho_i(x_i) = 1$ implies $\varrho_i(r(x_i)) = 1$ and so $s(\varrho_i) \leq r(x_i)$. Thus, combining (9.3) and (9.4), we conclude

$$r(h) = \left(e - \sum_{i=1}^{n} s(\varrho_i)\right) r(x).$$

As u commutes with h we see that u commutes with

$$r(h) = \left(e - \sum_{i=1}^{n} s(\varrho_i)\right) r(x)$$

and so it commutes with $\sum_{i=1}^{n} s(\varrho_i)$. Let us now put

$$v = u - \sum_{i=1}^{n} \varrho_i(u)\, b_i .$$

The elements v and $\sum_{i=1}^{n} s(\varrho_i)$ commute. Moreover,

$$\varrho_j(v) = \varrho_j(s(\varrho_j)\, v\, s(\varrho_j)) = \varrho_j(u) - \varrho_j(u) = 0 \qquad (j = 1, \dots, n)$$

implies immediately that

$$\left(\sum_{i=1}^{n} s(\varrho_i)\right) v \left(\sum_{i=1}^{n} s(\varrho_i)\right) = 0$$

and so

$$v \sum_{i=1}^{n} s(\varrho_i) = 0.$$

In summary,

$$v \in r(x) \left(1 - \sum_{i=1}^{n} s(\varrho_i)\right) A^{**}\, r(x) \left(1 - \sum_{i=1}^{n} s(\varrho_i)\right) \cap A .$$

Therefore $v \in \mathcal{C}$ and, immediately $u \in \mathcal{B}$ which shows that $\mathcal{B}$ is a maximal abelian subalgebra of $r(x)\, A^{**}\, r(x) \cap A$. Further, since $\varrho_i(b_i) = \|b_i\| = 1$ the Cauchy-Schwarz inequality entails that states $\varrho_1, \varrho_2, \dots, \varrho_n$ are definite at b_i and therefore multiplicative on $\mathcal{C}$.

Finally, let us extend $\mathcal{B}$ to a maximal abelian subalgebra B of A and show that B satisfies all statements of the Theorem 9.1.8. Since $B \cap r(x)\, A^{**}\, r(x) = \mathcal{B}$, $\mathcal{B}$ is a hereditary subalgebra of B. Indeed, whenever $0 \le f \le g$ with $f \in B$ and $g \in \mathcal{B}$, $\|g\| = 1$, then $f \le r(x)$ and so $f \in \mathcal{B}$. It implies that all states $\varrho_1, \varrho_2, \dots, \varrho_n$ are pure on B, which completes the proof. □

The following example shows that the previous theorem cannot be generalized to infinite sequences of orthogonal pure states and it is a negative answer to Problem 3.1.5 in its strongest form. The counterexample can be constructed on any irreducible separable C^*-algebra of infinite dimension that is not a commutative extension of compact operators.

9.1.9. Example. *Let A be a separable unital infinitely dimensional C^*-algebra acting irreducibly on a Hilbert space H. Let $K(H)$ be the algebra of all compact operators acting on H. Suppose that $A/A \cap K(H)$ is non-commutative. Then there is a sequence of pure, mutually orthogonal states on A, which do not restrict simultaneously to pure states on any maximal abelian subalgebra of A.*

Proof: Since A acts irreducibly on H, either $A \cap K(H) = \{0\}$ or $K(H) \subset A$. Therefore our discussion falls into two cases.
(a) Suppose that $A \cap K = \{0\}$. The separability of H allows one to find a dense sequence (x_n) in its unit sphere. Then the convex hull of the sequence (ω_{x_n}) of corresponding vector states is weak*-dense in the state space of A because the set ω_{x_n} is order determining for A_{sa}. Put

$$f = \sum_n \frac{1}{2^n} \omega_{x_n} .$$

By the Glimm Lemma there is an orthonormal basis (ξ_n) of H such that

$$f(a) = \lim_{n\to\infty} (a\,\xi_n, \xi_n) \qquad \text{for all } a \in A .$$

By taking the corresponding pure states (ω_{ξ_n}) we construct a sequence of orthogonal pure states on A (we make use of the fact that A is irreducible). Suppose that C is maximal abelian subalgebra of A such that all states (ω_{ξ_n}) are multiplicative on C. Then f is pure on C and so $f = \omega_{x_n}$ for all n on C. By the density of the convex hull of (ω_{x_n}) in the state space, we see that all states on C are multiples of f. By the Gelfand Theorem, C has to be one-dimensional and so $\dim A = 1$, which is a contradiction with the irreducibility of A on the infinite dimensional Hilbert space H.

(b) Let us now consider the case of $K(H) \subset A$. It requires a different technique. The quotient algebra $A/K(H)$ is non-abelian by the assumption, hence we can find a self-adjoint element $\hat{a}$ in A/K and a pure state $\hat{\varrho}$ of A/K such that $\hat{\varrho}(\hat{a})^2 \neq \hat{\varrho}(\hat{a}^2)$. We can lift $\hat{a}$ and $\hat{\varrho}$ to the self-adjoint element a in A and the pure state ϱ on A with $\varrho(a)^2 \neq \varrho(a^2)$. Making use of the Weyl-von Neumann-Berg theorem (Chapter 2, Compact and trace class operators), we can find for any $\varepsilon > 0$ an orthonormal basis (ξ_n) of H and a compact operator $k \in K$, $\|k\| \leq \varepsilon$, such that

$$a = a_d + k,$$

where $a_d \in A$ is a diagonal self-adjoint operator with respect to the basis (ξ_n) of H. By taking k sufficiently small we can arrange for $\varrho(a_d^2) \neq \varrho(a_d)^2$. Let us observe that a pure state ϱ is not equivalent to any pure state ω_{ξ_n} and so ϱ is orthogonal to any ω_{ξ_n}. Suppose that there is a maximal abelian subalgebra B of A such that the sequence $(\omega_{\xi_n}) \cup \{\varrho\}$ of orthogonal pure states restricts to the sequence of pure states on B and try to reach a contradiction. Take a self-adjoint element $b \in B$. The multiplicativity of ω_{ξ_n} on B implies that $b - \omega_{\xi_n}(b)\,1 = b - (b\xi_n, \xi_n)\,1$ belongs to the left kernel $\{x \in A \mid \omega_{\xi_n}(x^*x) = 0\}$. Hence, $b\xi_n = (b\xi_n, \xi_n)\,\xi_n$ and we have derived that each vector ξ_n is an eigenvector of b. Alternatively, b is diagonal with respect to the basis (ξ_n) and so B is the algebra of all such operators lying in A by maximality. As $a_d \in B$ and ϱ is not definite on a_d, we obtain a contradiction. The proof is complete. □

The previous example says that if A is an irreducible infinite-dimensional separable unital C^*-algebra acting on a Hilbert space H and not containing compact operators, then $A/K(H) \simeq A$ is a separable subalgebra of the Calkin algebra $B(H)/K(H)$ that admits a sequence of orthogonal pure states which do not restrict to a sequence of multiplicative states on any maximal abelian subalgebra. In contrast to this, Anderson proved in [24] that *every* sequence of pure states on the Calkin algebra admits a maximal abelian subalgebra such that all members of the given sequence are pure on it. It witnesses a delicate interplay between the "size" of the C^*-algebra and the "size" of its maximal abelian subalgebras.

9.2. Gleason type theorems for quantum logics

Extension principles play an important role in many fields of mathematics. As examples one can take the Hahn-Banach Theorem on extending linear functionals and the Gleason Theorem allowing one to extend probability measures from projections to linear functionals on a given algebra. The former is a basic principle of functional analysis, the latter is a central principle of quantum measure theory built on operators. In this section we show the interplay between these two deep theorems that enables us further strengthening of Gleason Theorem. First we demonstrate that the Mackey-Gleason problem can be reformulated as a problem of extending measures between projection lattices. As seen in Chapter 5, the transition from Gleason Theorem for separable Hilbert spaces and σ-additive probability measures to the Generalized Gleason Theorem involving vector finitely additive measures on projections lattices of von Neumann algebras is highly nontrivial. In fact, the effort needed is equivalent to the following statement: Let M be a von Neumann algebra without Type I_2 direct summand acting on a Hilbert space H. Then any bounded finitely additive measure on $P(M)$ extends to a probability measure on $P(H)$. Indeed, having such an extension at our disposal, we can use Gleason Theorem for finitely additive states on Hilbert space logic and describe in such a manner all measures on the projection lattices. This form of restating the Generalized Gleason Theorem enables one its generalization to the context of general orthomodular structures. The main goal of the present section is to show that the extension results for measures between projection lattices $P(M)$ and $P(H)$ hold even if we replace the enveloping Hilbert space logic $P(H)$ with any ordered structure consistent with very general features of quantum system. Moreover, we show that this type of extension can be obtained for any vector measure with values in boundedly complete vector lattice. The basic tool of our analysis will be the interplay between the following ordered structures: (1) the generalization of projection structures of C^*-algebras given by the generalized orthomodular posets, (2) the generalization of the state spaces of C^*-algebras given by compact convex sets, (3) the generalization of the linear structure of self-adjoint parts of C^*-algebras given by the order unit Banach spaces of affine functions on compact convex sets. We will now specify those structures precisely. The following concept generalizes the notion of quantum logic in order to accommodate the order structures of projections in nonunital C^*-algebras.

9.2.1. Definition. Let $(L, \leq, 0)$ be a poset with the order $\leq$ and a least element 0. For each $a \in L$ let $\perp_a \colon [0,a] \to [0,a]$ be an operation on the interval $[0,a] = \{x \in L \mid 0 \leq x \leq a\}$. The structure $(L, \leq, 0, (\perp_a)_{a \in L})$ is called the *generalized orthomodular poset* or *generalized quantum logic* if the following conditions are satisfied

(i) For each $a \in L$, the structure $([0,a], \leq, 0, a, \perp_a)$ is an orthomodular poset.

(ii) If $a \leq b$ then, for all $x \in [0,a]$, $x^{\perp_a} = x^{\perp_b} \wedge a$.

We say that elements $a, b \in L$ are *orthogonal* if there is an element $c \geq a, b$ such that a and b are orthogonal in the orthomodular poset $[0, c]$ (i.e. if $a \leq b^{\perp_c}$).

If A is a C^*-algebra, then it is easy to verify that the projection structure $P(A)$ is a generalized orthomodular poset where a least element is the zero projection, the order is inherited from the self-adjoint part of A, and for $p \leq q$ in $P(A)$,

$$p^{\perp_q} = q - p\,.$$

Moreover, two projections p, q in A are orthogonal as elements of $P(A)$ exactly when they are orthogonal in the usual sense, i.e. if $pq = 0$. For this observe that if $pq = 0$, then $p+q$ is a projection and hence p and q are orthogonal in the interval $[0, p+q]$. Let us notice that the orthomodular posets are generalized orthomodular posets that have a largest element. We say that a generalized orthomodular poset L is a *substructure* of a generalized orthomodular poset K if every interval in L is a sublogic of the corresponding interval in K. The concepts of measure theory extend to generalized orthomodular posets canonically. Let L be a generalized orthomodular poset and X a normed space. We say that a map $\mu\colon L \to X$ is a *finitely additive measure* if μ is a finitely additive measure when restricted to each interval $[0, a]$, $a \in L$. Other attributes of measures (complete additivity, etc.) are defined analogously. A nonnegative measure μ on L is called a *state* (or probability measure) if $\sup_{a \in L} \mu(a) = 1$. If, in addition, L has the largest element, 1_L, then μ is a state if, and only if, $\mu(1_L) = 1$. We shall denote by $S(L)$ the set of all states on L and call it the *state space* of L. If L has the largest element, then the state space is a compact convex set when endowed with the topology of pointwise convergence. If A is a nonzero C^*-algebra, then $S(P(A))$ is a set of probability measures on $P(A)$. The generalized orthomodular poset L is said to be *unital* if for every nonzero $a \in L$ there is $\varrho \in S(L)$ such that $\varrho(a) = 1$. Further, we say that the state space $S(L)$ is *order determining* (resp. *separating*) if $s(a) \leq s(b)$ for each $s \in S(L)$ implies $a \leq b$ (resp. if $s(a) = s(b)$ for each $s \in S(L)$ implies that $a = b$). Of course, every order determining state space is separating. In our basic example $P(A)$, where A is a C^*-algebra (not required to be unital), the state space is always order determining, which is a consequence of the fact that (linear) states determine the order on A. In general, the state space $S(L)$ need not be either compact or convex. For that reason we introduce the concept of quasi-state space. By a *quasi-state space*, $Q(L)$, we shall mean the set of all nonnegative measures on L with values in the interval $[0,1]$. The set $Q(L)$, equipped with the topology of pointwise convergence, is a nonempty convex set. Let us note that if A is a von Neumann algebra without Type I_2 direct summand, then the state spaces $S(P(A))$ and $Q(P(A))$ are affinely homeomorphic to the (compact) state space and quasi-state space of A, respectively, (the Generalized Gleason Theorem). In order to use the powerful apparatus of functional analysis we shall consider the following linear structure associated with a generalized orthomodular poset, L. Let $A(L)$ be the set of all bounded affine real-valued functions on $Q(L)$ vanishing at 0. (By 0 we mean a zero measure.) Endowed with the usual arithmetic operations, supremum norm, and pointwise order, $(A(L), \leq)$ becomes an ordered Banach space. Let u_L be the function on $Q(L)$ defined by

$$u_L(s) = \sup_{a \in L} s(a)\,.$$

Then u_L is an affine function on condition that L is an upwards directed poset, i.e. if for all $a, b \in L$ there exist $c \in L$ such that $c \geq a, b$. (This is of course satisfied if L is a lattice or a quantum logic.) Indeed, suppose that $s_1, s_2 \in Q(L)$ and take elements $a_1, a_2 \in L$ such that

$$s_1(a_1) > u_L(s_1) - \varepsilon/2\,, \qquad s_2(a_2) > u_L(s_2) - \varepsilon/2\,.$$

For an element $a \geq a_1, a_2$ we have

$$s_1(a) > u_L(s_1) - \varepsilon/2 \qquad s_2(a) > u_L(s_2) - \varepsilon/2\,.$$

Now, for any convex combination $\alpha\, s_1 + (1-\alpha)\, s_2$, $\alpha \in [0,1]$, we see that

$$u_L(\alpha\, s_1 + (1-\alpha)\, s_2) > \alpha\, s_1(a) + (1-\alpha)\, s_2(a) > \alpha\, u_L(s_1) + (1-\alpha)\, u_L(s_2) - \varepsilon\,.$$

In other words

$$u_L(\alpha\, s_1 + (1-\alpha)\, s_2) \geq \alpha\, u_L(s_1) + (1-\alpha)\, u_L(s_2)\,.$$

Since the reverse inequality is obvious, we infer that $u_L(\alpha\, s_1 + (1-\alpha)\, s_2) = \alpha\, u_L(s_1) + (1-\alpha)\, u_L(s_2)$. It is easy to verify that if u_L is an affine function, then $(A(L), \|\cdot\|, \leq, u_L)$ becomes an order unit Banach space (see Chapter 2,

Ordered Banach spaces). The structure L can be mapped into $A(L)$ via evaluation mapping, e_L, given by

$$[e_L(a)](s) = s(a) \qquad \text{for all } s \in Q(L) \text{ and } a \in L\,.$$

Then, of course, $e_L(a \vee b) = e_L(a) + e_L(b)$ whenever $a \perp b$. It means that $e_L\colon\ L \to A(L)$ is a finitely additive vector measure. If L has an order determining state space, then e_L is injective and order preserving in both directions.

Let A be a C^*-algebra. We say hat A has the *Gleason property* if every state on $P(A)$ extends to a state on A. Suppose that A is the C^*-algebra with the Gleason property and such that A is a closed linear span of its projections. (This is of course the case of von Neumann algebras.) Then every state on $P(A)$ extends *uniquely* to a state on A. In this case $A(L)$ can be identified with the self-adjoint part of the double dual A^{**} of A. Moreover, the canonical embedding $e_{P(A)}$ of $P(A)$ into $A(L)$ is given by the restriction of the canonical embedding of the Banach space A into its second dual (see Chapter 2, State spaces of C^*-algebras). The Gleason property for generalized orthomodular posets will be formulated in the following way. The generalized orthomodular poset L, for which u_L is an affine function, is said to have the *Gleason property* if every state φ on L is of the form

$$\varphi = f \circ e_L\,, \tag{9.5}$$

where f is a state on the order unit space $(A(L), \leq, u_L)$. For any C^*-algebra enjoying the Gleason property whose projection structure spans densely A the projection structure $P(A)$ has the Gleason property. Indeed, if A is a C^*-algebra with the properties stated above, then there is state f_1 on A extending the given state φ on $P(A)$. Extending f_1 to a state f of A^{**} in a standard way we obtain immediately that φ is of the required form. On the other hand, it can be demonstrated easily that also the converse implication holds. We now pass to the key extension property considered in this section. We say that a generalized orthomodular poset L has the *universal state extension property* if for any unital generalized orthomodular poset K containing L as a substructure and for any state φ on L there is a state ψ of K extending L. Loosely speaking, generalized orthomodular posets with the state extension properties are perfectly compatible with respect to any enlargement. All classical orthomodular structures, i.e. Boolean algebras, enjoy the universal state extension property as the following proposition shows. We say that a generalized orthomodular poset is Boolean if it is a lattice and any interval in it is a Boolean algebra.

9.2.2. Proposition. *Any Boolean generalized orthomodular poset B has the universal state extension property.*

Proof: Suppose that K is a unital generalized orthomodular poset containing B as a substructure. Let φ be a state on B. As an initial step we show that for a given nonempty finite set $F \subset B$ there is a state ψ on K such that $\varphi|F = \psi|F$. We may suppose that $\sup F \neq 0$. Using the standard argument we can find a finite sequence $a_1, \ldots, a_n$ of orthogonal nonzero elements in B such that every element of F is a supremum of some elements from the set $\{a_1, \ldots, a_n\}$. It is therefore sufficient to find a measure $\mu \in Q(K)$ with $\psi(a_i) = \varphi(a_i)$ for all $i = 1, 2, \ldots, n$. By the hypothesis, for each i there is a state ψ_i on K with $\psi_i(a_i) = 1$. Then, of course, $\psi_i(a_j) = 0$ whenever $i \neq j$. Let us now put

$$\psi = \psi_1(a_1)\,\psi_1 + \psi_2(a_2)\,\psi_2 + \cdots + \psi_n(a_n)\,\psi_n\,.$$

As $0 \leq \psi(a_1) + \cdots + \psi_n(a_n) \leq 1$, we see that $\psi \in Q(K)$. Moreover, $\psi|\{a_1, \ldots, a_n\} = \varphi|\{a_1, \ldots, a_n\}$.

For each finite nonempty subset $F \subset B$ we define a set $P_F \subset Q(K)$ consisting of all measures on K that coincide with φ on F. By the previous argument $P_F \neq \emptyset$. It is also clear that F is closed in the pointwise convergence topology. Denote by $\mathcal{F}$ the set of all finite nonempty subsets of L and consider a finite collection $F_1, \ldots, F_m \in \mathcal{F}$. Again by the standard boolean argument one can show that $F_1, \ldots, F_m$ are contained in some finite subset F of L. Since $P_F \neq \emptyset$, we infer that $P_{F_1} \cap P_{F_2} \cap \cdots \cap P_{F_m} \supset P_{\bigcup_{i=1}^m F_i} \neq \emptyset$. In other words, the system of closed subsets $(P_F)_{F\in\mathcal{F}}$ has the finite intersection property and hence the intersection $\bigcap_{F\in\mathcal{F}} P_F$ is nonvoid by the compactness of $Q(K)$. Take an element $\psi \in \bigcap_{F\in\mathcal{F}} P_F$. Then, obviously, ψ extends φ. Moreover, since φ is a state, the same must hold for ψ. □

The next result shows that the Gleason property of C^*-algebras with sufficiently many projections is strong enough to imply the universal state extension property of their projection structures.

9.2.3. Theorem. *Let A be a C^*-algebra such that every nonnegative element x of A is contained in the closed linear span of some family of mutually commuting projections. Then A has the Gleason property if, and only if, $P(A)$ has the universal state extension property.*

Proof: Assume that A has the Gleason property and that L is a generalized orthomodular poset with unital set of states containing $P(A)$ as a substructure. Given a state φ of A we will seek an element of $Q(L)$ extending $\varphi|P(A)$.

Substantial part of the proof consists in proving that $\varphi|P(M)$ extends to an element of $Q(L)$ provided that M is a separable subalgebra of A. At first stage we will prove it under assumption that $\varphi|M$ is a pure state. According to Corollary 9.1.4 there is a nonnegative determining element $x \in M$ for the state $\varphi|M$. Let $B \subset M$ be an abelian C^*-subalgebra of M which contains x in the norm closed linear span of its projections. Since $P(B)$ is a Boolean generalized orthomodular poset, $P(B)$ enjoys the universal state extension property and $\varphi|P(B)$ extends to $\bar{\varphi} \in Q(L)$. Let $\hat{\varphi}$ be a linear extension of $\bar{\varphi}|P(A)$ to A. Then $\hat{\varphi}(x) = \varphi(x)$ and, in turn, $\hat{\varphi} = \varphi$ on M. Therefore $\bar{\varphi}$ is an extension of $\varphi|P(M)$.

Let us now consider more general case of $\varphi|M$ being an arbitrary element of $Q(M)$. Since every extreme point of the compact convex set $Q(M)$ is either a pure state of M or zero, the Krein-Milman theorem says that $\varphi|M$ is a weak*-limit point of a net (ψ_α), where each ψ_α is a linear combination of pure states of M whose coefficients are from the interval [0,1]. Every state $\psi_\alpha|P(M)$ extends to a state $\tilde{\psi}_\alpha$ of L. Employing now compactness of $Q(L)$ we may assume (passing to a subnet if necessary) that $\tilde{\psi}_\alpha \to \tilde{\psi}$. Then $\tilde{\psi}$ is the desired extension.

Let now $p_1, \dots, p_n$ be projections in A. Considering the C^*-algebra $C^*(p_1, \dots, p_n)$ generated by elements $p_1, \dots, p_n$, we can extend $\varphi|P(C^*(p_1, \dots, p_n))$ to a measure in $Q(L)$. In other words, for each finite subset $S \subset P(A)$, the set $F_S = \{\varrho \in Q(L) \mid \varrho|S = \varphi|S\}$ is a nonempty (compact) set in $Q(L)$. As $F_{\cup S_i} = \bigcap F_{S_i}$ for any finite collection $S_1, \dots, S_n$ of finite subsets of $P(A)$, it implies immediately that the system $(F_S)_{S \in \Gamma}$, where Γ is a collection of all finite nonempty subsets of $P(A)$, has the finite intersection property. The set $Q(L)$ being compact, the intersection $\bigcap_{S \in \Gamma} F_S$ has to be non-void, and any element $\tilde{\varphi} \in \bigcap_{S \in \Gamma} F_S$ serves as a state of L extending $\varphi|P(A)$. The proof is complete. □

As a corollary we obtain the main result telling us that nearly all von Neumann projection lattices possess the universal state extension property.

9.2.4. Corollary. *Let I be a norm closed ideal in a von Neumann algebra M. Suppose that M does not contain type I_2 direct summand. Then the quotient structure $P(M/I)$ enjoys the universal state extension property.*

Proof: By Chapter 5, M has the Gleason property and so does the quotient M/I. Let us take a self-adjoint $x \in M/I$ and denote by $\pi : M \to M/I$ the canonical quotient map. Then $x = \pi(y)$ for some self-adjoint $y \in M$. Let B be an abelian von Neumann subalgebra of M generated by y. Then $\pi(B)$ contains x and is a normed closed linear span of its projections. Thus the quotient M/I satisfies all assumptions of Theorem 9.2.3 and the proof is complete. □

So far we have obtained that any positive measure on the projection lattice of von Neumann algebra M with no Type I_2 direct summand extends to a positive measure on arbitrarily larger orthomodular structure. (Let us recall that we require only that the projection lattice lives in some subinterval of the superstructure.) As any measure on the projections in a C^*-algebra that extends to a bounded linear functional is a combination of four positive measures, we see that the Gleason property guarantees also the possibility to extend universally every bounded complex measure on the projections. In the subsequent development we show that the universal state extension property entails even the existence of extensions of vector measures with values in vector lattices and that, moreover, these extensions can be induced by linear functionals on the linear structures associated with orthomodular posets.

9.2.5. Lemma. *Let L be a generalized orthomodular poset with the universal state extension property contained as a substructure in a generalized orthomodular poset K. Suppose that u_L and u_K are ordered units of $A(L)$ and $A(K)$, respectively. Then there is a linear isometry, i, of $A(L)$ into $A(K)$ preserving order in both directions.*

Proof: Let $r\colon\ Q(K) \to Q(L)$ be the natural restriction mapping, i.e. $r(s) = s|L$ for every $s \in Q(K)$. By the universal state extension property, the mapping r is a surjective affine mapping. Let $f \in A(L)$ and let us define a function $i(f)$ on $Q(K)$ by the formula

$$i(f) = f(r(s)), \qquad s \in Q(K).$$

Of course, $i(f)$ is an affine function. Moreover,

$$\sup_{s\in Q(K)} |i(f)(s)| = \sup_{s\in Q(K)} |f(r(s))| = \sup_{t\in Q(K)} |f(t)| = \|f\|.$$

Thus, $i(f) \in A(K)$ and $\|i(f)\| = \|f\|$. Therefore the mapping i is a linear isometry. Since $i(f) \geq 0$ if, and only if, $f \geq 0$, i preserves the order. □

Let us recall that an order unit vector space $(S, 1_S)$ is called a *complete vector lattice* if it is *boundedly complete* meaning that every increasing net in S that is upper bounded has a supremum. Every complete vector lattice is isomorphic to the algebra $C(X)$ of continuous real functions on an extremely disconnected compact Hausdorff space X. The algebra $C(X)$ specified above is called the *Stone algebra.* According to Nachbin [242] any complete vector lattice has the following universal extension property in the category of normed spaces. Let X be a normed space. We say that X has the *universal extension property* if the following holds: Suppose that E is a closed subspace of a normed space F and that $T\colon E \to X$ is a bounded linear map. Then there is a bounded linear map $\bar{T}\colon\ F \to X$ extending T and having the same norm as T. On the other hand, it was proved in [242] that every normed vector space with the universal state extension property that has an extreme point e of its unit sphere can be organized into a vector lattice with the order unit e. Now we want to prove the result saying that Stone algebra valued positive map on a von Neumann algebra always extends to a positive map on any larger orthomodular structure. We say that such a measure is positive if it has values in the positive cone.

9.2.6. Theorem. *Let M be a quotient of a von Neumann algebra W, where W does not contain a Type I_2 direct summand. Let μ be a positive S-valued measure on $P(M)$, where $(S, \leq, 1_S)$ is a complete vector lattice such that $\mu(1_M) = 1_S$. Then μ extends to a positive S-valued measure on any larger orthomodular lattice K with a unital set of states containing $P(M)$ as a substructure.*

Proof: According to Corollary 9.2.4 the projection structure $P(M)$ enjoys the universal state extension property and so the mapping $i\colon A(P(M)) \to A(K)$ defined in the proof of Lemma 9.2.5 is an isometric embedding in the category of ordered vector structures. Denote by (V, v) the order unit space $(i(A(P(M)), \leq, v)$, where $v = (i \circ e_{P(M)})(1_M)$. By the Gleason property of M we can identify $A(P(M))$ with the self-adjoint part of the second dual M^{**}. Moreover, employing the Generalized Gleason Theorem, we can extend the S-valued measure μ to a bounded linear functional $\hat{\mu}\colon M \to S$. We show that $\hat{\mu}$ is positive. Suppose on the contrary that $\hat{\mu}(x)$ is not positive for some $0 \leq x \leq 1$. In that case there is a (pure) state φ on S such that $\varphi(\hat{\mu}(x)) < 0$. By the spectral theorem, there exist orthogonal projections $p_1, \ldots, p_n$ in M such that $\varphi\Big(\sum_{i=1}^n \lambda_i \, \hat{\mu}(p_i)\Big) < 0$ for some real $\lambda_1, \ldots, \lambda_n \geq 0$. Immediately, $\varphi(\hat{\mu}(p_i)) < 0$ for some i, and hence we have reached a contradiction. Therefore, $\|\hat{\mu}\| = 1$ (see Chapter 2, Ordered Banach spaces). By the extension property of complete vector lattices [242] we can further extend $\hat{\mu}$ to a linear mapping $\tilde{\mu}\colon M^{**}_{sa} \to S$ with $\|\tilde{\mu}\| = \|\hat{\mu}\| = 1$. Since $\tilde{\mu}(1_M) = 1_S$, we see that $\tilde{\mu}$ is positive. In summary, μ induces a positive linear mapping $F\colon V \to S$ such that

$$\mu = F \circ i \circ e_{P(M)} \,.$$

Since K is a lattice, the function 1_K is an order unit for $A(K)$. We can define a linear mapping $\overline{F}\colon V + \mathbb{R}1_K \to S$ extending F and satisfying $\overline{F}(1_K) = 1_S$. Let us show that $\overline{F}$ is also positive. Suppose that $x + \lambda\, 1_K \geq 0$ for $x \in V$, and $\lambda \in \mathbb{R}$. Take a state s of the lattice K satisfying $v(s) = 1$. Then

$$(x + \lambda\, 1_K)(s) = x(s) + \lambda \geq 0$$

and, in turn,

$$(x + \lambda\, v)(s) = x(s) + \lambda \geq 0 \,.$$

Thus the preimage $i^{-1}(x + \lambda\, v)$ is positive and so is $x + \lambda\, v$. Hence, due to positivity of F,

$$\overline{F}(x + \lambda\, 1_K) = F(x) + \lambda\, 1_S = F(x + \lambda\, v) \geq 0 \,.$$

It implies that $\|\overline{F}\| = 1$. Using the extension property of the complete vector lattice $(S, 1_S)$, we can further extend $\overline{F}$ to a continuous linear mapping $\tilde{F}\colon A(K) \to S$ such that $\|\tilde{F}\| = \|\overline{F}\| = 1$. Since $\tilde{F}(1_K) = \overline{F}(1_K) = 1_S$, we see that $\tilde{F}$ is positive. It is now clear that the S-valued function $\tilde{\mu}$ on K given by

$$\tilde{\mu} = \tilde{F} \circ e_K$$

is an S-valued measure. Using the fact that $e_K|P(M) = i \circ e_{P(M)}$, we obtain, for $p \in P(M)$,

$$\tilde{\mu}(p) = (\tilde{F} \circ i \circ e_{P(M)})(p) = (F \circ i \circ e_{P(M)})(p) = \mu(p)\,.$$

Hence $\tilde{\mu}$ is the desired extension. □

The technique of the previous proof gives us that the universal state extension property implies in fact the existence of "linear extensions" as it is formulated for states on von Neumann algebras in the following corollary.

9.2.7. Corollary. *Let M be a von Neumann algebra without Type I_2 direct summand and let K be a generalized orthomodular poset containing $P(M)$ as a substructure. Then any positive measure μ on $P(M)$ is of the form*

$$\mu = f \circ e_K\,|P(M)\,,$$

where f is a positive functional on $A(K)$.

Let us recall that a quantum logic L with the separating state space has the Gleason property if every state μ on L is of the form $\mu = f \circ e_L$, where f is a state of the order unit space $(A(L), u_L)$. In general, if L has the Gleason property and the universal state extension property, then every possible extension is given by the (linear) state on the corresponding order unit space. However, not many logics with the Gleason property different from Boolean algebras and projection lattices of von Neumann algebra are known.

9.3. Historical remarks and comments

The multiplicativity properties of pure states on maximal abelian subalgebras have been studied first by Kadison and Singer, and by Stormer [203, 299], who proved that a state φ on a C^*-algebra A is pure exactly when its definite set $D_\varphi = \{x \in A_{sa} \mid \varphi(x^2) = \varphi(x)^2\}$ is maximal in the set of all definite sets corresponding to states that is ordered by the set inclusion. Aarnes and Kadison showed in [2] that a pure state on a separable unital C^*-algebra is multiplicative on some maximal abelian subalgebra. Akemann then extended this result by proving Corollary 9.1.5 [11]. Then it was Barnes [30] who realized that essential for establishing multiplicativity on some maximal abelian subalgebra is the separability of the G.N.S. representation. He showed that a pure state ϱ on a C^*-algebra such that its G.N.S. representation π_ϱ acts on a separable Hilbert space is multiplicative on some maximal abelian subalgebra. He also proved, independently of Bunce [49], that a pure state on Type I C^*-algebra restricts to a pure state on some maximal abelian subalgebra. Theorem 9.1.8 generalizes simultaneously a result of Akemann and Barnes and has been established by Hamhalter [181, 179, 176]. Results concerning the determinacy of pure states that extend properties of pure states on abelian algebras in another direction (Proposition 9.1.2, Theorem 9.1.3, Corollary 9.1.4) are due to Hamhalter [178, 180, 179]. In the works [178, 180, 179] all results formulated here for C^*-algebras are proved in a more general context of JB algebras. Let us remark that in this case some parts of the proof are more complex because one has to overcome the lack of the representation of a JB algebra on a Hilbert space as well as its nonassociativity. Theorem 9.1.8 is a very special variant of the following result proved in [181] concerning a sequence of infinitely many orthogonal pure states approaching infinity. We say that a sequence (ϱ_n) of orthogonal pure states on a JB algebra A *approaches infinity* if $\lim_{n\to\infty} \varrho_n(a) = 0$ for all $a \in A_{sa}$ such that the spectrum of a contains zero. Orthogonal pure states on an abelian C^*-algebra $C_0(\mathbb{R})$ approach infinity precisely when the corresponding sequence of points in X converges to infinity.

9.3.1. Theorem. *Let (ϱ_n) be a sequence of orthogonal pure states on a JB algebra A approaching infinity and such that $c(\varrho_n)\, A^{**}$ is σ-finite for all n. Suppose that $\sum_{n=m}^{\infty} s(\varrho_n)$ is a closed projection for all m. Then there exists a maximal associative subalgebra B of A such that each state ϱ_n restricts to a pure state on B.*

The systems of orthogonal pure states approaching infinity have been extensively studied by Akemann, Anderson and Pedersen in [13, 14]. In an important paper [14] many results on the orthogonal pure states approaching infinity on separable algebras may be found. Among others, it is proved there that $\sum_{n=m}^{\infty} s(\varrho_n)$ is closed, whenever (ϱ_n) is a sequence of orthogonal pure states on a separable C^*-algebra that are nearly inequivalent in the sense that the cardinalities of equivalent classes of (ϱ_n) are uniformly bounded by some integer. Combined with Theorem 9.3.1 it implies that any sequence of orthogonal pure states on a separable C^*-algebra that are nearly inequivalent and approaching infinity restrict to a sequence of pure states on some maximal abelian subalgebra. The Example 9.1.9 showing limitation to possible improvements of these results has been obtained by Hamhalter [181].

The results concerning restriction properties will be applied in studying independence of operator algebras in Chapter 11. We want now to show partial multiplicativity of extremal completely positive maps that represent another possible line of the research in the future. We first recall a few concepts. From now on we shall suppose that all C^*-algebras are unital. Let A be a C^*-algebra. By the symbol $M_n(A)$ we continue to denote the C^*algebra of all $n \times n$ matrices over A. A linear map $\varphi\colon\ A \to B(H)$ is called *completely positive* if, for each n, the map $\varphi^{(n)}:\ M_n(A) \to M_n(B(H))$ given by

$$\varphi^{(n)}\left((a_{ij})\right)_{ij} = \left(\varphi(a_{ij})\right)_{ij}$$

is positive. According to the fundamental Stinespring Theorem [298] any completely positive map is similar to a representation π of A on a Hilbert space K in the sense of the following equality

$$\varphi(a) = V^* \pi(a)\,, \qquad a \in A\,.$$

where V is a bounded operator of H into K and $[\pi(A)V(H)] = K$. (The symbol $[X]$ denotes the norm closed linear span of the set X). Moreover, a completely positive map is called *unital* if it preserves the unit. In that case the Stinespring's decomposition simplifies to V being an isometric embedding of H into K. So we can (and shall) identify H with the subspace of K and write

$$\varphi(a) = P\pi(a)P|H,$$

where P is the orthogonal projection of K onto H. Then $[\pi(A)H] = K$. There are a few generalizations of the concept of pure state to the context of completely positive maps. At first, a completely positive map $\varphi : A \to B(H)$ is called *pure* if the only completely positive map ψ for which $\psi \leq \varphi$ is a multiple of φ. (We use the notation $\psi \leq \varphi$ if $\varphi - \psi$ is completely positive.) According to Arveson Theorem, a completely positive map φ is pure if, and only if, its representation π in the Stinespring's decomposition is irreducible [28, Cor. 1.4.3]. A unital completely positive map φ is called *C^*-extreme* if the following condition holds: If $\varphi = \sum_{i=1}^{n} t_i^* \varphi_i t_i$, where t_i's are invertible operators in $B(H)$, $\sum_{i=1}^{n} t_i^* t_i = 1$, and φ_i's are completely positive unital maps of A into $B(H)$, then all maps φ_i's are unitarily equivalent to φ. Any pure completely positive unital map is C^*-extreme. The converse does not hold. Recently, it was shown by Farenick and Zhou [133] that if $\dim H < \infty$ then every completely positive C^* extreme map $\varphi\colon A \to B(H)$ is a special direct sum of pure completely positive maps. More precisely, a sequence $\varphi_1^\pi, \varphi_2^\pi, \dots, \varphi_n^\pi$ of unital completely positive maps, each mapping A into $B(K_i)$, is called a *nested sequence corresponding to the representation* $\pi : A \to B(K_0)$ if

$$\varphi_1^\pi = w_1^* \, \pi \, w_1 \qquad \text{and} \qquad \varphi_i^\pi = w_i^* \, \varphi_{i-1}^\pi \, w_i \text{ for } i \geq 2 ,$$

where each w_i is an isometric embedding of K_i into K_{i-1}.

9.3.2. Theorem. *A completely positive map $\varphi\colon A \to B(H)$, $\dim H < \infty$, is C^*-extreme if, and only if, there is a sequence of irreducible inequivalent representations $\pi_1, \pi_2, \dots, \pi_k$ of A such that φ is unitarily equivalent to the direct sum*

$$\sum_{i=1}^{k} \oplus \left(\sum_{j=1}^{n_i} \oplus \, \varphi_j^{\pi_i} \right) , \tag{9.6}$$

where each sequence $\varphi_1^{\pi_i}, \dots, \varphi_{n_i}^{\pi_i}$ is a nested sequence of completely positive maps corresponding to π_i.

We shall call a completely positive map φ separable if the representation in its Stinespring's decomposition acts on a separable Hilbert space.

As opposed to pure states, pure completely positive map on a separable algebra need not restrict to pure completely positive map on any maximal abelian subalgebra. Indeed, let us consider an irreducible C^*-algebra A acting on a Hilbert space H with dim $H > 1$. Since the identity map π is an irreducible representation of A, π is a pure map. Any maximal abelian subalgebra B of A is of dimension at least two. Thus the restriction of π to B cannot be a pure map because any pure map on an abelian C^*-algebra has to be a complex-valued homomorphism. Nevertheless, π is obviously multiplicative on B. So an appropriate formulation of the restriction property for completely positive maps is the property of being multiplicative on some maximal abelian subalgebra. The following theorem proved by Hamhalter in [181] shows that any C^*-extreme, finite dimensional, completely positive map enjoys this property.

9.3.3. Theorem. *Let $\varphi\colon A \to B(H)$ be a C^*-extreme, completely positive and separable map, where A is a unital C^*-algebra and dim $H < \infty$. Then there is a maximal abelian subalgebra B of A such that $\varphi|B$ is multiplicative.*

Proof: Any C^*-extreme completely positive map with values in a finite-dimensional Hilbert space is unitarily equivalent to a sequence of compressions of irreducible representations. So we can assume that φ is of the form (9.6). Suppose that each irreducible representations π_i acts on a Hilbert space H_i. Under obvious identification we can find, for each i, a decreasing sequence of non-zero finite dimensional orthogonal projections $P_1^i \supseteq \ldots \supseteq P_{n_i}^i$ in $B(H_i)$ such that

$$\varphi_j^{\pi_i}(a) = P_j^i \pi_i(a) P_j^i | P_j(H_i) \quad i = 1, \ldots, k; \; j = 1, \ldots n_i. \tag{9.7}$$

Now, let us pick an orthonormal basis $(x_n)_{n=1}^l$ of finite dimensional Hilbert space

$$K = P_1^1(H_1) \oplus P_1^2(H_2) \oplus \cdots \oplus P_1^k(H_k)$$

such that each Hilbert space $P_j^i(H_i)$ (viewed as a subspace of K) has an orthonormal basis which is a subsequence of $(x_n)_{n=1}^l$. We define states $\varrho_1, \ldots, \varrho_l$ by setting

$$\varrho_m(a) = (\pi_s(a) x_m, x_m) \quad (a \in A),$$

where s is such that $x_m \in P_1^s(H_s)$. The system $\varrho_1, \ldots, \varrho_l$ is a finite family of mutually orthogonal pure states on A. Employing Theorem 9.1.8 we can find a maximal abelian subalgebra B of A such that the states $\varrho_1, \varrho_2, \ldots, \varrho_l$ restrict to pure states on B. We show that each direct summand in (9.6) is multiplicative on B. For this it suffices to consider the case $i = 1$. Fix $1 \le j \le n_1$. By construction, there is an orthonormal basis $\xi_1, \ldots, \xi_r$ of $P_j^1(H_1)$ such that the corresponding sequence of vector states $\psi_1, \ldots, \psi_r$ is a subsequence of $\varrho_1, \ldots, \varrho_l$. By the uniqueness of the G.N.S. representation the spaces $[\pi_1(B)\xi_h]$, where $h = 1, 2 \ldots, r$, have to be one-dimensional. Taking into account that $\pi_1(1)\xi_h = \xi_h$ we see that

$$[\pi_1(B)\xi_h] = [\xi_h].$$

Thus, $\pi_1(b)\xi_h = \psi_h(b)\xi_h$ for all $b \in B$. Fix now $b \in B$ and $z \in P_j^1(H_1)$. Then

$$\begin{aligned}
\varphi_j^{\pi_1}(b)\, z &= P_j^1\, \pi_1(b)\, P_j^1 z = P_j^1\, \pi_1(b)\, z = \sum_{h=1}^{r} (\pi_1(b)\, z, \xi_h)\, \xi_h \\
&= \sum_{h=1}^{r} (z, \pi_1(b^*)\xi_h)\, \xi_h = \sum_{h=1}^{r} (z, \psi_h(b^*)\, \xi_h)\, \xi_h \\
&= \sum_{h=1}^{r} \psi_h(b)\, (z, \xi_h)\, \xi_h.
\end{aligned}$$

In other words,

$$\varphi_j^{\pi_1}(b) = \sum_{h=1}^{r} \psi_h(b)\, P_{\xi_h}, \tag{9.8}$$

where P_{ξ_h} is the orthogonal projection on $[\xi_h]$. Therefore $\varphi_j^{\pi_1}$ is multiplicative on B. The proof is complete. □

Let us remark that completely positive maps play an important role in quantum information theory. Besides, the set of all completely positive map, called the *generalized state space*, and its operator convexity structure is related to the problem of quantization. For this reason the research of the extreme completely positive maps and their multiplicativity properties seems to be interesting.

Regarding the restriction properties, let us recall that the following question complementary to multiplicativity of pure states on maximal abelian subalgebras has been studied. What maximal abelian subalgebras in C^*-algebras have the *extension property* meaning that any pure state on a given maximal abelian subalgebra extends uniquely to a pure state on the whole algebra [203, 23, 24]. For example, it was shown in the classical paper by Kadison and Singer [203] that a maximal abelian subalgebra in $B(H)$ where H is separable that is isomorphic to $L^\infty(0,1)$ does not have the extension property. On the other hand, despite all quite deep partial results [23, 24, 26, 15] it is not known whether "discrete" maximal abelian subalgebra in $B(H)$ that is generated by atomic projections enjoys the extension property.

Very interesting line of the recent research is centred around the concept of *pure extension property.* A C^*-subalgebra B of a C^*-algebra A is said to have the *pure extension property* if every pure state on B extends *uniquely* to a pure state on A. Remarkable results on algebras of this type have been obtained by Bunce, Cho-Ho-Chu, Archbold and others (see e.g. [27, 68, 67]). It was shown among others in [68] that if B is a separable subalgebra of a von Neumann algebra A, then the pure extension property of B is equivalent to the fact that B is isomorphic to algebra of compact operators and minimal projections in B are minimal in in A. This is connected with the determinacy of pure states [180] which says that a pure state on a von Neumann algebra has a determining element if, and only if, it is normal. Thus, if B is a separable C^*-subalgebra of a von Neumann algebra A having the pure extension property, then by Corollary 9.1.5 any pure state ϱ on B has a determining element. Consequently, as ϱ extends uniquely to a pure state of A its extension has to be normal. In other words, all pure states on B are restrictions of pure states on A which is a typical property of the algebras of compact operators.

The material in Section 9.2 dealing with the Generalized Gleason Theorem is based on the papers by Hamhalter [173, 179, 177]. It has been inspired by the classical result of Horn and Tarski [193] to the effect that any probability measure on a Boolean algebra B extends to a probability measure on any larger Boolean algebra B' containing B. The first "quantum measure theoretic" version of this result was given by Pták [270] who showed that B' can be replaced by any quantum logic with the unital set of states. Pták is credited for introducing the concept of the universal state extension property and developing the extension technique based only on the compactness and convexity of the state space. He also established the universal state extension property of the finite-dimensional Hilbert space lattices.

In connection with Theorem 9.2.6, let us remark that according to result of Wright [323] precisely the Stone algebras are injective in the category of ordered unit spaces and unital positive maps. It indicates that Theorem 9.2.6 cannot be improved by taking measures more general than Stone algebra valued. The properties of generalized orthomodular posets should further be studied. For example, it is not known how close the generalized orthomodular posets with the Gleason property and the universal state extension property are to the projection lattices of JBW algebras.

10. JAUCH–PIRON STATES

In the first part of the book the extensions of classical principles of measure and probability theory to noncommutative projection structures were investigated. It was seen that basic tools of classical analysis can be established for the quantum measure spaces given by ordered structures of projections. One of the most essential achievements along this line is the Gleason Theorem that guarantees the existence of quantum integral and underlines the physical meaning of the basic quantum calculus. In the present chapter we focus on a completely new type of results that hold only in the presence of noncommutativity and thus demonstrate a considerable difference between classical and quantum measure theory. In general, it will be shown that topological properties of states can be derived from their algebraic lattice-theoretic properties, which is not the case for measures on boolean structures.

The central topic of this chapter is that of Jauch-Piron state. It was introduced by Jauch and Piron in the 60's in connection with the propositional calculus of quantum mechanics and the problem of hidden variables. Let μ be a state (i.e. a probability measure) on an orthomodular poset, L. We say that μ is a *Jauch-Piron state* if the following condition holds: If $\mu(a) = \mu(b) = 0$ for $a, b \in L$, then there is an element $c \in L$ such that $c \geq a, b$ and $\mu(c) = 0$. The physical motivation for introducing this concept is as follows. The elements of L are interpreted as propositions (yes-no observables or experiments) on the quantum system in a state μ. The value $\mu(a)$, $a \in L$, represents the probability that the proposition a is true. The Jauch-Piron property now means that there is a proposition, c, covering both a and b such that the probability of c being true is also zero. If L is a lattice, then one can simply take $c = a \vee b$. The proposition $a \vee b$ represents the proposition "a or b". In this case the Jauch-Piron condition rephrases in the following way. If the propositions a and b are false, then the proposition "a or b" is also false. From the probabilistic point of view, the Jauch-Piron property encodes the following largely adopted rule of Kolmogorovian probability theory:

If two events A and B have zero probability, then the probability of occurring A or B is also zero.

The introduction of Jauch-Piron states was motivated by the requirement that states on projection structures that qualify for having physical meaning should satisfy the natural Jauch-Piron property. This requirement is obviously fulfilled for all states on abelian algebras. However, as we shall see later, the states not having the Jauch-Piron property are ubiquitous in noncommutative algebras.

Another motivation for investigating Jauch-Piron states is purely mathematical. If ϱ is a state on a von Neumann algebra M, then its left kernel $L_\varrho = \{x \in M \mid \varrho(x^* x) = 0\}$ is a normed closed left ideal of M. The natural question arises of what are the properties of states ϱ for which $L_\varrho \cap P(M)$ is a lattice ideal in the projection lattice $P(M)$. Such states show a nice interplay with the projection structure. Let us recall that a set $I \subset P(M)$ is an ideal in $P(M)$ if, for all projection $p, q \in P(M)$, the following conditions hold true: (i) $p \vee q \in I$ whenever $p, q \in I$ (ii) if $p \leq q$ and $q \in I$, then $p \in I$. Since the condition (ii) is automatically satisfied for every set $L_\varrho \cap P(M)$ by the monotonicity of $\varrho|P(M)$, we see that it is exactly the Jauch-Piron states that generate lattice ideals in von Neumann projection lattices. So the concept of Jauch-Piron state has also natural mathematical motivation.

We will show that the "discrete" Jauch-Piron property is surprisingly strong to imply continuity properties of states. Our discussion of this noncommutative phenomenon will be divided into a few steps. In the first paragraph (Section 10.1) examples of states enjoying resp. lacking the Jauch-Piron property will be exhibited. We will introduce the concept of regularity that can be viewed as a countable form of the Jauch-Piron property. Formally, we say that a state ϱ on a von Neumann algebra M is *regular* if the kernel of ϱ is closed under forming countable suprema of projections. Regularity is an idealized form of the Jauch-Piron property It is adopted for technical reasons because the infinite operations have no direct physical meaning. It is therefore desirable to know the extent of the equivalence between regularity and the Jauch-Piron property and the extent of their disparity when it occurs. This question will be addressed in Section 10.2 where we conduct a through investigation of the properties of general Jauch-Piron states. It turns out, surprisingly, that the relation between the Jauch-Piron property, regularity, and nonsingularity is very dependent upon the Type of von Neumann algebra and the size of its centre. Among others, we show that all Jauch-Piron states on factors or Type *III* σ-finite algebras are automatically regular. We also show that the regularity of Jauch-Piron states depends on the centres of finite hereditary von Neumann subalgebras.

In Section 10.3 we shall investigate the σ-additivity of important types of Jauch-Piron states. The most striking result in this part is the fact that pure Jauch-Piron states on all von Neumann algebras as well as Jauch-Piron factor states on properly infinite algebras have to be σ-additive. Moreover, the following characterization of σ-additivity explaining its physical content will be proved: A state on an essentially noncommutative von Neumann algebra (i.e. algebra having no nonzero abelian direct summand) is σ-additive if, and only if, all transforms of the state ϱ corresponding to time development of the system in irreversible dynamics are Jauch-Piron. In concluding section we discuss perspectives of the recent development and some of the open problems in this area.

In summary, the results on automatic continuity of Jauch-Piron states have both mathematical and physical consequences. From the point of view of axiomatic quantum theory we will see that topological and algebraic aspects of operator-algebraic quantum formalism are closely connected. Nothing like that holds for commutative algebras of functions. The requirement of σ-additivity of measures in classical Kolmogorovian probability model, which is necessary for one-sided continuity of the distribution function, is an indispensable axiom that cannot be derived from the order structure of events. In contrast to this, in "proper" quantum measure theory one need not postulate σ-additivity because it is already a consequence of natural algebraic axioms involving only finitely many operations. This is more realistic because a physical observation can only refer to finitely many propositions. In this regard the axiomatics of quantum theory may paradoxically seem to be more transparent and natural than that of classical physics.

10.1. Basic properties of Jauch-Piron states

In the present section we collect basic properties of Jauch-Piron states on von Neumann algebras. Besides, by exhibiting important examples of states enjoying resp. lacking the Jauch-Piron property, we indicate possible limitations of the interplay between continuity of functionals and their lattice-theoretic character. The first example shows that all tracial states are Jauch-Piron.

10.1.1. Example. Any tracial state on a von Neumann algebra is Jauch-Piron. In particular, any state on an abelian von Neumann algebra is Jauch-Piron.

Proof: Let ϱ be a tracial state on a von Neumann algebra M. Then for all projections $e, f \in M$ we have $e \vee f - f \sim e - e \wedge f$. Therefore, by the tracial property, $\varrho(e \vee f) = \varrho(e) + \varrho(f) - \varrho(e \wedge f)$, this gives us $\varrho(e \vee f) = 0$ whenever $\varrho(e) = \varrho(f) = 0$. □

Besides the tracial properties, the Jauch-Piron property is also implied by continuity or faithfulness of states.

10.1.2. Proposition. *Let ϱ be a state on a von Neumann algebra M.*

(i) *If ϱ is σ-additive, then ϱ is Jauch-Piron.*

(ii) *If ϱ has a support projection, then ϱ is Jauch-Piron.*

Proof: (i) If ϱ is σ-additive, then its kernel is closed under taking suprema of projections. Indeed, let $e, f \in P(M)$ with $\varrho(e) = \varrho(f) = 0$. Then $e \vee f = r(e + f)$ can be written as a supremum of an increasing sequence (p_n) of spectral projections of $e + f$ such that $p_n \nearrow e \vee f$ and $p_n(e + f) \geq \frac{1}{n} p_n$. Thus $\varrho(p_n) = 0$ for all n and hence, by the σ-additivity, $\varrho(e \vee f) = 0$.

(ii) It follows directly from the definition of Jauch-Piron property. □

Let us note that while any σ-additive state is Jauch-Piron, the converse is far from being true. Manifestly, this applies to infinite-dimensional abelian algebras, where all states are Jauch-Piron but not all of them are σ-additive or nonsingular. Nevertheless, the nonequivalence of σ-additivity and the Jauch-Piron property is typical for all infinite dimensional algebras. Indeed, let M be a σ-finite von Neumann algebra of infinite dimension. Then M admits a faithful normal state ϱ and a singular state φ. Set $\psi = \frac{1}{2}(\varphi + \varrho)$. Then ψ is faithful and therefore Jauch-Piron, giving an example of a Jauch-Piron state on a σ-finite algebra that is not σ-additive (normal). This example gives rise to the question of whether, at least, any Jauch-Piron state on a noncommutative algebra is always nonsingular. Even this is not true. Let us consider a finite von Neumann algebra M with the infinite-dimensional centre $Z(M)$. Then $Z(M)$ admits a singular state φ. Using the canonical trace $T\colon\ M \to Z(M)$, we can extend φ to a tracial state $\psi = \varphi \circ T$ that gives us an example of a singular Jauch-Piron state.

So far we have met more or less obvious examples of Jauch-Piron states. To obtain a state that is not Jauch-Piron is not as straightforward. However, let us recall that such states have already been constructed in Proposition 8.2.2. According to this construction any algebra $l^\infty \otimes M_n(\mathbb{C})$, $n \geq 2$, admits a (product) state not having the Jauch-Piron property. As a consequence we show that states that are not Jauch-Piron are ubiquitous in essentially noncommutative algebras.

10.1.3. Proposition. *Let M a von Neumann algebra. Then every state of M is Jauch-Piron if, and only if, $M = A \oplus B$, where A is abelian and B has finite dimension.*

Proof: If M has a finite-dimensional nonabelian part, then any algebra $W^*(e, f)$ generated by two projections e, f is finite-dimensional. Hence, any state on $W^*(e, f)$ is normal and therefore it is Jauch-Piron.

Conversely, if the nonabelian part of M has infinite dimension, we can find a von Neumann subalgebra W of M that is isomorphic to the sum of infinitely many copies of $M_2(\mathbb{C})$. Then, according to Proposition 8.2.2, W admits a state which is not Jauch-Piron, and so does M. ❑

Now we introduce the concept of regularity of states and explore its basic structure.

10.1.4. Definition. A nonnegative functional ϱ on a von Neumann algebra M is called *regular* if $\varrho(\bigvee_{n=1}^{\infty} e_n) = 0$ whenever (e_n) is a sequence of projections in M with $\varrho(e_n) = 0$ for all n.

Regularity can be viewed as the σ-Jauch-Piron property. The next proposition lists several conditions equivalent to regularity. Among others, it turns out that regularity is a weak form of σ-additivity in the sense that regular states are characterized as being σ-additive with respect to their zero projections. Also, the kernels of regular functionals can be recognized by their strong sequential closedness. Let us recall that, for a given a state ϱ, $L_\varrho = \{x \in M \mid \varrho(x^* x) = 0\}$ denotes its left kernel.

10.1.5. Proposition. *Let ϱ be a state on a von Neumann algebra, M. Then the following conditions are equivalent:*

(i) *ϱ is regular,*

(ii) *if (e_n) is a sequence of orthogonal projections and $\varrho(e_n) = 0$ for all n, then $\varrho\left(\sum_n e_n\right) = 0$,*

(iii) *if (e_n) is an increasing sequence of projections such that $e_n \nearrow e$ and $\varrho(e_n) = 0$ for all n, then $\varrho(e) = 0$,*

(iv) *if x is a nonzero element of M and $\varrho(x) = 0$, then $\varrho(r(x)) = 0$,*

(v) *the self-adjoint part of L_ϱ is sequentially closed in the strong operator topology.*

If M is σ-additive, then the above conditions are equivalent to

(vi) *ϱ has a support projection in M.*

Proof: The implications (i)$\Rightarrow$ (ii) $\Leftrightarrow$ (iii)$\Leftarrow$ (v) are immediate. In order to prove the implication (iii)$\Rightarrow$ (iv) we have to observe that if $x \geq 0$, then there is a sequence of spectral projections (e_n) in M such that $x \geq (\frac{1}{n})\, e_n$, for all n, and $e_n \nearrow r(x)$. Then, immediately, $\varrho(x) = 0$ implies $\varrho(e_n) = 0$ and condition (iii) yields $\varrho(r(x)) = 0$.

(iv)$\Rightarrow$ (i) Assume (iv) and suppose that (e_n) is a sequence of projections in M such that $\varrho(e_n) = 0$ for all n. Set $x = \sum_{n=1}^{\infty} \frac{1}{2^n}\, e_n$. Then $\varrho(x) = 0$ and $r(x) = \bigvee_{n=1}^{\infty} e_n$. For the latter equality we can notice that, obviously, $\bigvee_{n=1}^{\infty} e_n \geq r(x)$. On the other hand, $\frac{1}{2^n}\, e_n \leq x$ gives $r(\frac{1}{2^n}\, e_n) = e_n \leq r(x)$, and hence $\bigvee_{n=1}^{\infty} e_n \leq r(x)$.

Suppose now that any one of the equivalent conditions (i)-(iv) holds and let (x_n) be a sequence of self-adjoint elements of L_ϱ such that

$x_n \to x \in M$ in the strong operator topology. By the Uniform Boundedness Principle, the sequence (x_n) is bounded, and hence $y_n = x_n^2 \to x^2$ strongly. Further, for each n, we have that $\varrho(y_n) = 0$, and hence $\varrho(e) = 0$, where $e = \bigvee_{n=1}^{\infty} r(y_n)$. But $(1-e)\,y_n = 0$ for all n, and hence $(1-e)\,x^2 = 0$, which implies that $\varrho(x^2) = 0$, by the Cauchy-Schwarz inequality. Therefore, $x \in L_\varrho$, and the equivalence of conditions (i)-(v) is established.

Suppose now that M is σ-finite and ϱ is regular.
Set $p = \sup\{e \in P(M) \mid \varrho(e) = 0\}$. By the σ-finiteness, there is a sequence $(e_n) \subset P(M) \cap \operatorname{Ker} \varrho$ such that $p = \bigvee e_n$. The regularity of ϱ now implies that $\varrho(p) = 0$, proving that $(1 - p)$ is a support of ϱ. □

The following corollary says that regularity is a strong form of nonsingularity for σ-finite algebras.

10.1.6. Corollary. *Let ϱ be a positive functional on a von Neumann algebra M. The following conditions are equivalent:*

(i) *ϱ is regular,*

(ii) *ϱ is nonsingular on all σ-finite von Neumann subalgebras of M.*

If M is σ-finite, then the above conditions are equivalent to

(iii) *ϱ is nonsingular on $e\,M\,e$, for all projections e of M.*

Proof: (i)⇒ (ii) If ϱ is regular, then it has a support when restricted to any σ-finite subalgebra and has to be therefore nonsingular on all σ-finite subalgebras of M.

(ii)⇒ (i) Suppose that (ii) holds. Let (e_n) be an orthogonal sequence of projections in M with $\varrho(e_n) = 0$ for all n. Then $W^*(e_n) \simeq l^\infty$ is σ-finite and so $\varrho\left(\sum_n e_n\right) = 0$. By Proposition 10.1.5, ϱ is regular. The equivalence of the above conditions to the condition (iii) for σ-finite algebras follows from Proposition 10.1.5. □

In summary, we know that the following implications hold true:

$$\sigma\text{-additivity} \implies \text{regularity} \implies \text{Jauch-Piron property}$$

By the examples presented above, none of these implications can be reversed. In preparation for a more detailed analysis in the next section we expose now some of the limitations of the natural conjectures:

(A) The Jauch-Piron property implies regularity.

(B) The Jauch-Piron property implies nonsingularity.

A singular tracial state on a σ-finite von Neumann algebra is a Jauch-Piron state for which both (A) and (B) fail. In this case the nonsingularity is due to the misbehaviour of the state on the centre. But the next example shows that singularities of Jauch-Piron states are not confined strictly to the centre. It shows that also centres of hereditary subalgebras cannot be ignored in this regard.

10.1.7. Example. Let M be a semifinite and locally σ-finite von Neumann algebra without abelian part but with infinite-dimensional centre. Then M has a Jauch-Piron state that is both nonsingular and regular on $Z(M)$ but singular and nonregular on $Z(e\,M\,e)$ for some nonzero projection e in M for which $\varrho(e) \neq 0$.

Proof: It follows from the structure theory for locally σ-finite algebras that M has the form $N_1 \oplus (N_2 \overline{\otimes} B(H))$, where N_1 and N_2 are finite, one of which must possess a σ-finite direct summand with infinite-dimensional centre. So in order to establish the statement offered by example we can suppose that M is σ-finite and finite.

First we find a finite projection e in M such that for the following central supports we have

$$c(e) = c(1-e) = 1\,.$$

To show this suppose first that M is of Type I. Let us take an abelian projection e such that $c(e) = 1$. As M does not have any abelian direct summand, e is subequivalent to $1-e$. It gives $c(1-e) = 1$. If M is of Type II, we can choose e as being any finite projection that halves M. The general case follows from this.

For a central projection $z \in Z(M)$ we now have $c(z\,e) = z\,c(e) = z$. Therefore the equality $z\,e = 0$ implies $z = 0$, and hence $Z(M)$ is isomorphic to $e\,Z(M) = Z(e\,M\,e)$ via the map $z \to z\,e$. It yields that $e\,M\,e$ is a finite hereditary subalgebra with infinite dimensional centre. By the above example we can choose a Jauch-Piron state σ on $e\,M\,e$ that is both singular and nonregular on $Z(e\,M\,e)$. Let τ be any state with support equal to $1-e$. Now consider the state, ϱ, on M defined by

$$\varrho(x) = \frac{1}{2}\Big(\sigma(e\,x\,e) + \tau(x)\Big) \qquad x \in M\,.$$

We shall verify that ϱ has the required properties. Let p and q be projections in M with $\varrho(p) = \varrho(q) = 0$. Since τ is faithful on $(1-e)\,M\,(1-e)$ and $\tau(p) = \tau(q) = 0$, we infer that $(1-e)\,p\,(1-e) = (1-e)\,q\,(1-e) = 0$, implying that $(1-e)\,p = (1-e)\,q = 0$. Thus, $p, q \le e$ and, by the Jauch-Piron property of σ, $\sigma(p \vee q) = 0$. Hence, $\varrho(p \vee q) = 0$ and ϱ is Jauch-Piron.

Further, ϱ is faithful (hence regular) on $Z(M)$. Indeed, if $\varrho(z) = 0$ where $z \in Z(M)$, then $z\,(1-e) = 0$. As $c(1-e) = 1$, we obtain that $z = 0$. Finally, as ϱ agrees with $\frac{1}{2}\sigma$ on $e\,M\,e$, it is singular on $Z(e\,M\,e)$. □

As we shall see in the next chapter, the misbehaviour of the Jauch-Piron states on centres of hereditary subalgebras described in the previous example is the only obstacle for Jauch-Piron states being regular or nonsingular. We also show that (A) holds for all algebras without nonzero σ-finite direct summand. In that case the validity of (B) is not likely. The reason is that if (B) is true, then all σ-additive states are normal. Indeed, suppose that (B) holds and that ϱ is a σ-additive state on a von Neumann algebra M. Then ϱ is Jauch-Piron (on all hereditary subalgebras) and so, by (B), it is nonsingular on any hereditary subalgebra of M. Let $\varrho = \varrho_n + \varrho_s$, where ϱ_n and ϱ_s is the normal and the singular part of ϱ, respectively. Let e be a support projection of ϱ_n. Since ϱ_s is both singular and nonsingular on $(1-e)\,M\,(1-e)$, we have $\varrho_s(1-e) = 0$. As ϱ is σ-additive and therefore normal on $e\,M\,e$ (e is a σ-finite projection), we infer that $\varrho_s(e) = 0$. Therefore $\varrho_s = 0$ and ϱ is normal.

The following well known technical lemma will be useful in the sequel.

10.1.8. Lemma. *Let ϱ be a state on M and (e_n) be a sequence of orthogonal projections in M with $\varrho(e_n) = 0$ for all n. Then $\varrho\left(\sum_{i=1}^{\infty} e_{n_i}\right) = 0$ for some strictly increasing sequence (n_i).*

Proof: Let us identify $W^*(e_n)$ with l^∞ and consider the canonical map $\pi\colon l^\infty \to l^\infty/c_0$. We know that $c_0 \subset \operatorname{Ker}\varrho$. Let $\overline{\varrho}$ be the induced functional on l^∞/c_0 so that $\varrho = \overline{\varrho} \circ \pi$ on l^∞. Given an enumeration $\{r_n \mid n \in \mathbb{N}\}$ of the rationals, let us choose, for each real number α, a strictly increasing sequence (n_i) in $\mathbb{N}$ for which $r_{n_i} \to \alpha$ and put $e_\alpha = \sum_{i=1}^{\infty} e_{n_i}$. Then $(\pi\, e_\alpha)$ is a family of orthogonal projections in l^∞/c_0 indexed by a set continuum cardinality and consequently for uncountably many α we have $\varrho(e_\alpha) = 0$. □

10.2. Nonsingularity of Jauch-Piron states

In this section we exhibit most of the principal results concerning the regularity, nonsingularity, and the Jauch-Piron property of states. The main finding is that algebraic properties of states can imply in many cases strong continuity properties. Among others, we show that a Jauch-Piron state on a von Neumann algebra is regular if, and only if, it is nonsingular on the centres of hereditary von Neumann subalgebras – a condition that cannot be in general removed in view of Example 10.1.7.

We prepare the way to these results by a few auxiliary lemmas.

10.2.1. Lemma. *Let ϱ be a Jauch-Piron state on a von Neumann algebra M and let $e \in P(M)$. Either of the following conditions implies that ϱ is regular:*

(i) *ϱ is regular on $e\,M\,e$ and $\varrho(e) = 1$*

(ii) *$e \precsim 1 - e$ and $\varrho(e) = 1$.*

Proof: (i) Let e satisfy the stated condition and let (x_n) be a sequence of self-adjoint elements of L_ϱ such that $x_n \to x$ strongly. Then $e\,x_n^2\,e \to e\,x^2\,e$ strongly. Since $\varrho(e\,x_n^2\,e) = \varrho(x_n^2) = 0$ for all n and ϱ is regular on $e\,M\,e$, it follows from Proposition 10.1.5 that $\varrho(e\,x^2\,e) = 0$. Hence, $\varrho(x^2) = 0$ and consequently $x \in L_\varrho$. Proposition 10.1.5 now implies that ϱ is regular.

(ii) Let us suppose first that $e \sim 1 - e$, where $\varrho(e) = 1$. By choosing suitable matrix units we may suppose that $M = N \otimes M_2(\mathbb{C}) \simeq M_2(N)$ with

$$e = \begin{pmatrix} 1_N & 0 \\ 0 & 0 \end{pmatrix}.$$

Let p be an atomic projection in $M_2(\mathbb{C})$ that is the support of the pure state ϱ_2 on $M_2(\mathbb{C})$ given by $\varrho_2(y) = \varrho(1 \otimes y)$ for all $y \in M_2(\mathbb{C})$. Observe that $e = 1_N \otimes p$. As

$$(1 \otimes p)\,(1 \otimes y)\,(1 \otimes p) = (1 \otimes \varrho_2(y)\,p) = \varrho_2(y)(1 \otimes p)\,,$$

for each $y \in M_2(\mathbb{C})$, we see that

$$\varrho(x \otimes y) = \varrho((1 \otimes p)\,(x \otimes y)\,(1 \otimes p)) = \varrho(x \otimes\, p\,y\,p) = \varrho_2(y)\,\varrho(x \otimes 1)\,.$$

It means that ϱ is a product state, $\varrho = \varrho_1 \otimes \varrho_2$, where ϱ_2 is a pure state on $M_2(\mathbb{C})$. We establish the regularity of ϱ_1 on N. Seeking for a contradiction, suppose that there is a sequence (f_n) of orthogonal projections in N such that $\varrho_1(f_n) = 0$ for all n and $\varrho_1(f) \neq 0$, where $f = \sum_{n=1}^{\infty} f_n$. By Proposition 8.2.2, $\varrho_1(f)^{-1}\varrho = (\varrho_1(f)^{-1}\varrho_1) \otimes \varrho_2$ is not a Jauch-Piron state of $f\,N\,f \otimes M_2(\mathbb{C})$. This is a contradiction. Since

$$e\,M\,e = N \otimes \begin{pmatrix} 1 & 0 \\ 0 & 0 \end{pmatrix},$$

it implies that ϱ is regular on $e\,M\,e$, and hence on M, by (i).

Finally, if the condition (ii) holds as stated, then $e \sim f$ for some projection $f \leq 1 - e$. The above argument then implies that ϱ is regular on $(e+f)\,M\,(e+f)$. Now (i) implies that ϱ is regular on M. □

10.2.2. Lemma. *Let ϱ be a Jauch-Piron state on a von Neumann algebra M.*

(i) *If e and f are projections in M such that $f \precsim e$ and $\varrho(e) = 0$, then ϱ is regular on $(e \vee f)\,M\,(e \vee f)$.*

(ii) *If $e_1, \ldots, e_n$ are equivalent projections in M such that $\varrho(e_1) = 0$ and $e = \bigvee_{i=1}^{n} e_i$ is σ-finite, then ϱ is nonsingular on $e\,M\,e$.*

Proof: (i) The case $\varrho(e \vee f) = 0$ is trivial. We may suppose that $\varrho(e \vee f) = 1$. Then the assertion follows from Lemma 10.2.1 (ii), because in the given situation

$$e \vee f - e \sim f - e \wedge f \leq f \precsim e \qquad \text{and} \qquad \varrho(e \vee f - e) = 1\,.$$

(ii) Suppose, for a contradiction, that ϱ_i is singular on $e\,M\,e$. Put $f_i = e_1 \vee e_i$, $i = 1, \ldots, n$. By (i) ϱ is regular on $f_i\,M\,f_i$. Therefore $\varrho(f_i) = 0$, $i = 1, \ldots, n$. Since $e = \bigvee_{i=1}^{n} f_i$ and ϱ is a Jauch-Piron state, this means that $\varrho(e) = 0$. Hence, in any case ϱ is nonsingular on $e\,M\,e$. □

10.2.3. Proposition. *Let M be a finite von Neumann algebra without abelian part, and let ϱ be a Jauch-Piron state on M. Then there is a family (z_α) of central, σ-finite and orthogonal projections in M such that $\sum z_\alpha = 1$ and ϱ is nonsingular on $z_\alpha M$ for all z_α.*

Proof: We can assume that ϱ is not faithful for otherwise ϱ is trivially nonsingular. Moreover, as M is locally σ-finite we can suppose that M is σ-finite. Let e be a nonzero projection in M with $\varrho(e) = 0$. Extend $\{e\}$ to a (necessarily finite) maximal family $\{e = e_1, e_2, \ldots, e_n\}$ of orthogonal equivalent projections in M. By the Comparability Theorem (see Chapter 2, Projection structures of von Neumann algebras) together with maximality, there is a nonzero central projection z in M such that

$$f = z\left(1 - \sum_{i=1}^{n} e_i\right) \lesssim z\,e\,.$$

It immediately follows from Lemma 10.2.2 (i) that ϱ is regular and hence nonsingular on $f\,M\,f$. In addition, by Lemma 10.2.2 (ii) ϱ is nonsingular on $p\,M\,p$, where

$$p = \sum_{i=1}^{n} z\,e_i = z - f\,.$$

Therefore ϱ must be nonsingular on $z\,M$. In other words, M has a nonzero central summand such that ϱ is nonsingular on it. Since the same argument applies to any hereditary subalgebra of M, the assertion follows from Zorn's lemma. □

10.2.4. Theorem. *A Jauch-Piron state ϱ on a finite von Neumann algebra M is nonsingular if, and only if, it is nonsingular on $Z(M)$.*

Proof: Let ϱ be a Jauch-Piron state on M that is nonsingular on $Z(M)$. By Proposition 10.2.3, $M = \sum_\alpha z_\alpha M$, $z_\alpha \in Z(M)$, such that $\varrho | z_\alpha M$ is nonsingular for each α. Suppose for a contradiction that ϱ is singular. Thus, suppose that there are projections (e_β) such that $\varrho(e_\beta) = 0$ for all β and $\sum_\beta e_\beta = 1$. Then, for each z_α, $z_\alpha = \sum_\beta z_\alpha\, e_\beta$ with $\varrho(z_\alpha\, e_\beta) = 0$ for all β. The nonsingularity of ϱ on $z_\alpha M$ implies now that $\varrho(z_\alpha) = 0$ but this is a contradiction with the nonsingularity of ϱ on $Z(M)$. □

In the next theorem, the position of the Jauch-Piron property, regularity and nonsingularity is clarified for the σ-finite algebras.

10.2.5. Theorem. *Let M be a σ-finite von Neumann algebra. The following statements hold:*

(i) *If M is properly infinite, then all Jauch-Piron states on M are nonsingular.*

(ii) *If M is of Type III, then all Jauch-Piron states on M are regular.*

(iii) *If M is semifinite and ϱ is a Jauch-Piron state on M, then ϱ is regular, if and only if, ϱ is nonsingular on $Z(e\,M\,e)$ for all finite projections e of M.*

Proof: (i) Let ϱ be a Jauch-Piron state of M, where M is properly infinite. Choose an infinite orthogonal sequence (f_n) of projections in M such that $f_n \sim 1$ for all n. Let v_n be a partial isometry in M such that

$$v_n^* \, v_n = 1 \qquad \text{and} \qquad v_n \, v_n^* = f_n \,.$$

(That is, v_n has an initial projection 1 and as final projection f_n.) Suppose for a contradiction that ϱ is singular. Then there is a sequence (e_n) of projections in M such that

$$\sum_{n=1}^{\infty} e_n = 1 \qquad \text{and} \qquad \varrho(e_n) = 0 \qquad \text{for all } n \,.$$

In which case, let us put, for each n,

$$p_n = v_n \left(\sum_{i=1}^{n} e_i \right) v_n^* \,.$$

Then, we have, for $n \neq m$,

$$p_n \, p_m = v_n \left(\sum_{i=1}^{n} e_i \right) v_n^* \, v_m \left(\sum_{i=1}^{n} e_i \right) v_m^* = 0 \,,$$

as $v_n^* \, v_m = 0$ by the orthogonality of (f_n). Hence (p_n) is an orthogonal sequence of projections. Moreover,

$$p_n \sim \sum_{i=1}^{n} e_i \qquad \text{for all } n \,.$$

Since $\varrho\left(\sum_{i=1}^n e_i\right) = 0$, it follows from Lemma 10.2.2 (i) that ϱ is regular on $p_n M p_n$ and hence nonsingular on $p_n M p_n$ there by σ-finiteness. But ϱ is singular and hence $\varrho(p_n) = 0$ for all n. Therefore (Lemma 10.1.8) we have

$$\varrho\left(\sum_{i=1}^{\infty} p_{n_i}\right) = 0$$

for some strictly increasing sequence (n_i) of natural numbers. Since we must also have $\sum_{i=1}^{\infty} p_{n_i} \sim 1$, ϱ is regular and hence nonsingular on M by Lemma 10.2.2 (i), which is a contradiction.

(ii) As any nonzero projection in M is properly infinite, it follows from (i) that ϱ is nonsingular on any hereditary subalgebra and therefore regular therein by σ-finiteness of M.

(iii) Let M be semifinite and let ϱ be a Jauch-Piron state on M such that ϱ is nonsingular on $Z(e\,M\,e)$ for all finite projections of M. By Theorem 10.2.4, ϱ is nonsingular on any finite von Neumann subalgebra of M. Let p be a projection in M. By the structure theory there are projections $p_1, p_2 \in Z(p\,M\,p)$ such that $p = p_1 + p_2$, where p_1 is finite and p_2 is properly infinite or zero. By the reasoning above, ϱ is nonsingular on both $p_1\,M\,p_1$ and $p_2\,M\,p_2$, and hence it is nonsingular on $p\,M\,p$. In other words, ϱ is regular and it establishes (iii). □

The previous theorem shows that the more noncommutative the algebra considered is, the stronger the interplay between algebraic and topological properties of states is. By Theorem 10.2.5 we see, e.g., that Jauch-Piron states on σ-finite Type *III* algebras are just states having support projections. The following theorem records, for the sake of completeness, a straightforward generalization of Theorem 10.2.5 to locally σ-finite algebras.

10.2.6. Theorem. *Let M be a locally σ-finite von Neumann algebra and let ϱ be a Jauch-Piron state on M. The following statements hold:*

(i) *ϱ is nonsingular if, and only if, ϱ is nonsingular on $Z(M)$.*

(ii) *If ϱ is nonsingular on $Z(e\,M\,e)$ for all projections e, then ϱ is regular.*

(iii) *If M is of Type III and ϱ is nonsingular on $Z(M)$, then ϱ is regular.*

In case of properly infinite algebras we obtain the regularity of Jauch-Piron states on the centre no matter how large the algebra in question is.

10.2.7. Proposition. *Let M be a properly infinite von Neumann algebra. Then all Jauch-Piron states of M are regular on $Z(M)$.*

Proof: Let ϱ be a Jauch-Piron state and let (z_n) be a sequence of orthogonal projections in $Z(M)$ such that $\varrho(z_n) = 0$ for all n. As all nontrivial central projections in M are properly infinite we may as well suppose that $\sum_{n=1}^{\infty} z_n = 1$ and proceed towards a contradiction. Let (f_n) be a sequence of orthogonal projections in M such that $f_n \sim 1$ for all n. Put $p_n = \left(\sum_{i=1}^{n} z_i\right) f_n$, $n \in \mathbb{N}$. Then p_n are mutually orthogonal, $p_n \sim \sum_{i=1}^{n} z_i$ and $\varrho(p_n) = 0$, for all n. Proceeding now in a similar way as in the proof of Theorem 10.2.5, we find a projection $p \in M$ such that $p \sim 1$ and $\varrho(p) = 0$. This yields, by Lemma 10.2.2, (i) that ϱ is regular, which is a contradiction. □

Having established the case of σ-finite von Neumann algebras, we concentrate now on the von Neumann algebras that have no σ-finite direct summand. Such algebras have systems of orthogonal projections of large cardinalities. Previous examples showed that a Jauch-Piron state on a σ-finite algebra need not be regular. However, in the next theorem we prove that Jauch-Piron states are regular for all von Neumann algebras with trivial locally σ-finite parts. This result may seem surprising at first sight. Loosely speaking, it reflects the fact that a bounded measures always has to vanish at some member of uncountable collection of pairwise orthogonal projections. So the kernels of linear functionals on large abelian algebras are big. This observation is employed in the following lemma.

10.2.8. Lemma. *Let M be a von Neumann algebra. Let (e_α) be an uncountable family of pairwise orthogonal projections in M such that $e_\alpha \sim 1$ for all α. Then any Jauch-Piron state on M is regular.*

Proof: It follows immediately from the fact that $\varrho(e_{\alpha_C}) = 0$ for some α_0 and from Lemma 10.2.2 (i). □

As a consequence, we obtain regularity of Jauch-Piron states on any algebra $B(H)$, where H is a nonseparable Hilbert space. A more general variant of this observation is shown in the next theorem.

10.2.9. Theorem. *Let M be a von Neumann algebra with zero locally σ-finite part. Then all Jauch-Piron states on M are regular.*

Proof: By the structure theory of non-locally σ-finite von Neumann algebras discussed in Chapter 2 (Classification of von Neumann algebras), there is an uncountable set Γ and a family $(z_\lambda)_{\lambda\in\Gamma}$ of pairwise orthogonal central projections such that, for each $\lambda \in \Gamma$, $z_\lambda M$ is of uniform Type λ and

$$M = \sum_{\lambda\in\Gamma} M\, z_\lambda .$$

Let λ_0 be a least member of Γ. Now, for each $\lambda \in \Gamma$, z_λ can be written as a sum of locally σ-finite projections,

$$z_\lambda = \sum_{\alpha\in J_\lambda} f_{\lambda,\alpha} ,$$

with $\mathrm{card}(J_\lambda) = \lambda$ and $f_{\lambda,\alpha} \sim f_{\lambda,\beta}$ for all $\alpha, \beta \in J_\lambda$. For each $\lambda \in \Gamma$, let us take a partition

$$J_\lambda = \bigcup_{i\in I} J_{\lambda,i}$$

with $\mathrm{card}(I) = \lambda_0$ and $\mathrm{card}(J_{\lambda,i}) = \lambda$ for all $i \in I$. Put

$$e_{\lambda,i} = \sum_{\alpha\in J_{\lambda,i}} f_{\lambda,\alpha} , \quad \text{for each } i \in I .$$

Then

$$z_\lambda = \sum_{i\in I} e_{\lambda,i} \quad \text{and} \quad e_{\lambda,i} \sim e_{\lambda,j} \quad \text{for all } i, j .$$

Now, setting $f_i = \sum_{\lambda\in\Gamma} e_{\lambda,i}$ for each $i \in I$, we have $\sum_{i\in I} f_i = 1$ and $f_i \sim f_j$ for all $i, j \in I$. Finally, given a partition $I = \bigcup_{j\in I} I_j$, where $\mathrm{card}(I) = \mathrm{card}(I_j) = \lambda_0$ for all j, the projections

$$h_j = \sum_{i\in I_j} f_i \sim \sum_{i\in I} f_i = 1$$

form an uncountable family. Now, by virtue of Lemma 10.2.8, we obtain that all Jauch-Piron states on M are regular. □

By combining results on "small" and "large" algebras, we can now summarize our discussion as follows.

10.2.10. Theorem. *Let ϱ be a Jauch-Piron state on a von Neumann algebra M. Then ϱ is regular on M if ϱ is nonsingular on $Z(e\,M\,e)$ for all projections e of M for which the central support $c(e)$ is locally σ-finite.*

A transparent description of Jauch-Piron states is consequently obtained for factors.

10.2.11. Theorem. *Every Jauch-Piron state on a factor is regular.*

Proof: It is an easy consequence of Theorem 10.2.10 because all hereditary subalgebras of factors are factors again. □

10.3. Countable additivity of states

In the previous section the regularity properties of general Jauch-Piron states have been established. In the present section we focus on the relationship between σ-additivity and the Jauch-Piron property. The usual assumption on states in quantum mechanics is that there are normal. In case of σ-finite algebras which are mostly used in physics, this postulate is equivalent to σ-additivity. The σ-additivity itself has no a priori physical justification and was adopted originally solely for the reason of mathematical convenience. However, the main results of the present part allow one to deduce the σ-additivity from the natural Jauch-Piron property. We obtain thus an explanation of the σ-additivity in terms of physically plausible Jauch-Piron property. For example, we show that if all transformed states corresponding to a given state ϱ are Jauch-Piron, then ϱ is σ-additive. In other words, σ-additivity is a consequence of the fact that all states evolving from a given state in irreversible dynamics are Jauch-Piron. In particular, we obtain that pure Jauch-Piron states are σ-additive on essentially noncommutative algebras and that the same holds for factor states on properly infinite von Neumann algebras. From the mathematical point of view we show that σ-additivity can be derived from the inner structure of the projection lattice. This phenomenon is completely noncommutative and has no analogy in classical measure and probability theory. Our main tool will be a geometrical analysis of the facial structure of the state spaces. We first recall some notations (see Section 6.2). Let $S(M)$, $S_n(M)$, $S_\sigma(M)$, and $S_{cs}(M)$ be the sets of all states, normal states, σ-additive states, and countably singular states on M, correspondingly. By $S_{JP}(M)$ let us denote the set of all Jauch-Piron states on M. It is clear that $S_{JP}(M)$ is a convex set. The following property of the Jauch-Piron state space $S_{JP}(M)$ indicating that $S_{JP}(M)$ is "nearly" a face will often be used in the sequel. If $\alpha\,\varrho_1 + (1-\alpha)\,\varrho_2 \in S_{JP}(M)$ for states $\varrho_1, \varrho_2 \in S(M)$, and $\alpha \in (0,1)$, then a state $\beta\,\varrho_1 + (1-\beta)\,\varrho_2$ is in $S_{JP}(M)$ for all $\beta \in (0,1)$. It follows from the equality $\mathrm{Ker}(\alpha\,\varrho_1 + (1-\alpha)\,\varrho_2) \cap P(M) = \mathrm{Ker}(\varrho_1) \cap \mathrm{Ker}(\varrho_2) \cap P(M)$ for all $\alpha \in (0,1)$. However, as we shall see later, the convex set $S_{JP}(M)$ is neither a face nor a norm closed set for essentially noncommutative von Neumann algebras. Despite this, we show in the next theorem that the norm closure of $S_{JP}(M)$ is always a split face that contains all reasonably continuous states. Like in the Section 6.2, given a state ϱ on a von Neumann algebra M and $a \in M$ such that $\varrho(a^*\,a) \neq 0$, the symbol ϱ_a denotes the transformed state $\varrho_a(x) = \varrho(a^*\,x\,a)/\varrho(a^*\,a)$ $(x \in M)$.

10.3.1. Theorem. *Let M be a von Neumann algebra. The norm closure $\overline{S_{JP}(M)}$ of $S_{JP}(M)$ is a split face of $S(M)$. The facial complement $\overline{S_{JP}(M)}^{\#}$ satisfies the inclusion*

$$\overline{S_{JP}(M)}^{\#} \subset \{\varrho \in S(M) \mid \varrho(e) = 0, e \in P(M) \text{ is } \sigma\text{-finite}\}\,.$$

Proof: Let τ lie in the face F of $S(M)$ generated by $S_{JP}(M)$. It means that for some $0 < \lambda \leq 1$ and some Jauch-Piron state ϱ we have $\lambda\tau \leq \varrho$. If $\lambda = 1$ then $\tau = \varrho$. For $\lambda \neq 1$ we obtain

$$\varrho = \lambda\tau + (1-\lambda)\sigma\,, \qquad \text{where } \sigma = (1-\lambda)^{-1}(\varrho - \lambda\tau) \in S(M)\,.$$

It implies that $\alpha\,\tau + (1-\alpha)\,\sigma$ is a Jauch-Piron state for all $\alpha \in (0,1)$. Letting $\alpha \to 1$ we have $\tau \in \overline{S_{JP}(M)}$. By the geometry of the state space the norm closure $\overline{F}$ of F is a face again. Further, it is clear that $S_{JP}(M)$ is unitarily invariant. Therefore, $\overline{S_{JP}(M)}$ is unitarily invariant and so it is a split face by Proposition 6.2.1.

Let $\varrho \in S(M)$ and suppose that $\varrho(e) \neq 0$ for some σ-finite projection e of M. Let τ be a normal state with its support equal to e and put $\varphi = 1/2\,(\varrho + \tau)$. If p and q are projections in M with $\varphi_e(p) = \varphi_e(q) = 0$, then $e\,p = e\,q = 0$ as τ is faithful on $e\,M\,e$ so that $\varphi_e(p \vee q) = 0$. Therefore, $\varphi(e)^{-1}\varphi_e$ is a Jauch-Piron state. Hence, $\varrho(e)^{-1}\varrho_e \in \overline{S_{JP}(M)}$, by the first part of the proof.

Let J be the norm closed ideal of M generated by the σ-finite projections of M. Since the set of all σ-finite projections is closed under passing to subprojections and under forming of finite suprema and Murray-von Neumann equivalence classes, it follows from [311, Theorem 19] that every projection in J is σ-finite. The state space, $S(J)$, of J identifies with $\{\varrho \in S(M) \mid \varrho(z) = 1\}$ where z is the central projection in M^{**} with $J^{**} = M^{**}\,z$. It is enough to show that $S(J) \subset \overline{S_{JP}(M)}$. Suppose that x with $0 \leq x \leq 1$ is an element of J. Let $x = \sum_n \frac{1}{2^n}\,h_n$ where h_n are spectral projections of x. Then $\frac{1}{2^n}\,h_n \leq x$ and so $h_n \in J$ for all n. Therefore, for each k, $p_n = \bigvee_{n=1}^{k} h_n \in J$ thanks to the commutativity of p_n's. It is easy to verify that $p_n\,x \to x$ as $n \to \infty$. We have shown that the system (f_λ) of σ-finite projections forms an approximate unit for J. Let ϱ be a state of J. As $\varrho(z) = 1$ we have $\tilde{\pi}_\varrho(z) = 1$ where $\tilde{\pi}_\varrho\colon\ M^{**} \to B(H_\varrho)$ is the normal extension of the G.N.S. representation π_ϱ of ϱ. Hence, as $f_\lambda \to z$ strongly in M^{**}, there exists a sequence (e_n) of projections in J such that $\pi_\varrho(e_n)x_\varrho \to x_\varrho$ and $\varrho(e_n) \neq 0$ for all n. It implies that $\|\varrho(e_n)^{-1}\varrho_{e_n} - \varrho\| \to 0$ (see Chapter 2, The Gelfand-Naimark-Segal construction). Hence, by the preceeding paragraph, $\varrho \in \overline{S_{JP}(M)}$. □

Let us recall that any projection in a von Neumann algebra is a sum of cyclic (and thereby σ-finite) projections. Therefore, a state vanishing on all σ-finite projections is highly nonsingular. In the light of this we can interpret the previous theorem as the observation that states on non-σ-finite algebras that cannot be approximated by Jauch-Piron states are very discontinuous. When specifying Theorem 10.3.1 to σ-finite algebra, we obtain immediately that Jauch-Piron states are norm dense in the state space.

As a corollary to Theorem 10.3.1 we can extend the characterization of the von Neumann algebras for which all states are Jauch-Piron (Proposition 10.1.3). We can also show that only in this case the Jauch-Piron state space enjoys nice topological and convexity properties.

10.3.2. Corollary. *Let M be a von Neumann algebra. The following statements are equivalent:*

(i) $S_{JP}(M)$ *is a face of* $S(M)$.

(ii) $S_{JP}(M)$ *is norm closed.*

(iii) $S_{JP}(M) = S(M)$.

(iv) *The nonabelian part of M is finite-dimensional.*

Proof: The implication (ii)⇒(i) follows from Theorem 10.3.1. Equivalence (iii)⇔(iv) is Proposition 10.1.3. The implication (iii)⇒(i) is obvious. It remains to prove (i)⇒ (iv).

Let $S_{JP}(M)$ form a face in $S(M)$. We may suppose that M has zero abelian part. Let e be a σ-finite projection in M and ϱ be a state on $e\,M\,e$. Let us form a convex combination $\varphi = \frac{1}{2}(\overline{\varrho} + \tau)$, where $\overline{\varrho}$ is the state extension of ϱ to M and τ is a faithful state on $e\,M\,e$. Like in the proof of Theorem 10.3.1 we infer that φ_e is a Jauch-Piron state. Therefore ϱ is Jauch-Piron by assumption. By equivalence of (iii) and (iv) we therefore infer that $e\,M\,e$ has finite dimensional nonabelian part whenever e is a σ-finite projection. As a consequence, every nonzero projection in M dominates a nonzero abelian projection and so M is of Type I. By considering the decomposition of M into homogeneous direct summands, we see that M has to be finite-dimensional. □

Let us continue to denote by E_ϱ the norm closure of the set of all transforms of the state ϱ on a given von Neumann algebra.

10.3.3. Theorem. *Let M be a von Neumann algebra with no nonzero central abelian direct summand. Let ϱ be a state of M and F_ϱ be the split face generated by ϱ. The following conditions are equivalent.*

(i) $F_\varrho \subset S_{JP}(M)$.

(ii) $E_\varrho \subset S_{JP}(M)$.

(iii) ϱ *is σ-additive.*

Proof: By Proposition 6.2.1, $E_\varrho \subset F_\varrho$, giving (i)⇒(ii). On the other hand if ϱ is σ-additive, then $F_\varrho \subset S_\sigma(M) \subset S_{JP}(M)$ (F_ϱ is an intersection of all split faces containing ϱ) and (iii)⇒(i) follows. It remains to prove (ii)⇒(iii).

Let $E_\varrho \subset S_{JP}(M)$. Denote by $(\pi_\varrho, H_\varrho, x_\varrho)$ the G.N.S triple resulting from ϱ. As $\omega_h \circ \pi_\varrho \in E_\varrho$, $\omega_h \circ \pi_\varrho$ is a Jauch-Piron state of M for every unit vector h of H_ϱ. Let e and f be projections of M. We have $\pi_\varrho(e) \vee \pi_\varrho(f) \leq \pi_\varrho(e \vee f)$. Suppose that $\pi_\varrho(e) \vee \pi_\varrho(f) \neq \pi_\varrho(e \vee f)$ and let h be a unit vector in

$$(\pi_\varrho(e \vee f) - \pi_\varrho(e) \vee \pi_\varrho(f))H_\varrho .$$

Then $\omega_h \circ \pi_\varrho(e) = \omega_h \circ \pi_\varrho(f) = 0$ and $\omega_h \circ \pi_\varrho(e \vee f) = 1$, which implies that $\omega_h \pi_\varrho$ is not a Jauch-Piron state of M. This violates the previous remark. Therefore, $\pi_\varrho(e \vee f) = \pi_\varrho(e) \vee \pi_\varrho(f)$. Now, by Theorem 8.2.4, ϱ is σ-additive. □

As seen in Section 10.1 the Jauch-Piron property is far from being equivalent to σ-additivity. However the previous result tells us that if the entire collection of states that can be approximated by transformations of a give state ϱ consists of Jauch-Piron states, then σ-additivity of ϱ follows automatically. This may be interpreted physically as follows. Reversible dynamics of a closed quantum mechanical system is given by the group of unitary operators (see the discussion in Chapter 7 and Chapter 8). On the other hand, when the system is coupled to an environment or subject to a measurement, its most general time evolution is irreversible (i.e. it does not preserve all structure properties). Mathematically, it is given by the transforms $x \to a^* x a$: $M \to M$, for a given element a (not necessarily unitary) of the algebra with $\|a\| \leq 1$. Thus, in the generalized Schrödinger picture, the time evolution of a system initially prepared at state ϱ is given by the transformation $\varrho \to \varrho_a$. In this setup, Theorem 10.3.3 says that the norm closure of the set of all possible states that can be obtained from the state ϱ by the time evolution of the system contains no pathological states if, and only if, ϱ is σ-additive.

Another form of the fact that the structure of Jauch-Piron states, $S_{JP}(M)$, uniquely determines the structure of σ-additive states is the following corollary.

10.3.4. Corollary. *Let M be a von Neumann algebra without abelian part. Then $S_\sigma(M)$ is the largest split face contained in $S_{JP}(M)$.*

Corollary 10.3.4 implies that if $S_{JP}(M_1)$ an $S_{JP}(M_2)$ are affinely isomorphic, where M_1 and M_2 are von Neumann algebras without nonzero abelian parts, then $S_\sigma(M_1)$ and $S_\sigma(M_2)$ are also affinely isomorphic. Further aspect of the interplay between the Jauch-Piron property and σ-additivity is the following result characterizing σ-additivity of homomorphisms in terms of the Jauch-Piron property.

10.3.5. Theorem. *Let M and N be von Neumann algebras, M having no nonzero abelian direct summand. Let $\pi: M \to N$ be a $*$-homomorphism. Then π is σ-additive if, and only if,*

$$\pi^*(S_{JP}(N)) \subset S_{JP}(M), \tag{10.1}$$

where π^* is the adjoint map of π.

Proof: It has been proved in Theorem 8.2.4 that π is σ-additive if, and only if, π is a lattice homomorphism, i.e. exactly when

$$\pi(e \vee f) = \pi(e) \vee \pi(f) \qquad \text{for all projections } e, f \in M. \tag{10.2}$$

We show that (10.2) is equivalent to (10.1). Suppose that π has the the property (10.1). It means that $\varphi \circ \pi$ is a Jauch-Piron state on M whenever φ is a Jauch-Piron state on N. Assume, for a contradiction, that

$$\pi(e \vee f) \neq \pi(e) \vee \pi(f)$$

for some projections $e, f \in P(M)$. Since $\pi(e) \vee \pi(f) \leq \pi(e \vee f)$ it means that the projection $p = \pi(e \vee f) - (\pi(e) \vee \pi(f))$ is nonzero. Therefore we can find a normal state ψ with $\psi(p) = 1$. It implies that $\psi(\pi(e) \vee \pi(f)) = 0$ and so $\psi(\pi(e)) = \psi(\pi(f)) = 0$. The state ψ is normal and therefore Jauch-Piron. Nevertheless, for the state $\psi \circ \pi$ we have

$$(\psi \circ \pi)(e \vee f) = 1 \text{ while } (\psi \circ \pi)(e) = (\psi \circ \pi)(f) = 0,$$

contradicting thus (10.1). The fact that (10.2) implies (10.1) is obvious. □

As a consequence, any $*$-homomorphism $\pi: M \to M$, where M is essentially noncommutative, that sends Jauch-Piron states to Jauch-Piron states is σ-additive. That is a certain compensation for the fact that individual Jauch-Piron state need not be σ-additive at all.

We now summarize our discussion on the relationship between the Jauch-Piron property and σ-additivity in the list of a few criteria of the σ-additivity of states.

10.3.6. Theorem. *Let ϱ be a state on a von Neumann algebra M having no nonzero abelian direct summand. Let $\pi_\varrho \colon M \to B(H_\varrho)$ be the G.N.S. representation of ϱ and let M_ϱ be the von Neumann algebra generated by $\pi_\varrho(M)$. The following conditions are equivalent.*

(i) *ϱ is σ-additive.*

(ii) *The norm closure of the set of all transformed states of ϱ consists of Jauch-Piron states.*

(iii) *$\varphi \circ \pi_\varrho$ is a Jauch-Piron state whenever φ is a Jauch-Piron state on M_ϱ.*

(iv) *$\varphi \circ \pi_\varrho$ is a Jauch-Piron state whenever φ is a normal state on M_ϱ.*

(v) *$\varphi \circ \pi_\varrho$ is a Jauch-Piron state whenever φ is a σ-additive state on M_ϱ.*

(vi) *$\varphi \circ \pi_\varrho$ is a Jauch-Piron state whenever φ is a vector state on M_ϱ.*

Proof: The equivalence of (i) and (ii) is Theorem 10.3.3. The equivalence of conditions (i), (iii)-(vi) follows from the fact that ϱ is σ-additive if, and only if, π_φ is σ-additive (Theorem 8.2.5) and from a reasoning fully analogous to Theorem 10.3.5.

In the next part we concentrate on the study of the Jauch-Piron property of important types of states. The first result brings a rather striking assertion that pure states on essentially noncommutative von Neumann algebras are always σ-additive.

10.3.7. Theorem. *Let M be a von Neumann algebra having no nonzero abelian direct summand. Then a pure state ϱ on M is Jauch-Piron if, and only if, it is σ-additive.*

Proof: Let ϱ be a pure Jauch-Piron state on M and let π_ϱ be the corresponding (irreducible) G.N.S. representation on the Hilbert space H_ϱ. As all states $\omega_h \circ \pi_\varrho$, where $h \in H_\varrho$ is a unit vector, are unitarily equivalent to ϱ we infer that

$$E_\varrho = \{\omega_h \circ \pi_\varrho \mid h \in H_\varrho \, , \|h\| = 1\} = \{\varrho(u^* \cdot u) \mid u \text{ is unitary in } M\} \, .$$

However, since the Jauch-Piron property is preserved under unitary transformations, we see that $E_\varrho \subset S_{JP}(M)$ and hence ϱ is σ-additive by Theorem 10.3.3. □

Let us remark that the Jauch-Piron states on M which are pure are precisely the extreme points of the convex set $S_{JP}(M)$. In fact, if ϱ is an extreme point of $S_{JP}(M)$ and $\varrho = \frac{1}{2}(\varrho_1 + \varrho_2)$, where ϱ_1, ϱ_2 are states of M, then $\tau_1 = \frac{1}{3}\varrho_1 + \frac{2}{3}\varrho_2$ and $\tau_2 = \frac{3}{5}\varrho_1 + \frac{2}{5}\varrho_2$ are Jauch-Piron states with $\varrho = \frac{3}{8}\tau_1 + \frac{5}{8}\tau_2$. Therefore $\tau_1 = \tau_2$, and hence $\varrho_1 = \varrho_2$, implying that ϱ is a pure state of M. Therefore, in a slightly stronger formulation of the previous theorem we can say that all extreme points of the Jauch-Piron state space are σ-additive, provided that the algebra has no nontrivial abelian direct summand.

In the usual σ-finite model of quantum mechanics the previous theorem reads as follows. Let ϱ be a pure state on a σ-finite von Neumann algebra M without nonzero abelian direct summand. Then ϱ is a vector state concentrated at some Type I factor direct summand of M. Indeed, by Theorem 10.3.7 ϱ is normal. As ϱ is pure (i.e. 0-1) on the centre of M, there is an atomic projection $z \in Z(M)$ that is the support of $\varrho|Z(M)$. Therefore $z\,M$ is a factor. Since Type II and Type III factors have no normal pure states, we see that $z\,M$ has to be a Type I factor which forces ϱ to be a vector state on $z\,M$. In other words, vector states used in early quantum mechanics are nothing but pure Jauch-Piron states on σ-finite algebras, which provides an explanation for the appropriateness of this model.

Besides pure states, more general factor states play an important role both in the structure theory of operator algebras and quantum physics. Let us recall that a state ϱ on a C^*-algebra A is a *factor state* if $\pi_\varrho(A)''$ is a von Neumann factor, or which is the same, if $\varrho(z) = 1$ for some atomic central projection in the enveloping von Neumann algebra M^{**}. The factor states are much more general then pure states and they appear as ground states in quantum statistical mechanics and elsewhere. For example, the convex combinations of equivalent pure states are factor states. In the sequel we shall also often use the fact that factor state ϱ on a von Neumann algebra M is always pure on the centre $Z(M)$. For this, let us denote by $(\pi_\varrho, H_\varrho, x_\varrho)$ the G.N.S data of the factor state ϱ. Observe that $\pi_\varrho(Z(M)) \subset Z((\pi_\varrho(M)'') = \mathbb{C}1$. Hence, we obtain, for each $a \in Z(M)$, $\pi_\varrho(a)\,x_\varrho = \varrho(a)\,x_\varrho$, implying immediately that ϱ is multiplicative on $Z(M)$.

As a consequence $\varrho(a\,x) = \varrho(a)\,\varrho(x)$ whenever $a \in Z(M)$ and $x \in M$ (see Chapter 2, Representations and pure states). As seen before pure Jauch-Piron states on essentially noncommutative von Neumann algebras are always σ-additive. As opposed to this, we show that σ-additivity of factor states depends on the structure of the von Neumann algebra we work with. It turns out that for highly noncommutative von Neumann algebra every factor Jauch-Piron state is σ-additive while for finite algebras it may be or may not be true. We shall investigate Jauch-Piron factor states on a case by case basis. We start with the discussion of Type I algebras. For the first result on σ-additivity, however, we do not need the Jauch-Piron condition at all.

10.3.8. Theorem. *Let ϱ be a factor state on a Type I finite von Neumann algebra such that ϱ is σ-additive on $Z(M)$. Then ϱ is σ-additive.*

Proof: As ϱ is a σ-additive pure state of $Z(M)$ and M is a direct sum of countably many homogeneous summands, we may suppose that M has Type I_n for some $n < \infty$. Given a sequence (f_j) of orthogonal projections in M, let Ω be a maximal abelian von Neumann subalgebra of M containing all f_j's. By Kadison diagonalization result (see Chapter 2, Classification of von Neumann algebras), there exist orthogonal and equivalent abelian projections $e_1, \ldots, e_n$ of M contained in Ω and which sum up to 1. For each $i = 1, \ldots, n$ we have $e_i\,M\,e_i = Z(M)\,e_i$ and $\varrho(z\,e_i) = \varrho(z)\,\varrho(e_i)$ for all z in $Z(M)$. Further, if (z_k) is a sequence of projections in $Z(M)$ such that $(z_k\,e_i)$ is orthogonal for some i, then (z_k) itself is an orthogonal sequence. It follows from the fact that $c(e_i) = 1$, and hence $z \to e_i\,z$ is a $*$-isomorphism mapping $Z(M)$ onto $e_i\,M\,e_i$. (See also the arguments in the proof of Example 10.1.7.) Therefore ϱ is σ-additive on $e_i\,M\,e_i$ for each $i = 1, \ldots, n$, so that $\sum_{j=1}^{\infty} \varrho(e_i f_j) = \varrho(\sum_{j=1}^{\infty} e_i f_j)$. As $\sum_{i=1}^{n} e_i = 1$, we have $\sum_{j=1}^{\infty} \varrho(f_j) = \varrho(\sum_{j=1}^{\infty} f_j)$. □

Let us observe that the previous result is in fact valid for all states on finite Type I algebras that are pure on the centre.

10.3.9. Corollary. *A factor state on a Type I finite von Neumann algebra is either σ-additive or countably singular.*

Proof: Let ϱ be a factor state on a Type I finite von Neumann algebra. As $\varrho|Z(M)$ is pure (i.e. 0-1), it has to be either σ-additive or countably singular. In the former case, ϱ is σ-additive by Theorem 10.3.8. □

In the next theorem we analyze the Jauch-Piron property of factor states on Type I finite homogeneous algebras that are isomorphic to tensor products $C(X) \otimes M_n(\mathbb{C})$, where X is a hyperstonean space.

10.3.10. Theorem. *Let $M = C(X) \otimes M_n(\mathbb{C})$, where $2 \leq n < \infty$ and X is a hyperstonean space. Let ϱ be a state on M.*

(i) *If ϱ is a factor Jauch-Piron state that is not faithful on $1 \otimes M_n(\mathbb{C})$, then ϱ is σ-additive.*

(ii) *If ϱ is a product state that is faithful on $1 \otimes M_n(\mathbb{C})$, then ϱ is a Jauch-Piron state. In particular, if ϱ is a factor state that is faithful on $1 \otimes M_n(\mathbb{C})$, then ϱ is Jauch-Piron.*

Proof: (i) As $Z(M) = C(X) \otimes 1$, it follows from the remark preceeding Theorem 10.3.8 that $\varrho = \varrho_1 \otimes \varrho_2$. By the assumption ϱ_2 is not faithful. Let us choose minimal projections e and f of $M_n(\mathbb{C})$ with $\varrho_2(e) = 0$ and $\varrho_2(f) \neq 0$. Put $N = (e \vee f)\, M_n(\mathbb{C})\, (e \vee f)$, $\overline{e} = 1 \otimes e$, $\overline{f} = 1 \otimes f$ and $p = \overline{e} \vee \overline{f}$. As ϱ is a Jauch–Piron state with $\varrho(\overline{e}) = 0$ and $\overline{e} \sim \overline{f}$, we have that $\tau = \varrho(p)^{-1}\varrho$ is a regular state of $C(X) \otimes N = p\,M\,p$, by Lemma 10.2.2 (i). Therefore, ϱ is a regular pure state of $Z(M)$ and so is σ-additive on $Z(M)$. Hence, ϱ is σ-additive on M, by Theorem 10.3.8.

(ii) If ϱ satisfies the hypothesis, then $\varrho = \varrho_1 \otimes \varrho_2$, where ϱ_2 is faithful. By a compactness argument, there is an $\alpha > 0$ such that $\varrho_2 \geq \alpha$ on the positive part of the unit sphere of $M_n(\mathbb{C})$. Let e be a projection of M. Via the identification $C(X) \otimes M_n(\mathbb{C}) = C(X, M_n(\mathbb{C}))$, the set $E = \{t \in X : e(t) \neq 0\}$ is clopen in X and the central cover, $c(e)$, of e is given by $c(e) = \chi_E \otimes 1$, where χ_E denotes the characteristic function of the set E. To see this, it is clear that E is open. Suppose that $t_\alpha \to t \in X$, $t_\alpha \in E$. Then $e(t_\alpha) \to e(t)$ and, for some α_0, $\|e(t_{\alpha_0}) - e(t)\| < 1$, giving that $0 \neq e(t_\alpha) \sim e(t)$ and hence $t \in E$. Since the order on M is determined by the pointwise order on $C(X, M_n(\mathbb{C}))$ and $c(p) = 1$ for any nonzero projection $p \in M_n(\mathbb{C})$, we see immediately that $c(e) = \chi_E \otimes 1$. By the density of locally constant projections in the projection lattice (see Chapter 2, Classification of von Neumann algebras), given $0 < \varepsilon < 1$, there exists a projection f in M,

$f = \sum_{i=1}^{k} z_i \otimes f_i$, where $z_1, z_2, \ldots, z_k$ are orthogonal projections of $C(X)$ and $f_1, \ldots, f_k$ are non-zero projections of $M_n(\mathbb{C})$ such that $\|e - f\| < \varepsilon$. We have $\sum_{i=1}^{k} z_i \otimes 1 = c(f) = c(e)$. Therefore,

$$\varrho(e) > \varrho(f) - \varepsilon = \sum_{i=1}^{k} \varrho_1(z_i)\, \varrho_2(f_i) - \varepsilon \geq \alpha\, \varrho(c(e)) - \varepsilon.$$

Hence, $\varrho(e) \geq \alpha\, \varrho(c(e))$. Therefore, if p and q are projections in M with $\varrho(p) = \varrho(q) = 0$, then $\varrho(c(p)) = \varrho(c(q)) = 0$ so that

$$\varrho(p \vee q) \leq \varrho(c(q) \vee c(q)) = 0\,,$$

proving that ϱ is a Jauch-Piron state of M. □

We now exhibit two examples which show that the assumptions in the foregoing results on factor states are not superfluous and that we cannot expect factor Jauch-Piron states to be σ-additive or regular in general.

10.3.11. Example. (i) Let $M = Z \otimes M_n(\mathbb{C})$, where Z is an infinite dimensional abelian von Neumann algebra. Let $\varrho = \varrho_1 \otimes \tau$, where ϱ_1 is pure and countably singular and τ is the normalized trace. Then τ is tracial and therefore Jauch-Piron. Moreover, as $\tau = \sum_{i=1}^{n} \frac{1}{n} \psi_i$, where ψ_i are equivalent pure states on $M_n(\mathbb{C})$, we obtain that also

$$\varrho = \frac{1}{n} \sum_{i=1}^{n} \varrho_1 \otimes \varrho_n$$

is a convex combination of equivalent pure states and therefore a factor state.

(ii) Let $M = \sum_{n=2}^{\infty} M_n$ where each M_n is a homogeneous von Neumann algebra of Type I_n $(n \geq 2)$ or where each M_n is of Type II_1. Let J be a maximal ideal in M containing each M_n. Then M/J can be realized as a Type II_1 factor and so has a unique tracial state φ (see e.g. [307]). As φ is normal, the G.N.S. representation π_ϱ is normal too. Moreover, π_ϱ is faithful because M is a factor. So φ is a factor state and so is $\overline{\varphi} = \varphi \circ \pi$ where $\pi\colon\ M \to M/J$ is the canonical map. As $\overline{\varphi}$ is a trace of M it is a Jauch-Piron state. Therefore, $\overline{\varphi}$ is a countably singular factor Jauch-Piron state of M.

The previous example (i) shows that even Type I factor states on essentially noncommutative algebras need not be σ-additive. However, the criterion of σ-additivity in terms of the facial structure (Theorem 10.3.3) simplifies slightly for factor states as follows.

10.3.12. Proposition. *If ϱ is a factor state on the von Neumann algebra M such that the split face F_ϱ generated by ϱ contains at least one Jauch-Piron state which is an extreme point of the set $S_{JP}(M)$, then ϱ is σ-additive. The converse holds on condition that ϱ is of Type I.*

Proof: Suppose that $\varphi \in F_\varrho$ is an extreme point of $S_{JP}(M)$. Then φ is a pure state which implies that φ is σ-additive (Theorem 10.3.7). Thus, the split face F_ϱ has nonzero intersection with the split face $S_\sigma(M)$. Since the split faces are represented by central projections in M^{**}, we can write $F_\varrho = \{\psi \in S(M) \mid \psi(z_\varrho) = 1\}$ and $S_\sigma(M) = \{\psi \in S(M) \mid \psi(z_\sigma) = 1\}$, where z_ϱ and z_σ is a minimal central projection and and a central projection in M^{**}, respectively. So either $z_\varrho \leq z_\sigma$ or $z_\sigma z_\varrho = 0$. Since the second possibility is excluded by the assumption, we see that $\varrho \in F_\varrho \subset S_\sigma(M)$.

On the other hand, if ϱ is a σ-additive Type I factor state, then F_ϱ consists of σ-additive states and it contains a pure state because $\overline{\pi_\varrho(M)} \cong z_\varrho M^{**}$ is a Type I factor. □

As seen above, the relation between σ-additivity and the Jauch-Piron property of factor states is quite intriguing. As the concluding result of this section let us show (using the continuum hypothesis) that for properly infinite algebras the property of being factor state and the Jauch-Piron property give already σ-additivity.

10.3.13. Lemma. *Let ϱ be a Jauch-Piron state of M and let e and f be projections of M with $\varrho(e) = 0$ and $e \sim f$. Let ϱ be countably singular and let us suppose that either M is σ-finite or that the continuum hypothesis is true. Then $\varrho(f) = 0$.*

Proof: By Proposition 10.2.2, ϱ is regular on $(e \vee f)\, M\, (e \vee f)$ and hence is regular on $f\, M\, f$. If $\varrho(f) \neq 0$, then $\varrho(f)^{-1} \varrho_f$ is countably singular by Theorem 6.2.9. Consequently, there exists a sequence of projections (e_n) in M with $\sum e_n = 1$ and $\varrho(f\, e_n\, f) = 0$, for all n. But then $f(\sum_{i=1}^n e_i) f$ converges strongly to f as $n \to \infty$ so that $\varrho(f) = 0$ by Proposition 10.1.5. □

10.3.14. Theorem. *Let M be a properly infinite von Neumann algebra. Suppose that M is σ-finite or that the continuum hypothesis is true. Then all Jauch-Piron factor states of M are σ-additive.*

Proof: Let ϱ be a Jauch-Piron factor state of M. Suppose that ϱ is countably singular. Then $\sum e_n = 1$ and $\varrho(e_n) = 0$ for all n, for some sequence (e_n) of orthogonal projections in M. As M is properly infinite, there exists a sequence (p_n) of orthogonal projections in M such that $p_n \sim \sum_{i=1}^n e_i$ for each n. By Lemma 10.3.13, $\varrho(p_n) = 0$ for all n. Consequently, by Lemma 10.1.8 there exists a strictly increasing sequence (n_i) of natural numbers such that $\varrho(\sum_{i=1}^\infty p_{n_i}) = 0$. We therefore have $\sum_{i=1}^\infty p_{n_i} \sim 1$. Now Lemma 10.3.13 implies that $\varrho(1) = 0$, which is a contradiction. Therefore, ϱ cannot be countably singular and so it must be σ-additive since, by Theorem 6.2.9, the facial complement of the σ-additive state space is the set of all countably singular states. □

10.4. Historical remarks and comments

The concept of Jauch-Piron state was introduced and first studied by Jauch and Piron in the 60's and 70's [199, 200, 201, 263]. They proposed the Jauch-Piron states as being real physical states and by excluding all other states they succeeded in deriving no-go theorems for hidden variables on complete orthomodular lattices. Their research had strongly inspired the investigation of the relationship between regularity and the Jauch-Piron property and intensive ensuing investigation of Jauch-Piron states on abstract quantum logics [57, 268]. The study of Jauch-Piron states on operator algebras was initiated by Rütimann in [287] and, in connection with abstract quantum logics, by Bunce, Navara, Pták and Wright [57]. Further revival of the interest in Jauch-Piron states on von Neumann algebras was stimulated by a nice paper of Amann [21] in early 90's. He proved that pure Jauch-Piron states on von Neumann algebras acting on a separable Hilbert spaces are normal and derived some results on nonsingularity of Jauch-Piron states. The theory of Jauch-Pirons on von Neumann algebras was then developed by Bunce and Hamhalter in a series of papers [169, 50, 52, 53, 54]. The exposition in Section 10.1 and 10.2 is based mainly on papers [169, 50, 52], the material in Section 10.3 then follows [52, 53, 54, 186].

The results on automatic regularity and continuity of Jauch-Piron states have also been extended by Bunce and Hamhalter to Jordan operator algebras in [53]. Usual assumption on the physical system is that the bounded observables are given by a self-adjoint part of von Neumann algebra. Less pragmatic assumption is that the system of bounded observables forms a JW algebra. A *JW algebra* is a real weakly closed subspace of self-adjoint operators acting on some Hilbert space H that is closed under forming squares. It is equipped with the Jordan product $a \circ b = 1/2\,(a\,b + b\,a)$. This model is physically more plausible because it refers only to the Jordan product of self-adjoint operators. This product is self-adjoint again and so it represents an observable. (Note that the usual product of noncommuting self-adjoint operators does not seem to have any physical meaning.) Interpreting the Jauch-Piron states as the only states relevant to physics a natural question arises of whether the results presented in this chapter hold also for Jauch-Piron states on JW algebras. (The definitions of Jauch-Piron states as well as all concepts in the present chapter canonically generalize to JW algebras.) It has turned out that the answer is in the positive for nearly all JW algebras. This can be obtained from the principle that Jauch-Piron states on JW algebras usually extend to Jauch-Piron states on a von Neumann algebras that they generate. We briefly outline the results of this type and its physical meaning (for full details see the original paper [53]). Let us observe first that Jauch-Piron states do not always extend to Jauch-Piron states on larger systems. For this consider the infinite-dimensional spin factor V faithfully embedded into Type III von Neumann factor N that is generated as a von Neumann algebra by V. Let us pick up a tracial singular state ϱ on the tensor product $l^\infty_{\mathbb{R}} \otimes V$, where $l^\infty_{\mathbb{R}}$ is the real part of l^∞. Then ϱ is a Jauch-Piron state. Nevertheless, it has no Jauch-Piron extension to the von Neumann algebra $l^\infty \otimes M$, because this algebra is σ-finite and of Type III, and hence all Jauch-Piron states on M are regular. However, if we exclude the Type I_2 JW algebras, then the extension principle for Jauch-Piron states holds.

10.4.1. Theorem. *Let M be a JW algebra having no Type I_2 direct summand acting on a Hilbert space H. Let W be the von Neumann algebra generated by M. Then every Jauch-Piron state on M extends to a Jauch-Piron state on W.*

This result has a transparent physical meaning. All physically relevant states on almost all JW algebras are induced by physically relevant states on von Neumann algebras. Thus, von Neumann algebras have state space structure as rich as JW algebras. This contributes to the axiomatics of quantum mechanics by explaining the use of von Neumann algebras in terms of physically more plausible Jordan structures. Theorem 10.4.1 together with results of the present chapter now gives the following theorem.

10.4.2. Theorem. *Let ϱ be a Jauch-Piron state on a JW algebra M not having Type I_2 direct summand. If M is σ-additive, then*

(i) *ϱ is nonsingular if M is properly infinite,*

(ii) *ϱ is regular if M is of Type III,*

(iii) *when M is semifinite, ϱ is regular if, and only if, it is nonsingular on $Z(eMe)$ for all finite projections e of M.*

Let ϱ be a Jauch-Piron state on a JW algebra M, then

(i) *if M is properly infinite, then ϱ is regular on $Z(M)$,*

(ii) *if M has no σ-finite direct summand, then ϱ is regular.*

10.4.3. Corollary. *Every Jauch-Piron state on a JW factor is regular.*

We conclude the present chapter by challenging the reader with the following open problems.

10.4.4. Problem. *Let M be a von Neumann algebra. Are the Jauch-Piron states on M norm dense in the state space of M?*

The positive answer is known for σ-finite algebras (see Theorem 10.3.1).

10.4.5. Problem. *Are all factor Jauch-Piron states on properly infinite von Neumann algebras σ-additive?*

This result was proved under the continuum hypothesis in Theorem 10.3.14. Let us note that the positive answer to Problem 6.4.1 would imply the positive answer to Problem 10.4.5.

11. INDEPENDENCE OF QUANTUM SYSTEMS

The aim of this chapter is to discuss classical results in connection with the recent progress concerning independence of operator algebras. It has turn out that independence conditions imposed on operators that coincide in the standard probability theory are far from being equivalent in case of noncommutative algebras. The study of this special phenomenon of quantum measure theory has produced many significant results on the structure of subalgebras. The problem of independence of operator algebras arises primarily in the formalism of local algebras in quantum field theory where it embodies Einstein's causality principle. Nevertheless, we want to investigate the independence mainly on an abstract level, that is independently of special axioms of quantum field theory. We believe this approach is useful for the following reasons. First, the question of independence of systems is more general than the specific problem of locality in quantum field theory. Secondly, as we shall show, the main results on independence hold for general pairs of operator algebras and, thus, they are not dependent on all properties of local algebras in quantum field theory formalism. Consequently, the interplay between independence conditions is more or less unrelated to further axioms in quantum field theory. This may shed new light on the systems of postulates in relativistic quantum field theory.

Our exposition will follow the following plan. In the first preparatory Section 11.1. we show that for all abelian C^*-algebras all relevant conditions of independence coincide. We also recall the basic formalism of quantum field theory. In Section 11.2 we carefully investigate the independence of C^*-algebra. In its first part we deal with the independence of not necessarily commuting algebras. We provide a lucid characterization of C^*-independence (the existence of simultaneous extensions of states on C^*-algebras) by showing that although C^*-independent algebras may not be mutually commuting, they still enjoy many properties typical for tensor products. Further, we shall show that C^*-independence is always implied by the independence in the category of quantum logics and construct an example which shows that logical independence is nearly always stronger. Then we take up to commuting pairs of C^*-algebras. We reprove the classical Roos Theorem on the equivalence of C^*-independence and the Schlieder property by using only the compactness of the state space and show that the problem of independence of commuting algebras can be entirely reduced to the independence of their centres, i.e. it can be reduced to the classical case. We conclude Section 11.2 by discussing tensor product aspects of independence and product state extensions.

Section 11.3 deals with independence conditions arising in the category of von Neumann algebras. The position of W^*-independence (the existence of simultaneous normal extensions of normal states), C^*-independence and strict locality conditions is discussed. We shall characterize W^*-independence in terms of physically relevant Jauch-Piron states. On the one hand we show that W^*-independence is seldom equivalent to C^*-independence, on the other hand we prove that C^*-independence can be viewed as approximative form of W^*-independence. Hence we show that both concepts are equivalent from the physical point of view. Finally, we show that W^*-independence is equivalent to C^*-independence for commuting subalgebras of a σ-finite von Neumann algebra.

In concluding Section 11.4 a historical overview is offered. Moreover, the strict locality in the sense of vector states and the split property together with their relevance to structure of von Neumann factors is commented on. Finally, it is shown the relation of independence conditions to Bell inequalities and discussed further perspectives of the research in this area.

11.1. Independence in classical and quantum theory

In physics the concept of independence describes the degree of noninteraction between two or more subsystems of a global system. In probability theory it codifies the fact that random experiments do not affect each other. The independence is a well established notion in statistical sciences. Unlike this, as we shall see in the present chapter, there are many possible nonequivalent generalizations of the independence in the quantum context. In order to compare independence in classical and quantum probability theory we shall first examine equivalent forms of independence of probability spaces. Let $(X, \mathcal{P}_1, \mu_1)$ and $(Y, \mathcal{P}_2, \mu_2)$ be two probability spaces living, roughly speaking, in a global probability space $(Z, \mathcal{P}, \mu)$ that is generated by X and Y. We say that subspaces X and Y are *independent* in Z if Z is isomorphic to the product space $(X \times Y, \mathcal{P}_1 \times \mathcal{P}_2, \mu_1 \times \mu_2)$. Further, we say that two random variables f and g on the space $(Z, \mathcal{P}, \mu)$ are independent if μ is a product measure with respect to the fields generated by f and g, i.e. if

$$\mu(f^{-1}(-\infty, a]) \cap g^{-1}(-\infty, b]) = \mu(f^{-1}(-\infty, a]) \cdot \mu(g^{-1}(-\infty, b])$$

for all $a, b \in \mathbb{R}$. As a consequence, for the expectation value $E\colon h \to \int h(x)\, d\mu(x)$, we have $E(f\,g) = E(f)\,E(g)$, whenever f and g are independent. In applications the probability space is usually represented by the structure of random variables endowed with the expectation functional. This leads to a slightly more general description of the probability model given by a complex associative unital $*$ algebra A (representing random variables) equipped with a state φ (representing probability measure). We shall now provide a few characterizations of independence in case when A is an abelian C^*-algebra. For this, let us recall that the unital abelian C^*-algebra is isomorphic to the algebra $C(X)$ of continuous functions on its spectrum, X. Assume that D is a unital abelian C^*-algebra generated by two C^*-subalgebras A and B. Then the algebras A and B are independent in D if the spectrum of D is homeomorphic to the cartesian product of the spectra of A and B. This can be expressed in a few equivalent forms.

11.1.1. Theorem. *Let A and B be unital C^*-subalgebras of an abelian C^*-algebra D and let D be generated by A and B. Let X, Y, Z be the spectra of the algebras A, B, and D, respectively. Then the following conditions are equivalent:*

(i) *Z is homeomorphic to $X \times Y$.*

(ii) *D is isomorphic to the tensor product $A \otimes B$.*

(iii) *For any state φ_1 on A and for any state φ_2 on B there is a state φ on D such that $\varphi(a\,b) = \varphi(a)\,\varphi(b)$ for all $a \in A$, $b \in B$.*

(iv) *For any state φ_1 on A and for any state φ_2 on B there is a state φ on D extending both φ_1 and φ_2.*

(v) *If $a \in A$ and $b \in B$ are both nonzero, then $a\,b \neq 0$.*

Proof: The equivalence of (i) and (ii) is a well known fact. The implication (ii)$\Rightarrow$(iii) is a consequence of the fact that any pair of states φ_1 and φ_2 on the factors of the tensor product has a product extension. The implication (iii)$\Rightarrow$(iv) is trivial. Suppose that (iv) holds and consider two nonzero elements $a \in A$ and $b \in B$. In order to prove that $a\,b \neq 0$ it is clearly enough to assume that $0 \leq a, b \leq 1$ and $\|a\| = \|b\| = 1$, for otherwise we can replace a and b by $a^*\,a$ and $b^*\,b$, respectively, and scale them after it appropriately. There are states φ_1 on A and φ_2 on B, respectively, such that $\varphi_1(a) = \varphi_2(b) = 1$. Let φ be a common extension of φ_1 and φ_2 to D. We obtain $\varphi(a\,b) = \varphi(a)\,\varphi(b) = 1$, giving us $a\,b \neq 0$.

It remains to prove the implication (v)$\Rightarrow$(ii). Let Φ be a $*$-homomorphism mapping $A \otimes B$ into D, uniquely determined by the condition $\Phi(a \otimes b) = a\,b$. Our goal is to prove that Φ is a $*$-isomorphism. Let us suppose that this is not the case and try to reach a contradiction. Let h be a nonzero nonnegative element in the ideal Ker Φ. Let us canonically identify $A \otimes B$ with $C(X \times Y)$, where X and Y are the spectra of A and B, respectively, via the map

$$f \otimes g \to u(x, y) = f(x)\,g(y) \;:\; A \otimes B \to C(X \times Y)\,.$$

In this interpretation, h is a nonzero nonnegative function on $X \times Y$, and we can therefore find a point $(x, y) \in X \times Y$ with $h(x, y) > 0$. By the continuity there are compact neighbourhoods $U(x)$ and $U(y)$ of x and y, respectively, such that h is strictly positive at any point of $U(x) \times U(y)$. By the compactness of $U(x) \times U(y)$ we infer that, for some $c > 0$, $h(z) > c$ for all $z \in U(x) \times U(y)$. Employing now the compactness of X and Y, we can find nonzero nonnegative continuous functions f and g on X and Y, respectively, such that f vanishes outside $U(x)$ and g vanishes outside $U(y)$. In addition, by scaling the functions appropriately, we can arrange for $0 \leq f, g \leq \sqrt{c}$. It is now clear that the function $\overline{h}(x, y) = f(x)\,g(y)$ is an element of $C(X \times Y)$ satisfying $\overline{h} \leq h$, and hence $\Phi(\overline{h}) = 0$. However, since $\overline{h} = f \otimes g$, we obtain that the product of nonzero elements f and g in D is zero, which is a contradiction. □

Theorem 11.1.1 characterizes the C^*-tensor products in the category of abelian C^*-algebras. The condition (iii) explains why the tensor product embodies the independence of local algebras. Indeed, only in this situation the states on local algebras can be chosen arbitrarily because they always extend to a state on the global algebra. The condition (iii) is called the C^*-independence. It will be thoroughly analyzed in the next section. The condition (v), called the Schlieder property, is a surprisingly simple tool to force a very specific tensor product organization.

Each of the conditions in Theorem 11.1.1 can be immediately generalized to nonabelian algebras. The discussion in the next section will show that conditions (i)–(v) in Theorem 11.1.1 are all inequivalent in general when the global algebra D is noncommutative. Each of them can therefore express certain aspects of the independence of two quantum systems. In quantum mechanics the standard definition of independence is the strongest one in the hierarchy of conditions in Theorem 11.1.1. Two quantum mechanical subsystems A and B of a system D are supposed to be independent if $D = A \otimes B$ for the corresponding algebras. This seems to be a reasonable definition since forming the tensor products is parallel to forming the products of probability spaces. However, there are many important pairs of subalgebras in quantum field theory that are independent but that do not generate tensor product. For this reason many alternative independence conditions have been introduced and studied in the context of quantum field theory. Since all the concepts treated in this chapter have their origin in the algebraic approach to quantum field theory, let us briefly outline the basic formalism.

The essential feature of the quantum field theory is the locality principle that has two aspects. First, according to Einstein relativity theory, no observation has its meaning for all spacetime. It is therefore necessary to define observables as only attached to bounded regions in the Minkowski spacetime. Secondly, according to the Einstein principle of causality, the observables associated to causally disjoint regions should be independent. There are many mathematical models implementing these requirements. Without going into the details, we shall single out basic axioms fulfilled by all of the models. Let $\mathcal{M}$ be the *Minkowski spacetime*, i.e. $\mathcal{M} = \{\vec{x} \mid \vec{x} = (x_0, x_1, x_2, x_3) \in \mathbb{R}^4\}$ (x_0 is the time coordinate) endowed with the indefinite inner product

$$\langle \vec{x}, \vec{y} \rangle = c^2 \, x_0 \, y_0 - x_1 \, y_1 - x_2 \, y_2 - x_3 \, y_3 \, ,$$

where c denotes the speed of the light. We say that $\vec{x}$ and $\vec{y}$ in M are *spacelike separated* if

$$\varrho(\vec{x},\vec{y}) = \langle \vec{x}-\vec{y}, \vec{x}-\vec{y} \rangle < 0 .$$

In other words, because the speed of the light is the largest parameter, no signal can get from $\vec{x}$ to $\vec{y}$ and vice versa. Two bounded regions O_1 and O_2 in M are said to be *causally independent* (or *spacelike separated*) if $\varrho(\vec{x},\vec{y}) < 0$ for all $\vec{x} \in O_1$ and $\vec{y} \in O_2$. Quantum field theory assigns to each bounded open set O in $\mathcal{M}$ the C^*-algebra $A(O)$ representing all observables in the region O. For obtaining the physical content of this approach, the following axioms for the assignment $O \to A(O)$ seem to be indispensable:

(i) (Isotony) If $O_1 \subset O_2$, then $A(O_1)$ is a subalgebra of $A(O_2)$.

(ii) (Local commutativity) If O_1 and O_2 are causally independent, then $A(O_1)$ and $A(O_2)$ mutually commute, i.e. $a\,b = b\,a$ whenever $a \in A(O_1)$ and $b \in A(O_2)$.

(iii) (Lorenz covariance) There exists a representation of the Poincare group $\mathcal{P}$ by automorphism $\alpha(g)$, $g \in \mathcal{P}$, acting on the inductive limit A of the algebras $A(O)$, O varies over all subregions of M, such that

$$\alpha(g)\,A(O) = A(g(O)) .$$

The axiom (i) says that what can be measured in O_1 can also be measured in O_2. The postulate (iii) encodes the invariance of the physical laws with respect to the Lorenz transformations. The principle axiom that is behind all independence conditions studied in the present chapter is the condition (iii). If O_1 and O_2 are causally independent bounded open sets in M, then no measurement performed in the region O_1 should disturb or be disturbed by any experiment performed in the region O_2. That is, algebras $A(O_1)$ and $A(O_2)$ should be independent. In the axiom (iii) the independence is expressed by the commutativity of the corresponding algebras. This requirement is sometimes called the *kinematical independence* and it codifies the fact that observables from regions O_1 and O_2 are not connected be the uncertainty relations. However, in view of the Theorem 11.1.1, the commutativity itself does not embody real statistical independence. For this reason the axiom (iii) is often replaced by another independence condition that is either postulated or derived from other axioms. For example, Haag and Kastler suggested C^*-independence as the most natural independence condition having physical content. Besides this, many other independence conditions have been developed in the framework of the quantum field theory. The aim of the present chapter is to present very recent results on logical interrelations between them. In particular, we focus on the C^*-independence that is central for the Haag-Kastler theory in which the local algebras are C^*-algebras and on the W^*-independence that is principal for the Haag-Araki theory in which all local algebras are von Neumann algebras.

11.2. Independence of C^*-algebras

Throughout this section, all C^*-algebras will be assumed to be unital. Moreover, by saying that A is a C^*-subalgebra of B we shall imply that A and B have the same unit. The following concept is principal in the present section.

11.2.1. Definition. Let A and B be C^*-subalgebras of a C^*-algebra D. Then A and B are said to be C^**-independent* in D if the following condition holds. If φ_1 and φ_2 are states on A_1 and A_2, respectively, then there is a state φ on D extending both φ_1 and φ_2.

The C^*-independence was introduced by Haag and Kastler in [155] in the context of the algebraic quantum field theory. According to their axiom of independence, the local algebras corresponding to causally independent regions should be "completely uncoupled", i.e. C^*-independent. The C^*-independence embodies the fact that any preparation (i.e. any state) of one subsystem cannot affect any preparation of another subsystem. In other words, no restrictions on the given subsystems are imposed by viewing them as a part of the larger one. In addition to its physical interest, C^*-independence has a natural mathematical meaning for it is the property of simultaneous extension. Let us note that the Hahn-Banach Theorem guarantees that any state can be extended from subalgebra to a larger algebra. However, for having the simultaneous extension the special position of the algebras is required. For example, if A and B are C^*-independent in the algebra D, then necessarily $A \cap B = \mathbb{C}1$ (such a condition is called the *extended locality*). Indeed, suppose that $a \in A \cap B$ is not a multiple of the unit. Then there is a positive norm-one element $c \in A \cap B$ such that $1 - c \neq 0$. Take now a state φ_1 on A with $\varphi_1(c) = 1$ and any state φ_2 on B such that $\varphi_2(1-c) > 0$. If it were a state extension φ of φ_1 and φ_2, then $1 = \varphi(1) = \varphi(c) + \varphi(1-c) > 1$, which is absurd. The C^*-independence is therefore interesting for the structural properties of subalgebras. An obvious example of an algebra in which C^*-algebras A and B are independent is the tensor product $A \otimes B$. In this case all pairs of states have product extensions. However, as the following example shows, there is an abundance of noncommuting C^*-independent algebras.

11.2.2. Example. (i) Noncommutative abelian algebras that are C^*-independent:

Let $H = \mathbb{C}^6$. Define projections p and q by

$$p = \begin{pmatrix} 1 & 0 & 0 & 0 & 0 & 0 \\ 0 & 1 & 0 & 0 & 0 & 0 \\ 0 & 0 & 0 & 0 & 0 & 0 \\ 0 & 0 & 0 & 0 & 0 & 0 \\ 0 & 0 & 0 & 0 & 1/2 & 1/2 \\ 0 & 0 & 0 & 0 & 1/2 & 1/2 \end{pmatrix} \qquad p = \begin{pmatrix} 1 & 0 & 0 & 0 & 0 & 0 \\ 0 & 0 & 0 & 0 & 0 & 0 \\ 0 & 0 & 1 & 0 & 0 & 0 \\ 0 & 0 & 0 & 0 & 0 & 0 \\ 0 & 0 & 0 & 0 & 1 & 0 \\ 0 & 0 & 0 & 0 & 0 & 0 \end{pmatrix}$$

and put $A = C^*(p) = \mathrm{sp}\{p, 1-p\}$, $B = C^*(q) = \mathrm{sp}\{q, 1-q\}$. Evidently, A and B do not commute. Let φ_1 and φ_2 be states on A and B, respectively. They are uniquely determined by the parameters $a = \varphi_1(p)$ and $b = \varphi_2(q)$. Define a diagonal matrix with nonnegative diagonal entries

$$\varrho = \mathrm{diag}(a\,b, a\,(1-b), (1-a)\,b, (1-a)\,(1-b), 0, 0)\,.$$

Then ϱ has trace 1 and so it defines a density matrix of a state φ on $B(H)$. It can be verified easily that $\varphi(p) = a$ and $\varphi(q) = b$. Let us remark that the state φ constructed above is not the only extension of φ_1 and φ_2. For example, the state given by the diagonal density matrix

$$\begin{aligned} \varrho' &= \mathrm{diag}\Bigg(\frac{b\,(a+a^2)}{2}, \frac{(1-b)\,(a+a^2)}{2}, \frac{b\,(2-3\,a+a^2)}{2}, \\ &\qquad \frac{(1-b)\,(2-3\,a+a^2)}{2}, b\,(a-a^2), (1-b)\,(a-a^2)\Bigg) \end{aligned}$$

is a product extension of φ_1 and φ_2 that is faithful if $a, b \in (0,1)$.

(ii) Noncommuting C^*-independent algebras one of which is not abelian: We shall again consider the algebra $B(H)$, where $H = \mathbb{C}^6$. Let A be the C^*-algebra generated by the diagonal matrices $p = \text{diag}(1,1,1,0,0,0)$ $1-p = (0,0,0,1,1,1)$, and let B be the C^*-algebra generated (as a linear space) by the matrices p_α given by

$$\begin{pmatrix} c_\alpha^2 & c_\alpha\, s_\alpha & 0 & 0 & 0 & 0 \\ c_\alpha\, s_\alpha & s_\alpha^2 & 0 & 0 & 0 & 0 \\ 0 & 0 & c_\alpha^2 & c_\alpha\, s_\alpha & 0 & 0 \\ 0 & 0 & c_\alpha\, s_\alpha & s_\alpha^2 & 0 & 0 \\ 0 & 0 & 0 & 0 & c_\alpha^2 & c_\alpha\, s_\alpha \\ 0 & 0 & 0 & 0 & c_\alpha\, s_\alpha & s_\alpha^2 \end{pmatrix},$$

where $c_\alpha = \cos\alpha$, $s_\alpha = \sin\alpha$, $\alpha \in [0, \pi/2]$. The algebras A and B do not commute. (For instance, the diagonal matrix $\text{diag}(1,1,1,0,0,0)$ does not commute with any p_α, where $\alpha \in (0, \frac{\pi}{2})$.) It is also clear that B is not commutative. Consider now the states φ_1 and φ_2 on A and B, respectively. Let ϱ_2 be the density matrix corresponding to the state φ_2. Without loss of generality we can suppose that ϱ_2 has nonzero entries only in the 2×2 upper left corner. The state φ_1 is uniquely determined by the parameter $\gamma = \varphi_1(p)$. Since any state on A is a convex combination of states concentrated either at p or at $1-p$, there is no loss of generality in assuming that $\gamma = 1$. Put

$$\varrho_1 = \text{diag}(1/2, 1/2, 0, 0, 0, 0)$$

and set $\varrho = 2\,\varrho_1\varrho_2$. We shall verify that the functional φ with the density matrix ϱ is a common extension of φ_1 and φ_2. Indeed, ϱ_1 and ϱ_2 commute, and hence ϱ is nonnegative. Moreover, $\text{tr}\,\varrho_1\,\varrho_2 = \frac{1}{2}\,\text{tr}\,\varrho_2 = \frac{1}{2}$, showing that ϱ is a state. Further,

$$\varphi(p) = \text{tr}(2\,\varrho_1\,\varrho_2\,p) = 2\,\text{tr}(p\,\varrho_1\,\varrho_2) = 2\,\text{tr}(1/2\,\varrho_2) = 2\,\frac{1}{2} = 1\,.$$

Similarly, for p_α we obtain

$$\varphi(p_\alpha) = 2\,\text{tr}(\varrho_1\,\varrho_2\,p_\alpha) = 2\,\text{tr}(1/2\,\varrho_2\,p_\alpha) = \text{tr}(\varrho_2\,p_\alpha) = \varphi_2(p_\alpha)\,.$$

We will now exhibit an example of abelian C^*-algebras that are manifestly not C^*-independent.

11.2.3. Example. Let $\mathcal{A}_\theta$ be the irrational rotation algebra (see Chapter 2, Projection structure of C^*-algebras) generated by two unitaries u and v satisfying the relations

$$u\,v = e^{2\pi i\theta}\,v\,u\,, \tag{11.1}$$

where θ is irrational. Let $A = C^*(u)$, $B = C^*(v)$ be unital C^*-subalgebras of $\mathcal{A}_\theta$. (Let us remark that u and v are transformed forms of the position and the momentum operator.) Suppose that φ_1 and φ_2 are pure states on A and B, respectively. We will show that there is no common state extension of φ_1 and φ_2 to $\mathcal{A}_\theta$. Suppose, for a contradiction, that φ is such an extension. Then φ is pure on A and B and therefore multiplicative on these algebras. It implies (see Chapter 2, The Gelfand-Naimark-Segal construction)

$$\varphi(a_1\,b_1\,a_2\,b_2\cdots a_n\,b_n) = \varphi(a_1)\,\varphi(a_2)\,\cdots\varphi(a_n)\,\varphi(b_1)\,\varphi(b_2)\cdots\varphi(b_n)$$

for all $a_1, a_2, \cdots a_n \in A$ and $b_1, b_2, \cdots, b_n \in B$. In other words, φ is multiplicative on $\mathcal{A}_\theta$. However, $\mathcal{A}_\theta$ is simple (and infinite-dimensional) and so it admits no multiplicative state, which is a contradiction.

The pair of abelian algebras in the previous example is far from being C^*-independent in the sense that that no pair of pure states extends to a global state. On the other hand, the existence of simultaneous extensions for pure states is equivalent to C^*-independence as the following proposition shows.

11.2.4. Proposition. *Let A and B be C^*-subalgebras of a C^*-algebra D. Then A and B are C^*-independent if, and only if, for any pure state φ_1 on A and for any pure state φ_2 on B there is a state φ on D such that $\varphi|A = \varphi_1$ and $\varphi|B = \varphi_2$.*

Proof: Let φ_1 and φ_2 be prescribed states on A and B, respectively. Assume first that both φ_1 and φ_2 is a convex combination of pure states. Let us write

$$\varphi_1 = \sum_{i=1}^{n} \alpha_i\,\varrho_i\,, \qquad \varphi_2 = \sum_{j=1}^{m} \beta_j\,\psi_j\,,$$

where $\varrho_1, \ldots, \varrho_n$ are pure states of A and $\psi_1, \ldots, \psi_m$ are pure states of B ($\alpha_i, \beta_j \geq 0$, $\sum_{i=1}^m \alpha_i = \sum_{j=1}^m \beta_j = 1$). By the hypothesis, we can find states $\tilde{\varrho}_1, \ldots, \tilde{\varrho}_n$ of A such that

$$\tilde{\varrho}_i|A = \varrho_i\,, \qquad \tilde{\varrho}|B = \psi_1 \qquad (i = 1, \ldots, n)\,.$$

Putting $\tilde{\psi}_1 = \sum_{i=1}^n \alpha_i\, \tilde{\varrho}_i$, we obtain a state of A with

$$\tilde{\psi}_1|A = \varphi_1\,, \qquad \tilde{\psi}_1|B = \psi_1\,.$$

Similarly we can find states $\tilde{\psi}_2, \ldots, \tilde{\psi}_m$ of A such that

$$\tilde{\psi}_j|A = \varphi_1\,, \qquad \tilde{\psi}_j|B = \psi_j\,.$$

Letting now $\varphi = \sum_{j=1}^m \beta_j\, \tilde{\psi}_j$, we see that φ is the desired extension of both φ_1 and φ_2. Finally, let φ_1 and φ_2 be arbitrary states. By the Krein-Milman Theorem both φ_1 and φ_2 are weak*-limit point of nets $(\varphi_\alpha)_{\alpha\in I}$ and $(\psi_\beta)_{\beta\in J}$ consisting of convex combinations of pure states of A and B, respectively. Considering now a net $(\varphi_{\alpha,\beta})_{\alpha\in I,\beta\in J}$ consisting of states $\varphi_{\alpha,\beta}$ extending simultaneously φ_α and φ_β, we can employ the compactness of the state space of A and obtain so the weak*-limit point, φ, of the net $(\varphi_{\alpha,\beta})$. It is clear that φ is the desired extension of φ_1 and φ_2. □

Next theorem provides a lucid characterization of C^*-independence.

11.2.5. Theorem. *Let A and B be C^*-subalgebras of a C^*-algebra D. The following conditions are equivalent.*

(i) *A and B are C^*-independent.*

(ii) *For every positive norm-one elements $a \in A$ and $b \in B$ there is a state φ on D such that $\varphi(a) = \varphi(b) = 1$.*

(iii) $\|a\,b\| = \|a\| \cdot \|b\|$ *for all $a \in A$ and $b \in B$.*

Proof: First we prove the equivalence (i)⇔(ii). Since for any $0 \leq a \in A$, $\|a\| = 1$, and any $0 \leq b \in B$, $\|b\| = 1$, there are states φ_1 and φ_2 on A and B such that $\varphi_1(a) = \varphi_2(b) = 1$, we conclude that (i)⇒(ii). Suppose that (ii) holds. Let φ_1 and φ_2 be assigned states on the algebras A and B, respectively. We shall establish a common extension of φ_1

and φ_2, first we shall do it on condition that D is separable. According to Proposition 11.2.4, we can consider the case of φ_1 and φ_2 being pure states. By the virtue of Theorem 9.1.3 there are determining positive elements a and b of φ_1 and φ_2, respectively. Let us now take a state φ on D such that $\varphi(a) = \varphi(b) = 1$. Then the determinacy of φ_1 and φ_2 by a and b gives us $\varphi|A = \varphi_1$ and $\varphi|B = \varphi_2$. Having proved the separable case, let us now proceed to the case when D is arbitrary. Denote by $\mathcal{S}$ the system of all finite nonempty subsets of $A \cup B$. For any $S \in \mathcal{S}$ we put

$$F_S = \{\varphi \in S(D) \mid \varphi|S \cap A = \varphi_1 \,, \varphi|S \cap B = \varphi_2\} \,.$$

Since any C^*-algebra D_S generated by the set S, $(S \in \mathcal{S})$, is separable, there is a state of D_S extending both $\varphi_1|D_S \cap A$ and $\varphi_2|D_S \cap B$. In other words, the sets F_S, $S \in \mathcal{S}$, are nonempty closed sets in the compact space $S(D)$. Moreover, the inclusion

$$F_{\bigcup_{i=1}^n S_i} \subset \bigcap_{i=1}^n F_{S_i}$$

says that the system $(F_S)_{S \in \mathcal{S}}$ enjoys the finite intersection property. Hence, $F = \bigcap_{S \in \mathcal{S}} F_S \neq \emptyset$ by compactness and any state $\varphi \in F$ is the required extension of φ_1 and φ_2.

(ii)$\Rightarrow$ (iii) Suppose that (ii) holds and that $a \in A$, $b \in B$. For proving the property (iii) there is no loss of generality involved in assuming that $\|a\| = \|b\| = 1$. We first prove the validity of (iii) for all $a, b \geq 0$. By (ii), there is a state φ on D with $\varphi(a) = \varphi(b) = 1$. Then φ is definite at a and b, and hence $\varphi(a\,b) = \varphi(a)\varphi(b) = 1$, giving immediately that $\|a\,b\| = 1$. Suppose now that a and b are arbitrary and write $|a| = \sqrt{a^*\,a}$, $|b^*| = \sqrt{b\,b^*}$. Then

$$\|a\,b\,\|^2 = \|b^*\,a^*\,a\,b\,\| = \|b^*\,|a|^2\,b\,\| = \|\,|a|\,b\|^2 \,.$$

Further,

$$\|\,|a|\,b\|^2 = \|b^*\,|a|\,\|^2 = \|\,|a|\,b\,b^*\,|a|\,\| = \|\,|a|\,|b^*|\,\|^2 \,.$$

Hence, $\|a\,b\| = \|\,|a|\,|b^*|\,\|$, and thus by the previous reasoning $\|a\,b\| = \|\,|a|\,\| \cdot \|\,|b^*|\,\| = \|a\| \cdot \|b\|$.

(iii) $\Rightarrow$ (ii) Suppose that (iii) holds and consider norm one nonnegative elements $a \in A$, $b \in B$. Then $\|a\,b\| = 1$, and hence $\|b\,a^2\,b\| = 1$. Let us take a state φ on D with $\varphi(b\,a^2\,b) = 1$. Then

$$1 = \varphi(b\,a^2\,b) \leq \|a\|^2\,\varphi(b^2) = \varphi(b^2) \,.$$

However, since $0 \leq b^2 \leq b$, we see that $\varphi(b) = 1$. Using now definiteness of φ at b we infer that

$$1 = \varphi(b\, a^2\, b) = \varphi(b)\, \varphi(a^2)\, \varphi(b) = \varphi(a^2)\,,$$

implying $\varphi(a) = 1$. □

The condition (ii) in the previous theorem means that for establishing simultaneous extensions it is enough to expose two positive normalized elements by a common positive linear functional on the global algebra. In terms of probability theory, the C^*-independence means that given two normalized positive observables from local algebras, one can always prepare the global system in such a state φ that both observables have the expectation value simultaneously one. The condition (iii) characterizes the C^*-independence only by referring to the norm on the global algebra. It says that this norm is multiplicative with respect to local algebras, and hence it has a typical property of the tensor product norm. (Compare the abelian case in Theorem 11.1.1.)

Many independence conditions in quantum field theory are formulated as so-called strict locality principles. They postulate that any preparation of one system does not disturb the statistical properties of prescribed observable from another system. As an immediate corollary of Theorem 11.2.5 we obtain that the C^*-independence and the principle of strict locality coincide in the category of C^*-algebras.

11.2.6. Corollary. *Let A and B be C^*-subalgebras in a C^*-algebra D. The following conditions are equivalent.*

(i) *A and B are C^*-independent.*

(ii) *For any nonnegative $a \in A$ with $\|a\| = 1$ and any state ψ on B there is a state φ on D such that $\varphi(a) = 1$ and $\varphi|B = \psi$.*

Moreover, when considered A in its universal representation as an algebra acting on a Hilbert space H_u, then the above stated conditions are equivalent to

(iii) *For all positive norm-one elements $a \in A$ and $b \in B$ there is a common eigenvector $\xi \in H_u$ of a and b corresponding to eigenvalue one.*

Proof: The equivalence of (i) and (ii) follows from Theorem 11.2.5. The equivalence of (i) and (iii) is a consequence of the fact that every state in the universal representation is a vector state ω_ξ. Now, the condition $(a\,\xi,\xi) = (b\,\xi,\xi) = 1$ implies that $a\,\xi = \xi$ and $b\xi = \xi$ by the Cauchy-Schwarz inequality. □

Another direct consequence of Theorem 11.2.5 is the fact that the product of two nonzero elements from C^*-independent local algebras is always nonzero. This condition is called the *Schlieder property* and it is recorded in the following definition.

11.2.7. Definition. Let A and B be C^*-subalgebras of a C^*-algebra D. The algebras A and B are called *Schlieder independent* (S-independent in the abbreviation) if $a\,b \neq 0$ whenever $0 \neq a \in A$ and $0 \neq b \in B$.

The S-independence is more or less technical condition that can be verified sometimes more easily than extension type properties. It is a necessary condition for C^*-independence but not sufficient as the following example demonstrates.

11.2.8. Example. Let $p, q \in M_2(\mathbb{C})$ be one-dimensional distinct and nonorthogonal projections. Let $A = C^*(p,1)$, $B = C^*(q,1)$. The algebras A and B are S-independent. Indeed, without loss of generality it is possible to assume that $a = p + \lambda_1\,(1-p) \neq 0$ and $b = q + \lambda_2\,(1-q) \neq 0$ ($\lambda_1, \lambda_2 \in \mathbb{C}$) and to prove that $a\,b \neq 0$. Suppose on the contrary that

$$(p + \lambda_1\,(1-p))\,(q + \lambda_2(1-q)) = 0\,.$$

This rewrites as

$$p\,q = c_1\,(1-p)\,q + c_2\,(1-p)\,(1-q) + c_3\,p\,(1-q)\,,$$

for some $c_1, c_2, c_3 \in \mathbb{C}$. By multiplying the above equation by p and q from the left and right, successively, we obtain $p\,q = 0$, which is a contradiction. Nevertheless, we shall show that A and B are not C^*-independent by demonstrating that there is no state φ on $M_2(\mathbb{C})$ such that $\varphi(p) = \varphi(q) = 0$. (Such states of course exist on partial algebras.) The reason is that for any unit vector $\xi \in \mathbb{C}^2$

$$\Big((p+q)\xi, \xi\Big) = \|p\,\xi\|^2 + \|q\,\xi\|^2 > 0\,.$$

By the compactness of the unit sphere there is a constant $c > 0$ with

$$\Big((p+q)\xi, \xi\Big) \geq c > 0 \text{ for all unit vectors } \xi \in \mathbb{C}^2 .$$

In other words,

$$p + q \geq cI ,$$

where I is the identity operator, and hence,

$$\varphi(p+q) \geq c > 0 \qquad \text{for every state } \varphi .$$

Further natural independence condition that has been introduced by Redei in the context of quantum logic is that of logical independence [274].

11.2.9. Definition. Let A and B be C^*-subalgebras of a C^*-algebra D. We say that A and B are *logically independent* if for all nonzero projections $p \in A$ and $q \in B$ there is a nonzero projection $r \in D$ such that $r \leq p$ and $r \leq q$.

Of course, the condition of logical independence is strong only for C^*-algebras having sufficiently many projections. For example, if A and B are von Neumann subalgebras in a von Neumann algebra D, then A and B are logically independent if nonzero projections from A and B have nonzero intersections in D. The meaning of logical independence can be explained as follows. If r and p are projections in a C^*-algebra A, then $r \leq p$ if, and only if, the following condition holds: $\varphi(p) = 1$ whenever φ is a state on A with $\varphi(r) = 1$. The projections r and p represent propositions (yes-no decision effects) on the system. In this light the relation $r \leq p$ reads: If r is true in a given state, then p is also true in this state. Therefore, A and B are logically independent if every pair of projections from local algebras is implied by some proposition on D. Loosely speaking, no nontrivial proposition on A implies or is implied by any proposition in B. Now we want to explain the probability content of the logical independence. The infimum $p \wedge q$ is nonzero for nonzero projections $p \in P(A)$ and $q \in P(B)$ if, and only if, there is a *normal* state ψ on D such that $\psi(p) = \psi(q) = 1$. Indeed, if $p \wedge q$ is nonzero, then any vector state concentrated at $p \wedge q$ is concentrated simultaneously at p and q. On the other hand, if ψ is a normal state with $\psi(p) = \psi(q) = 1$, then the Jauch-Piron property of normal states entails that $\psi(p \wedge q) = 1$, meaning that $p \wedge q \neq 0$. Suppose now that A and B are logically independent and that $p \in A$ and $q \in B$ are nontrivial projections. Given a parameter $a \in [0,1]$, we can find normal states ϱ_1 and ϱ_2 on the global algebra with $\varrho_1(p) = 1$, $\varrho_1(q) = 1$ and $\varrho_2(p) = 0$, $\varrho_2(q) = 1$. Then the normal state $\varrho = a\,\varrho_1 + (1-a)\,\varrho_2$ has the property $\varrho(p) = a$ and $\varrho(q) = 1$. Similarly, given a parameter $b \in [0,1]$, we can find a normal state φ such that $\varphi(p) = a$ and $\varphi(1-q) = 1$. Given a second parameter $b \in [0,1]$, we can set $\psi = b\,\varrho + (1-b)\,\varphi$ to obtain a normal state with

$$\psi(p) = a \qquad \text{and} \qquad \psi(q) = b\,.$$

Therefore, systems are logically independent if there is always a *normal* state realizing arbitrary probability of the propositions from local algebras.

We shall start the study of logical independence by stating that algebras generated by the position and the momentum operator are violating logical independence in nearly all projections. In a way this example is a W^*-version of Example 11.2.3.

11.2.10. Example. Let Q and P be the observables of the position and the momentum of the particle moving in one dimension. More specifically, Q and P are unbounded operators on the Hilbert space $H = L^2(\mathbb{R}, \mu)$, where μ is the Lebesgue measure, such that

$$(P\,f)(x) = x\,f(x) \text{ for all } f \in L^2(\mathbb{R}, \mu) \text{ with } x\,f(x) \in L^2(\mathbb{R}, \mu)\,,$$

and

$$Q = \mathcal{F}^{-1}\,P\,\mathcal{F}\,,$$

where $\mathcal{F}\colon\ L^2(\mathbb{R}, \mu) \to L^2(\mathbb{R}, \mu)$ is the Fourier-Plancherel transformation. We shall show that von Neumann algebras A and B generated by the spectral projections of Q and P, respectively, are not logically independent. For this let us take bounded Borel sets $B_1, B_2 \subset \mathbb{R}$. Let p and q be the spectral projections of Q and P, respectively, corresponding to B_1 and B_2. Then $f \in p(H)$ if, and only if, the support of f is in B_1 and $f \in q(H)$ if, and only if, the Plancherel-Fourier transform $\hat{f}$ of f has the support contained in B_2. However, according to the Payley-Wiener Theorem no function can be localized both in the time and the frequency domain. It means that $f = 0$, and thus $p \wedge q = 0$.

The next result shows that for C^*-algebras with rich projection structure logical independence implies C^*-independence.

11.2.11. Theorem. *Let D be a real rank zero algebra containing two real rank zero C^*-subalgebras A and B. If A and B are logically independent, then they are C^*-independent.*

Proof: Assume that A and B are logically independent. According to Theorem 11.2.5, in order to prove the C^*-independence we have to show that there is a state, φ, attaining value one on the prescribed norm-one positive elements $a \in A$ and $b \in B$. For a given $n \in \mathbb{N}$, there are self-adjoint elements $a_n \in A$, $b_n \in B$ with finite spectra such that

$$\|a - a_n\| \leq \frac{1}{n}, \qquad \|b - b_n\| \leq \frac{1}{n}. \tag{11.2}$$

Let us write $a_n = \sum_{j=1}^{k} \lambda_j p_j$, where $\lambda_j \in \mathbb{R}$ and p_j are orthogonal projections in A. As $\|a_n\| = \max_j |\lambda_j|$, we see by (11.2) that

$$\lambda_{j_0} \geq 1 - \frac{1}{n}$$

for some j_0. For an arbitrary state ϱ on D with $\varrho(p_{j_0}) = 1$ we have

$$\frac{1}{n} \geq |\varrho(a) - \varrho(a_n)| = |\varrho(a) - \lambda_{j_0}|.$$

Therefore,

$$\varrho(a) \geq \lambda_{j_0} - \frac{1}{n} \geq 1 - \frac{2}{n}.$$

Consequently, for each n there are projections p_n and q_n in A and B, respectively, such that whenever ϱ is a state on D with $\varrho(p_n) = \varrho(q_n) = 1$, then $\varrho(a) \geq 1 - \frac{2}{n}$ and $\varrho(b) \geq 1 - \frac{2}{n}$. The logical independence of A and B implies the existence of nonzero projection $r_n \leq p_n, q_n$ in D. Take now a state ϱ_n on D satisfying $\varrho_n(r_n) = 1$. We have $\varrho_n(a) \geq 1 - \frac{2}{n}$, $\varrho_n(b) \geq 1 - \frac{2}{n}$. Employing now the compactness of the state space, we obtain a weak*-limit point, φ, of the sequence (ϱ_n) that consequently satisfies $\varphi(a) = \varphi(b) = 1$.

□

The next theorem shows that logical independence is strictly stronger then C^*-independence for all essentially noncommutative von Neumann algebras.

11.2.12. Theorem. *Let M be a von Neumann algebra. The following conditions are equivalent*

(i) *There are finite-dimensional von Neumann subalgebras A and B of M which are C^*-independent but not logically independent.*

(ii) *M has an infinite-dimensional nonabelian central summand.*

Proof: (ii)$\Rightarrow$ (i). Supposing that the nonabelian central summand of M is infinite-dimensional, we are going to construct a pair of finite-dimensional subalgebras of M which are C^*-independent but not logically independent. As a preparation for this we shall prove that M contains a copy of the algebra

$$(l^\infty \overline{\otimes} M_2(\mathbb{C})) \oplus \mathbb{C} \oplus \mathbb{C} \oplus \mathbb{C}.$$

At first step we show that there is a copy of

$$l^\infty \overline{\otimes} M_2(\mathbb{C})$$

contained, as a subalgebra, in some hereditary von Neumann subalgebra of M. According to the assumption, the nonabelian Type I part or the non-Type I part of M has infinite dimension. Let us consider the first case, supposing that M has an infinite-dimensional direct summand consisting of Type I_{n_k} parts, each $n_k \geq 2$. Assume that one of these homogeneous Type I summands is infinite-dimensional. In that case there is an infinite-dimensional central summand of M isomorphic to $R \overline{\otimes} B(K)$, where R is abelian and $\dim K \geq 2$, which implies that R is infinite-dimensional or K is infinite-dimensional. In the former case R contains a copy of l^∞ and so $R \overline{\otimes} B(K)$ contains a copy of $l^\infty \overline{\otimes} M_2(\mathbb{C})$. In the latter case $B(K) \simeq B(L) \overline{\otimes} (M_2)(\mathbb{C})$ with $\dim L = \infty$, and we can find a copy of l^∞ in $B(L)$, and hence a copy of $l^\infty \overline{\otimes} M_2(\mathbb{C})$ in M. On the other hand, if each homogeneous Type I_{n_k} direct summand has finite dimension, then M has to contain infinitely many nonzero direct summands of this kind and so there is a copy of $\sum_{n=1}^{\infty} \oplus M_2(\mathbb{C})$ in M, which is an algebra isomorphic to $l^\infty \overline{\otimes} M_2(\mathbb{C})$.

Let us now discuss the case of M having nonzero central summand G which is of Type II or III. In both cases there is a projection r in G such that $r \sim 1 - r$ and $\dim r\,G\,r = \infty$. So, again, $r\,G\,r$ contains a subalgebra isomorphic to l^∞ and $G \simeq r\,G\,r \overline{\otimes} M_2(\mathbb{C})$ contains a copy of $l^\infty \overline{\otimes} M_2(\mathbb{C})$.

In summary, we have shown that M admits a projection p such that the algebra $p\,M\,p$ has a subalgebra isomorphic to $l^\infty \overline{\otimes} M_2(\mathbb{C})$. In the sequel we shall always identify $l^\infty \overline{\otimes} M_2(\mathbb{C})$ with $\sum_{n=1}^\infty \oplus M_2(\mathbb{C})$ via the map

$$(a_n)_{n=1}^\infty \otimes b \to \sum_{n=1}^\infty \oplus\, a_n\, b, \qquad (a_n)_{n=1}^\infty \in l^\infty,\ b \in M_2(\mathbb{C}).$$

Assume that $1 - p \neq 0$. Then

$$S = \mathbb{C} \oplus \mathbb{C} \oplus \sum_{n=3}^\infty \oplus M_2(\mathbb{C})$$

is a (unital) subalgebra of $\sum_{n=1}^\infty \oplus M_2(\mathbb{C})$. On taking $S \oplus \mathbb{C}\,(1 - p)$, we obtain the desired subalgebra. If $p = 1$ we can, in the same way, take the subalgebra

$$\mathbb{C} \oplus \mathbb{C} \oplus \mathbb{C} \oplus \sum_{n=4}^\infty \oplus M_2(\mathbb{C})$$

in $\sum_{n=1}^\infty \oplus M_2(\mathbb{C})$ to obtain the subalgebra required.

In the rest of the proof we shall assume that

$$M = (l^\infty \overline{\otimes} M_2(\mathbb{C})) \oplus \mathbb{C} \oplus \mathbb{C} \oplus \mathbb{C}.$$

Let $x, x_1, x_2, \ldots,$ be a sequence of distinct unit vectors in the two-dimensional Hilbert space such that $x_n \to x$ and $(x, x_n) \neq 0$ for all n. Denote by e and f_n, $n = 1, 2, \ldots,$ the atomic projections in $M_2(\mathbb{C})$ (identified with $B(H)$, $\dim H = 2$) with ranges spanned by the vectors $x, x_1, x_2, \ldots,$ correspondingly. We define two projections g and h in M by letting

$$\begin{aligned} g &= \left(\sum_{n=1}^\infty \oplus e\right) \oplus 1 \oplus 0 \oplus 0 \\ h &= \left(\sum_{n=1}^\infty \oplus f_n\right) \oplus 0 \oplus 1 \oplus 0. \end{aligned}$$

Put

$$\begin{aligned} A &= W^*(1, g) = sp\{g, 1 - g\} \\ B &= W^*(1, h) = sp\{h, 1 - h\}. \end{aligned}$$

It is clear that $g \wedge h = 0$, while

$$(1-g)\wedge h \ \geq \ \left(\sum_{n=1}^{\infty} \oplus 0\right) \oplus 0 \oplus 1 \oplus 0 \neq 0 \tag{11.3}$$

$$g \wedge (1-h) \ \geq \ \left(\sum_{n=1}^{\infty} \oplus 0\right) \oplus 1 \oplus 0 \oplus 0 \neq 0 \tag{11.4}$$

$$(1-g)\wedge (1-h) \ \geq \ \left(\sum_{n=1}^{\infty} \oplus 0\right) \oplus 0 \oplus 0 \oplus 1 \neq 0. \tag{11.5}$$

As $x_n \to x$, we infer that $||e\, f_n\, e|| \to 1$. Whence,

$$||g\, h\, g|| = \sup_n ||e\, f_n\, e|| = 1.$$

The element $g\,h\,g$ being positive, there is a state φ on M such that $\varphi(g\,h\,g) = 1$. The inequality $g\,h\,g \leq g$ entails $\varphi(g) = 1$. By employing the Cauchy-Schwarz inequality we obtain $\varphi(h) = \varphi(g\,h\,g) = 1$. Summing it up, there is a state on the global algebra taking value 1 at both g and h. The order relations (2.1), (2.2), (2.3) imply that the same holds for the pairs $(1-g, h), (g, 1-h), (1-g, 1-h)$. Since all states on A and B are convex combinations of states concentrated at atomic projections, we infer that A and B are C^*–independent. However, these algebras are not logically independent because $g \wedge h = 0$.

(i)$\Rightarrow$ (ii) We shall prove that if M has finite-dimensional nonabelian direct summand, then *all* C^*-independent von Neumann subalgebras are logically independent. To this end, take a C^*-independent pair of von Neumann subalgebras A and B and consider nonzero projections $p \in A$ and $q \in B$. The C^*-independence guarantees the existence of the state φ on M with $\varphi(p) = \varphi(q) = 1$. By Proposition 10.1.3 φ is Jauch-Piron, and hence $\varphi(p \wedge q) = 1$. Thus, $p \wedge q \neq 0$ and the proof is complete. □

By the proof technique of the previous theorem we see that in any von Neumann algebra with infinite-dimensional nonabelian part there is a pair of two-dimensional subalgebras that is C^*-independent but not logically independent. In other words, one can find two positive operators $a, b \in M$, each having two-point spectrum, such that $W^*(a)$ and $W^*(b)$ are C^*-independent but not logically independent. Let us remark that typical examples of such operators are given by the spin of electrons.

So far we have proved the implications:

logical independence $\Longrightarrow$ C^*-independence $\Longrightarrow$ S-independence

and shown that none of these implications can be reversed. However, if the local algebras commute then all independence conditions stated above coincide. Indeed, let A and B be mutually commuting von Neumann subalgebras of a von Neumann algebra M. If A and B are S-independent, then $pq = p \wedge q \neq 0$, whenever $p \in P(A)$ and $q \in P(B)$ are nonzero, and hence there must be logically independent. On the C^*-level, if A and B are arbitrary commuting S-independent C^*-algebras, then they are C^*-independent. This is the content of the classical Roos Theorem.

11.2.13. Theorem (Roos). *Let A and B be C^*-subalgebras of a C^*-algebra D. Then A and B are C^*-independent if, and only if, A and B are S-independent.*

Proof: Suppose that A and B are S-independent. Let $a \in A$, $b \in B$ be positive norm one elements. The abelian subalgebras $C^*(1, a)$ and $C^*(1, b)$ are S-independent in the abelian algebra $C^*(1, a, b)$. By virtue of Theorem 11.1.1, $C^*(1, a)$ and $C^*(1, b)$ are C^*-independent in $C^*(a, b)$, and hence there is a state φ on D with $\varphi(a) = \varphi(b) = 1$. Thus, A and B are C^*-independent by Theorem 11.2.5. □

In case of abelian algebras the usual independence conditions are equivalent with the fact that local algebras generate the tensor product. Our next goal is to show that this is to some extent true also for mutually commuting C^*-algebras. We shall first need the following auxiliary results concerning the structure of the algebraic tensor product.

11.2.14. Lemma. *Let D be a C^*-algebra generated by mutually commuting S-independent C^*-algebras A and B. Suppose that $\sum_{i=1}^{n} x_i y_i = 0$ with $x_i \in A$ and $y_i \in B$ $(i = 1, \ldots, n)$ and suppose that neither $x_1, \ldots, x_n$ nor $y_1, \ldots, y_n$ is a zero sequence. Then neither the set $\{x_i \mid i = 1, \ldots, n\}$ nor the set $\{y_i \mid i = 1, \ldots, n\}$ can be linearly independent.*

Proof: The core of the proof is to establish the required statement on condition that $x_1, \ldots, x_n$ commute and $y_1, \ldots, y_n$ commute. Moreover, taking into account that any element can be written as a linear combination of self-adjoint elements, we see that it is sufficient to prove the claim under additional assumption that $x_1, \ldots, x_n$ and $y_1, \ldots, y_n$ are self-adjoint. Suppose that it is the case and set $A_1 = C^*(1, x_1, \ldots, x_n)$, $B_1 = C^*(1, y_1, \ldots, y_n)$. Then A_1 and B_1 are abelian S-independent subalgebras in a commutative C^*-algebra $D = C^*(A_1 \cup B_1)$. By Theorem 11.1.1, every pair of states $(\varphi_1, \varphi_2) \in A_1^* \times B_1^*$ has a product extension $\varphi = \varphi_1 \otimes \varphi_2$ to D_1. We can find a state φ_2 on B such that not all $\varphi_2(y_i)$ are zero. For the element $x = \sum_{i=1}^n \varphi_2(y_i)\, x_i$ we have $\varphi_1(x) = (\varphi_1 \otimes \varphi_2)\left(\sum_{i=1}^n x_i\, y_i\right) = 0$. Since φ_1 can be an arbitrary state, we see that $x = 0$, and hence $x_1, \ldots, x_n$ are not linearly independent. The same argument applies to $y_1, \ldots, y_n$. As the next step, let us assume that there is an element, say y_1, such that not all $y_i' = [y_1, y_i]$ vanish. Then, of course,

$$\begin{aligned} \sum_{i=2}^n x_i\, y_i' &= \sum_{i=2}^n (x_i\, y_1\, y_i - x_i\, y_i\, y_1) = -y_1\, x_1\, y_1 - \sum_{i=2}^n x_i\, y_i\, y_1 \\ &= \left(-\sum_{i=1}^n x_i\, y_i\right) y_1 = 0\,. \end{aligned}$$

We shall proceed by induction on the number of summands. Trivially, the lemma is true for $n = 1$. Suppose it is true for all $k \leq n-1$. Because the sum above has less then n terms, the set $\{x_2, \ldots, x_n\}$ is linearly independent and, obviously, the same holds for the set $\{x_1, \ldots, x_n\}$. Without loss of generality we can assume that $x_1 = -\sum_{i=2}^n c_i\, x_i$, for some $c_i \in \mathbb{C}$, $i = 2, \ldots, n$. By the assumption $\sum_{i=2}^n x_i\,(y_i - c_i\, y_1) = 0$. Then either $y_i = c_i\, y_1$, for all i, or not all $y_i - c_i\, y_1$ vanish. The former case gives us already the linear dependence of $y_1, \ldots, y_n$. The latter case implies, by the induction hypothesis, that the set $\{y_i - c_i\, y_1 \mid i = 1, \ldots, n\}$ is linearly dependent, which implies the desired linear dependence of $y_1, \ldots, y_n$. □

11.2.15. Lemma. *Suppose that C^*-algebras A, B, D are as in Lemma* 11.2.14. *Assume that $\sum_{i=1}^n x_i\, y_i = 0$ for $x_i \in A$, $y_i \in B$, $i = 1, \ldots, n$, and that both $x_1, \ldots, x_n$ and $y_1, \ldots, y_n$ is not a zero sequence.*

Then there exists a complex $n \times n$ *matrix* (c_{ij}) *that is neither zero nor unit and that satisfies*

$$\begin{aligned} \sum_{i=1}^{n} c_{ij}\, x_i &= 0 \qquad j = 1, \ldots, n \\ \sum_{j=1}^{n} c_{ij}\, y_j &= y_i \qquad i = 1, \ldots, n\,. \end{aligned}$$

Proof: For $n = 1$ the statement holds because by S-independence the assumption is vacuous. We will proceed by induction on n. The case $n = 1$ follows from S-independence of A and B. Suppose that the statement holds true for all $k \leq n - 1$. By Lemma 11.2.14, the sequences $x_1, \ldots, x_n$ and $y_1, \ldots, y_n$ are linearly dependent. Suppose, without loss of generality, that

$$x_1 = -\sum_{i=2}^{n} \lambda_i\, x_i\,, \qquad \lambda_i \in \mathbb{C}\,.$$

It implies that

$$\sum_{i=2}^{n} x_i\,(y_i - \lambda_i\, y_1) = 0\,.$$

If $y_i - \lambda_i\, y_1$ are not all zero, there exists by the induction hypothesis an $(n-1) \times (n-1)$ matrix (c'_{ij}) such that

$$\begin{aligned} \sum_{i=2}^{n} c'_{ij}\, x_i &= 0 \qquad\qquad j = 2, \ldots, n \\ \sum_{j=2}^{n} c'_{ij}\,(y_j - \lambda_j\, y_1) &= y_i - \lambda_i\, y_1 \qquad i = 2, \ldots, n\,. \end{aligned}$$

Putting now

$$\begin{aligned} c_{11} &= 1 \\ c_{1j} &= 0 \qquad\qquad j = 2, \ldots, n \\ c_{i1} &= \lambda_i - \sum_{j=2}^{n} c'_{ij}\, \lambda_j \quad i = 2, \ldots, n \\ c_{ij} &= c'_{ij} \qquad\qquad i, j = 2, \ldots, n\,, \end{aligned}$$

one can verify by a direct calculation that the matrix (c_{ij}) satisfies the requirement. It remains to consider the case when $y_i = \lambda_i y_1$ for all $i = 2, \ldots, n$. In this case we can set $c_{11} = 1$, $c_{i1} = \lambda_i$ for $i = 2, \ldots, n$; and $c_{ij} = 0$ otherwise. We can then verify easily that $\sum_{i=1}^{n} c_{i1} x_i = x_1 + \sum_{i=2}^{n} \lambda_i x_i = 0$, $\sum_{i=1}^{n} c_{1j} x_i = 0$ for $j \geq 2$, and $\sum_{j=1}^{n} c_{ij} y_j = \lambda_i y_1 = y_i$ for $i = 1, \ldots, n$. □

11.2.16. Proposition. *Let A and B be mutually commuting C^*-subalgebras that are S-independent in a C^*-algebra D. Then the $*$-algebra C generated by A and B is isomorphic to the algebraic tensor product $A \otimes_{alg} B$ via the $*$-isomorphism*

$$\Psi : C \to A \otimes_{alg} B : \quad \sum_{i=1}^{n} a_i b_i \to \sum_{i=1}^{n} a_i \otimes b_i \, .$$

Proof: It is clear that Ψ is a surjective $*$-homomorphism. For proving that Ψ is injective we show that A and B are *algebraically independent* in the sense that if $\{x_1, \ldots, x_n\} \subset A$ and $\{y_1, \ldots, y_m\} \subset B$ are linearly independent sets, then the set $\{x_i y_j \mid i = 1 \ldots n, j = 1, \ldots m\}$ is linearly independent subset of C. To this end, let us consider a linear combination

$$\sum_{i=1}^{n} \sum_{j=1}^{m} \lambda_{ij} x_i y_j = 0 \qquad (\lambda_{ij} \in \mathbb{C}) \, .$$

This can be rewritten as

$$\sum_{j=1}^{m} x'_j y_j = 0 \, , \text{ where } x'_j = \sum_{i=1}^{n} \lambda_{ij} x_i \, .$$

By Lemma 11.2.15 there is a nontrivial matrix (c_{ij}) such that $\sum_{j=1}^{m} c_{ij} y_j = y_i$, $i = 1, \ldots, n$, which is in a contradiction with the linear independence of y_i's. Therefore, $\sum_{j=1}^{m} \lambda_{ij} x_i = 0$ for all j, meaning that $\lambda_{ij} = 0$ for all i, j.

Having established the algebraic independence, we see that the map Ψ sends the linear basis $\{x y \mid x \in B_1, y \in B_2\}$ of C, where B_1 and B_2 are (linear) bases of A and B, respectively, to the linear basis $\{x \otimes y \mid x \in B_1, y \in B_2\}$ of $A \otimes_{alg} B$ and it is therefore an isomorphism. □

The next theorem summarizes our discussion and shows that in the case of commuting algebras we obtain the same equivalences as in the abelian case (compare with Theorem 11.1.1) if we replace the minimal tensor product by general one.

11.2.17. Theorem. *Let D be a C^*-algebra generated by mutually commuting C^*-subalgebras A and B. The following conditions are equivalent*

(i) *D is $*$-isomorphic to a tensor product $A \otimes_\beta B$, via the natural map $\Psi : D \to A \otimes_\beta B$ satisfying*

$$\psi\left(\sum_{i=1}^{n} a_i\, b_i\right) = \sum_{i=1}^{n} a_i \otimes b_i \qquad a_i \in A\,, b_i \in B\,.$$

(ii) *For any state φ_1 on A and any state φ_2 on B there is a (product) state φ on D such that $\varphi(a\,b) = \varphi_1(a)\,\varphi_2(b)$ for all $a \in A$ and $b \in B$.*

(iii) *A and B are C^*-independent.*

(iv) *A and B are S-independent.*

Proof: The implication (i) $\Rightarrow$ (ii) follows from the existence of the product state on the tensor product. The implication (ii) $\Rightarrow$ (iii) is trivial. The implication (iii) $\Rightarrow$ (iv) is a special case of the fact that C^*-independence always implies S-independence. We shall establish the implication (iv) $\Rightarrow$ (i). By Proposition 11.2.16 we already know that the $*$-algebra C generated by A and B is (linearly) isomorphic to $A \otimes_{alg} B$. Define now the norm β on $A \otimes_{alg} B$ by setting

$$\left\| \sum_{i=1}^{n} a_i \otimes b_i \right\|_\beta = \left\| \sum_{i=1}^{n} a_i\, b_i \right\|.$$

Then β is a C^*-norm, and hence the completion of $(A \otimes_{alg} B, \beta)$ is the tensor product $A \otimes_\beta B$ that must be isomorphic to D. □

Finally, as the last result in this section we show that testing the independence of commuting von Neumann algebras can be reduced to the centres of the algebras considered and therefore to an entirely abelian situation.

11.2.18. Theorem. *Let A and B be mutually commuting von Neumann subalgebras of a von Neumann algebra M. Then A and B are C^*-independent if, and only if, the centres $Z(A)$ and $Z(B)$ are C^*-independent.*

Proof: Suppose that $Z(A)$ and $Z(B)$ are C^*-independent. Let us recall that all independence conditions coincide for mutually commuting algebras. Let us consider nonzero projections $e \in A$ and $f \in B$ and denote by $c(e)$ and $c(f)$ the central covers of projections e and f with respect to algebras A and B, respectively. Then $c(e) = \bigvee_{u\in U(A)} u^* e u$, where u runs through the unitary group $U(A)$ of A. Let r, s be projections in A. Then $(r \vee s) f = r f \vee s f$. (Indeed, the map $\pi\colon A \to B : a \to a f$ is a normal $*$-homomorphism that must preserve the suprema of projections.) Now,

$$c(e) f = \Big(\bigvee_{u\in U(A)} u^* e u \Big) f = \bigvee_{u\in U(A)} u^* e u f \neq 0 .$$

if, and only if, $u_0^* e u_0 f \neq 0$ for some $u_0 \in U(A)$, or what is the same, if and only if, $e f \neq 0$. By the symmetry

$$e f \neq 0 \qquad \text{if, and only if,} \qquad c(e)\, c(f) \neq 0 .$$

Nevertheless, $c(e)\, c(f) \neq 0$ for all nonzero $c(e)$ and $c(f)$ by S-independence of $Z(A)$ and $Z(B)$. Hence, we have shown that A and B are logically independent and the proof is complete. □

Let M be a von Neumann algebra containing two mutually commuting von Neumann subalgebras A and B. If A has a discrete centre (i.e. if it is a direct sum of factors), then A and B are C^*-independent if, and only if, $pq \neq 0$, whenever p is an atomic projection in $Z(A)$ and q is a nonzero projection in $Z(B)$. This generalizes the classical result of Murray and von Neumann to the effect that pair of commuting von Neumann algebras is S-independent on condition that at least one of them is a factor [238, Corollary of Theorem III].

11.3. Independence of von Neumann algebras

In the present part we shall examine the independence conditions in the category of von Neumann algebras. The central notion will be that of W^*-independence that is the simultaneous extension property in the category of von Neumann algebras.

11.3.1. Definition. Let A and B be von Neumann subalgebras of a von Neumann algebra M. We say that

(i) A and B are *W^*-independent* if for every normal state φ_1 on A and for every normal state φ_2 on B there is a normal state φ on M extending both φ_1 and φ_2;

(ii) the pair (A, B) is *W^*-strictly local* if for any nonzero projection $p \in P(A)$ and for any normal state φ on B there is a normal state ψ on M such that $\psi(p) = 1$ and and $\psi|B = \varphi$;

(iii) the pair (A, B) is *W^*-strictly local in the sense of vector states* if for any nonzero projection $p \in P(A)$ and for any vector state φ on B there is a vector state ψ on M such that $\psi(p) = 1$ and $\psi|B = \varphi$.

It is clear that if M has finite dimension, then the C^*-independence coincides with the W^*-independence and so examples in Example 11.2.2 provide automatically examples of noncommuting W^*-independent algebras. The W^*-strict locality is a weaker condition than W^*-independence. It is not known whether it is always equivalent to W^*-independence and whether it is symmetric with respect to the local algebras. The W^*-strict locality describes the situation when the preparation of one system in a given state does not exclude the preparation of the second system such that the given proposition is true. As far as the position of W^*-strict locality and W^*-strict locality in the sense of vector states is concerned, let us note that W^*-strict locality in the sense of vector states always implies W^*-strict locality because any normal state is a σ-combination of vector states. Both properties coincide in case when all normal states are vector states which occurs, e.g., when the global algebra M has a separating vector. However, W^*-strict locality in the sense of vector states is in many cases much stronger than other independence conditions. The main goal of the present discussion is to compare W^*-independence, C^*-independence and the related independence properties. The following easy observation will clarify the situation.

11.3.2. Proposition. *Let A and B be von Neumann subalgebras of a von Neumann algebra M. If the pair (A, B) is W^*-strictly local, then A and B are logically independent.*

Proof: Suppose that (A, B) is W^*-strictly local and consider nonzero projections $p \in P(A)$ and $q \in P(B)$. Since we can find a normal state φ on B with $\varphi(q) = 1$, we obtain by the strict locality a normal state ψ on M with $\psi(p) = \psi(q) = 1$. By the Jauch-Piron property of normal states we have $\psi(p \wedge q) = 1$, and hence $p \wedge q \neq 0$. □

Using the results of the previous section we have the following chain of implications for von Neumann algebras:

$$W^*\text{-independence} \Longrightarrow W^*\text{-strict locality} \Longrightarrow \text{logical independence}$$

$$\Longrightarrow C^*\text{-independence} \Longrightarrow S\text{-independence}$$

As a consequence of this and Theorem 11.2.12 and its proof, we see that in all essentially noncommutative von Neumann algebras one can find a pair of finite-dimensional subalgebras that are C^*-independent but not W^*-independent. More precisely, the following statement is true. A von Neumann algebra M has two finite-dimensional subalgebras A and B that are C^*-independent but not W^*-independent if, and only if, the nonabelian part of M has infinite dimension. Indeed, suppose that M has finite-dimensional nonabelian central summand and that A and B are finite-dimensional subalgebras that are C^*-independent. Then the C^*-algebra generated by A and B is finite-dimensional again, and thus A and B are W^*-independent. The reverse implication follows from Theorem 11.2.12.

Although W^*-independence and C^*-independence typically dispair, we show that C^*-independence is nothing but an approximate W^*-independence. As a preparation for this let us prove the following characterization of W^*-independence.

11.3.3. Proposition. *Let A and B be von Neumann subalgebras of von Neumann algebra M acting on a Hilbert space H. The following conditions are equivalent:*

(i) *A and B are W^*-independent.*

(ii) *For any pair of normal states φ_1 and φ_2 on A and B, respectively, there is a normal, nonzero, positive functional φ on M such that $\varphi \leq \varphi_1$ on A and $\varphi \leq \varphi_2$ on B.*

(iii) *For any pair of unit vectors $\xi_1, \xi_2 \in H$ there is a nonzero vector $\xi \in H$ such that $\omega_\xi \leq \omega_{\xi_1}$ on A and $\omega_\xi \leq \omega_{\xi_2}$ on B.*

Proof: It is obvious that (i)⇒(ii). Let us prove the implication (ii)⇒ (i). Take two normal states φ_1 and φ_2 on A and B, respectively, and define the set

$$S = \{\varphi \in (M_*)^+ \mid \varphi|A \leq \varphi_1, \, \varphi|B \leq \varphi_2\}.$$

The condition (ii) guarantees that S contains a nonzero element. Moreover, S has a maximal element with respect to the order given by the positive cone $(M_*)^+$. Indeed, let us consider an increasing chain (φ_α) in S. Since $\|\varphi_\alpha\| \leq 1$, we can define a positive linear form ψ on M by

$$\psi(x) = \sup_\alpha \varphi_\alpha(x), \quad \text{for all } \; 0 \leq x \in M.$$

Then

$$\|\psi - \varphi_\alpha\| = (\psi - \varphi_\alpha)(1) \to 0,$$

implying that ψ is a normal fuctional. On applying Zorn's lemma, there is a maximal (nonzero) element, let us say τ, of S. We prove that τ has to be a common extension of φ_1 and φ_2. Suppose, for a contradiction, that $\tau \neq \varphi_1$ on A and $\tau \neq \varphi_2$ on B. Then $\|\varphi_1 - \tau\| = \|\varphi_2 - \tau\| = 1 - \tau(1) > 0$ and so by (ii) there is a nonzero normal positive functional τ' on M with $\tau' \leq \varphi_1 - \tau$ on A and $\tau' \leq \varphi_2 - \tau$ on B. Thus, $\tau \leq \tau + \tau' \in S$, which contradicts the maximality of τ. Hence, without loss of generality, we can assume that $\tau|A = \varphi_1$. Then immediately

$$\|\varphi_2 - \tau|B\| = \varphi_2(1) - \tau(1) = 0,$$

which concludes the proof of (ii)⇒ (i).

The equivalence of (ii) and (iii) follows immediately from the fact that any normal state is a σ-convex combination of vector states. □

The following corollary tells us that in order to obtain simultaneous normal extension it is enough to find a common nonsingular extension.

11.3.4. Corollary. *Let M be a von Neumann algebra containing von Neumann subalgebras A and B. The algebras A and B are W^*-independent if, and only if, for any normal state φ_1 on A and for any normal state φ_2 on B there is a nonsingular state φ on M with $\varphi|A = \varphi_1$ and $\varphi|B = \varphi_2$.*

Proof: Since any nonsingular state majorizes a nonzero normal functional, the statement follows from the condition (ii) of Proposition 11.3.3. □

As another consequence of Proposition 11.3.3 we obtain the characterization of W^*-independence in terms of Jauch-Piron states treated in the previous chapter.

11.3.5. Proposition. *Let M be a von Neumann algebra satisfying (a) or (b) below:*

(a) M is σ-finite and properly infinite.

(b) M is a σ-finite factor.

Suppose that A and B are von Neumann subalgebras of M. The following conditions are equivalent:

(i) *A and B are W^*-independent.*

(ii) *For any normal state φ_1 on A and for any normal state φ_2 on B there is a Jauch-Piron state φ on M such that $\varphi|A = \varphi_1$ and $\varphi|B = \varphi_2$.*

Proof: The analysis of the the continuity properties of the Jauch-Piron states carried out in the previous chapter yields that if M is of type (a) or (b), then any Jauch-Piron state on M is automatically nonsingular and the statement follows from Corollary 11.3.4. □

The following characterization of W^*-independence is an analog of Theorem 11.2.5. It says that C^*-algebras are independent if normalized positive elements from local algebras can be exposed by a common state. (Let us note that for any positive norm-one element in a von Neumann algebra M there is a state φ on M such that $\varphi(a) = 1$. However, it is not always possible to find φ to be normal.)

11.3.6. Proposition. *Two von Neumann subalgebras A and B of a von Neumann algebra M are C^*-independent if, and only if, for all self-adjoint elements $a \in A$ and $b \in B$, and for any $\varepsilon > 0$, there is a normal state ϱ on M with*

$$|\varrho(a)| > \|a\| - \varepsilon, \quad \textit{and} \quad |\varrho(b)| > \|b\| - \varepsilon.$$

Proof: Suppose that A and B are C^*-independent and fix self-adjoint elements $a \in A, b \in B$, respectively, and $\varepsilon > 0$. Then there are states φ_1 on A and φ_2 on B such that

$$|\varphi_1(a)| = \|a\|, \qquad |\varphi_2(b)| = \|b\|.$$

There is a common state extension of φ_1 and φ_2 over M. By the weak*-density of normal states in the state space, we can always find a normal state ϱ on M with

$$|\varrho(a) - \varphi_1(a)| < \varepsilon \qquad \text{and} \qquad |\varrho(b) - \varphi_2(b)| < \varepsilon\,.$$

Therefore

$$|\varrho(a)| > \|a\| - \varepsilon \qquad |\varrho(b)| > \|b\| - \varepsilon\,.$$

On the other hand, suppose that the statement of the Proposition 11.3.6 is fulfilled. Then for all norm one positive elements $a \in A$ and $b \in B$, there is a net (ϱ_n) of normal states on M with

$$\varrho_n(a) > 1 - \frac{1}{n}, \qquad \varrho_n(b) > 1 - \frac{1}{n}.$$

Employing the compactness of the state space, we can assume that the sequence (ϱ_n) has a limit ϱ in the weak*-topology. Then $\varrho(a) = \varrho(b) = 1$ and a straightforward application of Theorem 11.2.5 concludes the proof. □

By the symbol $S_n(M)$ we continue to denote the normal state space of a von Neumann algebra M.

11.3.7. Lemma. *Let A and B be C^*-independent von Neumann subalgebras of a von Neumann algebra M. For a fixed nonnegative element $b \in B$ and $\varepsilon > 0$ the set*

$$S_\varepsilon = \{\varphi|A \mid \varphi \in S_n(M) \text{ with } \varphi(b) > \|b\| - \varepsilon\}$$

is norm dense in $S_n(A)$.

Proof: The set S_ε is nonempty and convex. Therefore the norm closure $\overline{S_\varepsilon}$ is a closed convex subset of $S_n(A)$. Suppose that $\overline{S_\varepsilon} \neq S_n(A)$ and take a normal state $\psi \in S_n(A) \setminus \overline{S_\varepsilon}$. Using the Hahn-Banach Theorem for the real Banach space of all self-adjoint normal functionals on A the dual of which is the self-adjoint part of A, we can find a real number α and a self-adjoint element $c \in A$ such that

$$\psi(c) > \alpha > \psi'(c) \text{ for all } \psi' \in \overline{S_\varepsilon}\,.$$

By adding a suitable multiple of identity to c and modifying α accordingly, we can assume that $c \geq 0$. We have,

$$\|c\| \geq \psi(c) > \alpha > \psi'(c) \text{ for all } \psi' \in \overline{S_\varepsilon}.$$

However, by Proposition 11.3.6 there is a state $\overline{\psi} \in S_{\varepsilon'} \subset S_\varepsilon$, where $\varepsilon' = \min(\varepsilon, \frac{\|c\|-\alpha}{2})$ such that

$$\overline{\psi}(c) > \|c\| - \varepsilon'.$$

It leads to a contradiction

$$\overline{\psi}(c) > \|c\| - \varepsilon' > \alpha > \overline{\psi}(c).$$

Thus, we have $\overline{S_\varepsilon} = S_n(A)$. □

Now we are ready to prove one of the main results on the relationship between W^*-independence and C^*-independence.

11.3.8. Theorem. *Let A and B be C^*-independent von Neumann subalgebras of a von Neumann algebra M. Then for any $\varepsilon > 0$ and for all normal states φ_1 on A and φ_2 on B there is a normal state φ on M such that*

$$\|\varphi|A - \varphi_1\| < \varepsilon \text{ and } \|\varphi|B - \varphi_2\| < \varepsilon.$$

Proof: Let us consider the topological cartesian product $S_n(A) \times S_n(B)$ of the topological spaces $(S_n(A), \|\cdot\|)$ and $(S_n(B), \|\cdot\|)$. We have to show that the set

$$L = \{(\tau|A, \tau|B) \mid \tau \in S_n(M)\}$$

is dense in $S_n(A) \times S_n(B)$. Fix a state $\varphi \in S_n(A)$, $\varepsilon > 0$, and define the set

$$K_\varepsilon = \{\tau|B \mid \tau \in S_n(M), \text{ with } \|\tau|A - \varphi\| < \varepsilon\}.$$

The set K_ε is a nonempty convex set. We shall show that $\overline{K_\varepsilon} = S_n(B)$. Suppose the contrary and take $\psi \in S_n(B) \setminus \overline{K_\varepsilon}$. Employing the Hahn-Banach theorem in the way we did in the previous lemma, we can find a non-negative $c \in B$ and a real α such that

$$\psi(c) > \alpha > \psi'(c) \text{ for all } \psi' \in \overline{K_\varepsilon}.$$

It yields

$$\|c\| \geq \psi(c) > \alpha > \psi'(c) \text{ for all } \psi' \in \overline{K_\varepsilon}.$$

But by Lemma 11.3.7 there is a state $\tau \in S_n(M)$ such that

$$\tau(c) > \|c\| - \frac{\|c\| - \alpha}{2} = \frac{\|c\| + \alpha}{2} > \alpha \text{ and } \|\tau|A - \varphi\| < \varepsilon.$$

So $\tau|B \in K_\varepsilon$, which implies $\tau(c) > \alpha > \tau(c)$. That is a contradiction. Therefore, $\overline{K_\varepsilon} = S_n(B)$ for any $\varepsilon > 0$, proving that the set L is dense in $S_n(A) \times S_n(B)$. □

The previous theorem says that the C^*-independence of von Neumann algebras is in fact approximate W^*-independence that guarantees simultaneous extensions of normal states up to an arbitrarily small error. Since any physical experiment involves some uncertainty, it seems correct to say that C^*-independence and W^*-independence coincide from the physical point of view.

The last result on independence of algebras in general position says that in case when one of the algebras is discrete, then W^*-strict locality coincides with logical independence. (Let us recall that von Neumann algebra is called discrete if it is a sum of Type I factors.)

11.3.9. Proposition. *Let A and B be von Neumann subalgebras of a von Neumann algebra M. Suppose that B is discrete. Then A and B are logically independent if, and only if, (A, B) is W^*-strictly local.*

Proof: Let p be a nonzero projection in A and φ be a normal state on B. As B is discrete, φ is a mixture of pure normal states on B, i.e.

$$\varphi = \sum_{i=1}^{\infty} \lambda_i \varphi_i, \qquad 0 \le \lambda_i \le 1, \qquad \sum_{i=1}^{\infty} \lambda_i = 1,$$

where each φ_i is a pure normal state concentrated at atomic projection $p_i \in B$. If A and B are logically independent, then $p_i \wedge p \neq 0$ for all i. Therefore there is always a normal state ϱ_i on M such that $\varrho_i(p) = \varrho_i(p_i) = 1$. By setting

$$\varrho = \sum_{i=1}^{\infty} \lambda_i \varrho_i$$

we obtain a normal state on M with $\varrho(p) = 1$ and $\varrho|B = \varphi$. □

As a consequence of Proposition 11.3.9 we see that a pair $(A, W^*(b))$, where b is an operator with the discrete spectrum, is W^*-strictly local if, and only if, it is logically independent.

After discussing the interrelations between major independence conditions for not necessarily commuting algebras we now turn to the case when these algebras commute. Under this circumstance we obtain that all variants of independence are equivalent, provided that the global algebra is σ-finite. For explaining this result we need the followings elements of the modular theory. (For a more detailed exposition let us refer, e.g., to the monograph [39].) Let M be a von Neumann algebra acting on a Hilbert space H that admits a separating and generating vector $\xi \in H$. It means that $[M\,\xi] = H$ and $[M'\,\xi] = H$, where M' denotes the commutant of M. (We continue to use the standard notation for $M\,\xi = \{a\,\xi \mid a \in M\}$ and $[M\,\xi]$ for its norm closure in H.) Then the conjugate linear map

$$S_0 \colon\ M\,\xi \to H : a\,\xi \to a^*\xi, \qquad a \in M,$$

is closable with the closure $S = \overline{S_0}$. (Let us recall that conjugate linear map T on a complex vector space V is an additive map such that $T(\alpha\, x) = \overline{\alpha}\, T(x)$ for all $x \in V$ and $\alpha \in \mathbb{C}$.) Let Δ be the unique positive self-adjoint operator (possibly unbounded) and J the unique conjugate linear isometry occurring in the polar decomposition

$$S = J\,\Delta^{1/2}\,.$$

The operators Δ and J are called the *modular operator* and the *modular conjugation* associated with (M, ξ). It holds that $J = J^*$, $J^2 = 1$ and $J\xi = \Delta\xi = \xi$. By the celebrated Tomita-Takesaki Theorem $J\,M\,J = M'$, which defines the conjugate linear $*$-isomorphism j between M and its commutant M' by $j(a) = J\,a\,J$ $(a \in M)$. Moreover, define the *natural positive cone* $\mathcal{P}$ in H associated with (M, ξ) as the closure of the set $\{a\,j(a)\,\xi \mid a \in M\}$. The following property of the natural positive cone is important. For every normal positive functional ψ on M there is exactly one vector $\xi_\psi \in \mathcal{P}$ such that ψ is a vector functional given by the vector ξ_ψ. In addition, the map between M_*^+ and $\mathcal{P}$, $\psi \to \xi_\psi$ is a homeomorphism when both $\mathcal{P}$ and M_*^+ are equipped with the norm topology. We shall need the following technical lemmas.

11.3.10. Lemma. *Let M be a von Neumann algebra acting on a Hilbert space H. Suppose that ξ_1 and ξ_2 are unit vectors in H such that the vector states ω_{ξ_1} and ω_{ξ_2} coincide on M. Denote by p and q the orthogonal projections onto subspaces $[M\,\xi_1]$ and $[M\,\xi_2]$, respectively. Then the map v': $[M\,\xi_1] \to [M\,\xi_2]$ given by the condition*

$$v'\,a\,\xi_1 = a\,\xi_2\,, \qquad a \in M\,,$$

extends canonically to the partial isometry in M' with the initial projection p and the final projection q.

Proof: Let us observe at first that for each $a, b \in M$,

$$\begin{aligned}(a\,\xi_1, b\,\xi_1) &= (b^*\,a\,\xi_1, \xi_1) = \omega_{\xi_1}(b^*\,a) = \omega_{\xi_2}(b^*\,a) \\ &= (a\,\xi_2, b\,\xi_2)\,,\end{aligned}$$

giving that v' is a well defined isometry between $p(H)$ and $q(H)$. It can be extended (by putting $v' = 0$ on $(1-p)(H)$) to a partial isometry, $v' \in B(H)$, such that $v'^*\,v' = p$ and $v'\,v'^* = q$. It remains to prove that $v' \in M'$. For this, let us now take $a, b \in M$ and compute

$$a\,b\,\xi_2 = v'\,a\,b\,\xi_1 = v'\,a\,p\,b\,\xi_1 = v'\,a\,v'^*\,v'\,b\,\xi_1 = (v'\,a\,v'^*)\,b\,\xi_2\,.$$

Hence, $a\,q = v'\,a\,v'^*\,q$ and in turn $v'^*\,a\,q = p\,a\,v'^*\,q$. Using now the identities $v' = v'\,p = q\,v'$ and the fact that $p, q \in M'$, we conclude that

$$v'^*\,a\,q = v'^*\,q\,a = v'^*\,a$$

and

$$v'^* a q = p a v'^* q = a p v'^* q = a v'^* .$$

Hence, v'^* commutes with M and so does v'. □

11.3.11. Lemma. *Let M be a von Neumann algebra with a separating and cyclic vector ξ. Then for every nonzero vector $\nu \in H$ there are elements $a \in M$, $a' \in M'$ such that $a \xi = a' \nu \neq 0$.*

Proof: Consider normal functional l on M' defined by

$$l(a') = (a' \nu, \xi), \qquad a' \in M' .$$

The functional l is nonzero because ξ is cyclic for M'. By the polar decomposition of the normal functional there is a partial isometry $u' \in M'$ such that the functional ϱ on M' given by

$$\varrho(a') = (a' u' \nu, \xi) = (u' \nu, a'^* \xi), \qquad a' \in M',$$

is nonnegative. Since ϱ is nonzero, we have $u' \nu \neq 0$. Then there exists a self-adjoint operator Q affiliated with M such that $u' \nu = Q \xi$. (We say that a (possibly unbounded) operator S is affiliated with the von Neumann algebra M if $u^* S u = S$ for all unitary u from the commutant M' of M.) Indeed, define an operator T on $M' \xi$ by

$$T a' \xi = a' u' \nu, \qquad a' \in M' .$$

Taking a unitary $v' \in M'$ we compute

$$T v' a' \xi = v' a' u' \nu = v' T a' \xi, \qquad a' \in M' .$$

In other words,

$$v' T v'^* = T,$$

because ξ is generating for M'. Moreover,

$$(T a' \xi, a' \xi) = (a' u' \nu, a' \xi) = (a'^* a' u' \nu, \xi) = \varrho(a'^* a') \geq 0 .$$

This means that T is a positive, symmetric operator. Let Q be the Friedrich's extension of T. Then Q is a self-adjoint operator and, by the uniqueness of the Friedrich's extension,

$$v' Q v'^* = Q$$

for all $v' \in M'$. Hence, Q is affiliated with M and such that

$$Q\xi = u'\nu .$$

Consider now the positive linear functional ψ on M defined by this vector, i.e.

$$\psi(a) = (a\,Q\,\xi, Q\,\xi) = (a\,u'\,\nu, u'\,\nu)\,, \qquad a \in M\,.$$

There is a unit vector ξ_ψ in the natural cone $\mathcal{P}$ corresponding to (M, ξ) representing ψ, i.e.

$$(a\,u'\,\nu, u'\,\nu) = (a\,Q\,\xi, Q\,\xi) = (a\,\xi_\psi, \xi_\psi) \qquad \text{for all } a \in M\,.$$

By Lemma 11.3.10 there is a partial isometry $v' \in M'$ carrying $[M\,u'\,\nu]$ to $[M\,\xi_\psi]$ such that

$$v'\,u'\,\nu = v'\,Q\,\xi = \xi_\psi\,.$$

Since the modular conjugation J corresponding to (M, ξ) leaves the cone $\mathcal{P}$ pointwise invariant, one has

$$v'\,u'\,\nu = J\,v'\,u'\,\nu = J\,v'\,Q\,\xi = (J\,v'\,J)\,(J\,Q\,J)\,\xi\,. \tag{11.6}$$

By the spectral theorem we can choose a suitable spectral projection $p' \in M'$ of the self-adjoint operator $J\,Q\,J$ (that is affiliated with M') such that

$$p'\,(J\,Q\,J) \in M' \qquad \text{and} \qquad p'\,(J\,Q\,J)\,\xi \neq 0\,. \tag{11.7}$$

Then, using (11.6)

$$\begin{aligned} p'\,v'\,u'\,\nu &= p'\,(J\,v'\,J)\,(J\,Q\,J)\,\xi = (J\,v'\,J)\,p'\,(J\,Q\,J)\,\xi \\ &= J\,v'\,(J\,p'\,J)\,Q\,\xi\,. \end{aligned} \tag{11.8}$$

As v' is isometry when restricted to $[M\,Q\,\xi]$ we see from (11.8) that

$$p'\,v'\,u'\,\nu \neq 0\,.$$

Using now the arguments above for the nonzero normal functional induced by the vector $p'\,v'\,u'\,\nu$, we a obtain a partial isometry $w' \in M'$ such that

$$0 \neq w'\,p'\,v'\,u'\,\nu \in \mathcal{P}\,.$$

Hence, by applying (11.8)

$$\begin{aligned} 0 &\neq w'\,p'\,v'\,u'\,\nu = J\,w'\,(p'\,v'\,u'\,\nu) \\ &= J\,w'\,(J\,v'\,J)\,p'\,(J\,Q\,J)\xi \\ &= v'\,(J\,w'\,J)\,(J\,p'\,J)\,Q\,\xi\,. \end{aligned} \tag{11.9}$$

Noting that v' is a partial isometry the initial projection of which has range $[M\,Q\,\xi]$, we deduce from (11.9) that

$$v'^*\,w'\,p'\,v'\,u'\,\nu = v'^*\,v'\,(J\,w'\,J)\,(J\,p'\,J)\,Q\,\xi = (J\,w'\,J)\,(J\,p'\,J)\,Q\,\xi\,.$$

Let us put

$$a' = v'^*\,w'\,p'\,v'\,u' \in M'$$

and

$$a = (J\,w'\,J)\,(J\,p'\,J)\,Q\,.$$

Using the fact that $p'\,J\,Q\,J \in M'$, we have

$$a = (J\,w'\,J)\,J\,(p'\,J\,Q\,J)\,J \in M\,.$$

Hence, by (11.8), a and a' are the desired elements. □

We are now in the position to prove the equivalence of W^*-independence and C^*-independence in an important case of commuting σ-finite algebras.

11.3.12. Theorem (Florig-Summers). *Let M be a σ-finite von Neumann algebra containing two mutually commuting von Neumann subalgebras A and B. Then A and B are W^*-independent if, and only if, they are C^*-independent.*

Proof: Suppose that A and B are C^*-independent. Without loss of generality we can realize M as a von Neumann algebra acting on a Hilbert space H and having generating and separating vector $\xi \in H$. For proving W^*-independence of A and B it is enough to show, in view of Proposition 11.3.3, that given normal states ψ_1 and ψ_2 on A and B, respectively, one can find a nonzero normal positive functional ψ on M such that $\psi \le \psi_1$ on A and $\psi \le \psi_2$ on B. Since M has a separating vector, all normal states are vector states. Denote by p' and q' the projections projecting orthogonally onto $[A\,\xi]$ and $[B\,\xi]$, respectively. Then $p' \in A'$. The map $a \to p'\,a : A \to p'\,A$ is a $*$-isomorphism because ξ is separating for A. Hence, there is a normal state ψ_1' on $p'\,A$ such that

$$\psi_1(x) = \psi_1'(p'\,x) \qquad \text{for all } x \in A\,.$$

Any vector positive functional ω on $B(H)$ for which $\omega \leq \psi_1'$ on $p'\,A$ has to be concentrated at $p'\,M\,p'$, and hence $\omega(x) = \omega(p'\,x) \leq \psi_1'(p'\,x) = \psi_1(x)$. In summary, we can assume that ψ_1 and ψ_2 are vector states given by unit vectors $\xi_1 \in p'(H)$ and $\xi_2 \in q'(H)$, respectively. Employing Lemma 11.3.11 for the algebra $p'\,A$ acting on $p'(H)$ with a separating and cyclic vector $p'\xi = \xi$, we have elements $a \in A$ and $a' \in A'$ with

$$0 \neq a\,\xi = p'\,a\,p'\,\xi = p'\,a'\,p'\,\xi_1 = a'\,\xi_1\,.$$

Similarly, we can find $b \in B$ and $b' \in B'$ with

$$0 \neq b\,\xi = b'\,\xi_2\,.$$

Since $a, b \neq 0$, the S-independence of A and B implies that $a\,b \neq 0$. As ξ is cyclic for M' we deduce that there is an element $c' \in M'$ such that

$$a\,b\,c'\,\xi \neq 0\,.$$

Then

$$\begin{aligned} 0 &\neq b\,c'\,a'\,\xi_1 = b\,c'\,a\,\xi = a\,b\,c'\,\xi = a\,c'\,b\,\xi \\ &= a\,c'\,b'\,\xi_2\,. \end{aligned}$$

Write $x' = b\,c'\,a' \in A'$ and $y' = a\,c'\,b' \in B'$. Then

$$x'\,\xi_1 = y'\,\xi_2 \neq 0\,.$$

Consider now a nonzero a positive vector functional ϱ determined by

$$\varrho(a) = \omega_{x'\,\xi_1}(a) = (a\,x'\,\xi_1, x'\,\xi_1)\,, \qquad a \in M\,.$$

For any $a \in A$, $a \geq 0$, we have

$$\begin{aligned} \omega_{x'\,\xi_1}(a) &= (a\,x'\,\xi_1, x'\,\xi_1) = (a^{1/2}\,x'^*\,x'\,a^{1/2}\,\xi_1, \xi_1) \\ &\leq \|x'^*\,x'\| \cdot (a\,\xi_1, \xi_1) = \|x'^*\,x'\| \cdot \psi_1(a)\,. \end{aligned}$$

By the symmetry, we can find a nonzero multiple of ϱ that is dominated by both ψ_1 and ψ_2. □

As we saw, for commuting local algebras acting on a separable Hilbert space most of the independence conditions considered coincide. This applies to the most significant physical situations. However, it turns out that W^*-strict locality is equivalent to logical independence (and therefore to C^*-independence and S-independence) for *all* commuting algebras no matter how large they are. The arguments in this case are much simpler then in the proof of Florig-Summers Theorem.

11.3.13. Theorem. *Let A and B be commuting von Neumann subalgebras of a von Neumann algebra M. Then A and B are W^*-strictly local if, and only if, A and B are S-independent.*

Proof: Suppose that A and B are S-independent. Let $p \in A$ be a nonzero projection and φ be a normal state on B. By S-independence the map $\tau\colon B \to pBp = Bp : b \to pb$ is a $*$-isomorphism. Define a state ω on pB as $\omega = \varphi \circ \tau^{-1}$ and extend it to a normal state, again denoted by ω, on the whole algebra M. Then $\omega(p) = 1$ and, using the Cauchy-Schwarz inequality

$$\begin{aligned} \omega(b) &= \omega(bp) = \omega(\tau(b)) = (\varphi \circ \tau^{-1})(\tau(b)) \\ &= \varphi(b)\,, \end{aligned}$$

for all $b \in B$. It shows that (A, B) is W^*-strictly local. □

11.4. Historical remarks and comments

Although many independence conditions have been known from the 60's (see e.g. [92, 155, 216, 218]), only recently their interrelationship have been clarified (see [302] for the development till the beginning of the 90's, and [175, 180, 182, 136, 145] for new results). The concept of C^*-independence first appeared in the paper by Haag and Kastler [155], where it is called *statistical independence* and it is proposed to implement Einstein causality principle. The examples of noncommutative C^*-algebras (Example 11.2.2) can be found in [136, 145, 243, 302]. The characterization of C^*-independence in terms of pure states and exposing elements by functionals (Proposition 11.2.4, Theorem 11.2.5) was proved by Hamhalter in [182]. Result on the product property of norm on C^*-independent algebras (condition (ii) in Theorem 11.2.5) is due to Florig and Summers [136]. The logical independence was introduced and first investigated in case of von Neumann algebras by Redei [272, 273, 274]. He was motivated by semantic approach to quantum logics [274]. In our treatment we have formulated this property for all C^*-algebras. The fact that logical independence of real rank zero algebras implies C^*-independence (Theorem 11.2.11) was proved by Hamhalter in [175], where also examples of finite-dimensional subalgebras that are C^*-independent but not logically independent (Theorem 11.2.12) are provided. The results that C^*-independence implies S-independence and that strict locality in the category of C^*-algebras coincides with C^*-independence were proved in [175], too.

The Roos Theorem can be found in [277]. The original proof of Theorem 11.2.13 on the equivalence of S-independence and C^*-independence for commuting algebras is based on the tensor product technique. The proof presented here uses only the compactness of the state space and it is therefore extendable to Jordan algebras where we have not tensor product technique at our disposal. The following result generalizes Roos Theorem (a) by considering finitely many algebras (b) by replacing the C^*-algebras by the JB algebras. It was proved by Hamhalter in [180].

11.4.1. Theorem. *Let $A_1, \ldots, A_n$ be operator commuting JB subalgebras of a JB algebra A. The following conditions are equivalent*

(i) *If $a_i \in A_i$ $i = 1, \ldots, n$ are nonzero elements, then the product $a_1 \circ \cdots \circ a_n$ is a nonzero element in A.*

(ii) *Any n-tuple of states $\varphi_1 \in A_1^*$, $\varphi_2 \in A_2^*, \ldots, \varphi_n \in A_n^*$ has a common extension to A.*

Theorem 11.2.17 saying that commuting algebras are S-independence if, and only if, they generate a tensor product is due to Roos [277]. His argument is based on a result of Turumaru [312], who proved that S-independent commuting algebras are algebraically independent (Proposition 11.2.16).

The Roos Theorem implies that any pair of states on C^*-independent commuting C^*-algebras extends to a product state across the local algebras.

We expect the problem of the existence of various product type simultaneous extensions and their uniqueness to be subject to intensive research in the future.

The W^*-independence has been mainly studied in [136, 182]. It was proved by Florig and Summers that commuting von Neumann subalgebras in a σ-finite algebra are W^*-independent if, and only if, they are C^*-independent (Theorem 11.3.12). The fact that this equivalence is far from being true for noncommuting algebras was shown by Hamhalter in [182]. The equivalence of C^*-independence with approximative W^*-independence (Theorem 11.3.8) has been established by Hamhalter in [182].

The W^*-strict locality was introduced on physical ground by Kraus [216] and developed further by Licht [218]. The equivalence of W^*-strict locality and C^*-independence for commuting von Neumann algebras (Theorem 6.2.7) is due to Florig and Summers [136]. Hamhalter has proved that W^*-strict locality coincide with C^*-independence on condition that one local algebra is discrete. Although recent results have solved many problems in the classification of independence conditions, the following important questions remain open.

11.4.2. Problem.
(i) *Does logical independence imply W^*-independence?*
(ii) *Does W^*-strict locality imply W^*-independence?*

In the previous section we have not devoted much space to W^*-strict locality in the sense of vector states. The reason is that this property does not fit well in the hierarchy of other independence conditions because it has consequences for the structural properties of von Neumann algebras as the following result of Licht [218] shows

11.4.3. Theorem. *Let A and B be commuting von Neumann subalgebras of a von Neumann algebra M acting on a Hilbert space H. Suppose that there is a vector that is cyclic for B. Then the pair (A, B) is W^*-strictly local in the sense of vector states if, and only if, any nonzero projection $p \in A$ is equivalent to 1 with respect to the algebra B'.*

Proof: Suppose that $p \sim 1$ in B' for any nonzero $p \in A$. There is a partial isometry $w \in B'$ such that $w^* w = 1$ and $w w^* = p$. For any unit vector $\nu \in H$, let us put $\xi = w \nu \in p(H)$. Then the vector state ω_ξ is concentrated at p and, for any $b \in B$,

$$\omega_\xi(b) = (b w \nu, w \nu) = (w^* w b \nu, \nu) = (b \nu, \nu) .$$

Thus, (A, B) is W^*-strictly local in the sense of vector states.

Let us now prove the reverse implication. Denote by ξ the cyclic vector of B and fix a nonzero projection $p \in A$. If (A, B) is W^*-strictly local in the sense of vector states, then there is a unit vector $\nu \in p(H)$ such that the vector states ω_ν and ω_ξ coincide on B. Employing now Lemma 11.3.10, the projections p' and q' onto $[B \xi]$ and $[B \nu]$, respectively, are equivalent in B'. But $p' = 1$ by the cyclicity of ξ and $q' \leq p$ because B' leaves $p(H)$ invariant. Hence, p is equivalent to 1 in B'. □

The previous result shows that W^*-strict locality in the sense of vector states has bearing on the structure of von Neumann algebras. This is apparent from the following corollary [218].

11.4.4. Corollary. *Let M be a von Neumann factor acting on a separable Hilbert space H and having a cyclic vector. Then the pair (M', M) is W^*-strictly local in the sense of vector states if, and only if, M is of Type III.*

Proof: By Theorem 11.4.3 (M', M) is W^*-strictly local in the sense of vector states if, and only if, every nonzero projection in factor M' is equivalent to 1. Hence, M' and so M has to be of Type III.

On the other hand, Type III factor acting on a separable Hilbert space has the property that all nontrivial projections are equivalent to 1. □

By the Reech-Schlieder Theorem in quantum field theory the vacuum state is typically cyclic for all local algebras. From this point of view the assumption of cyclicity in the previous corollary is not so restricting. Corollary 11.4.4 provides an explanation of Type III structure of local algebras in quantum field theory in terms of physically meaningful independence condition. It does not mean of course that all pairs that are W^*-strictly local in the sense of vector states have to be of Type III. (Note that W^*-strict locality in the sense of vector states is inherited by subalgebras.) In

[205] Kadison constructed a pair (A, B) of commuting Type I infinite factors that have a joint cyclic and separating vector and therefore (A, B) is W^*-strictly local in the sense of vector states by Theorem 11.4.3 without being Type III. We shall now discuss the situation of when a pair of algebras is (M, M'), where M is a factor acting on a Hilbert space H. This case is a source of many important examples. First, let us note that by the classical result of Murray and von Neumann M and M' are always S-independent (and thus C^*-independent) (see also Theorem 11.2.18). However, if M is not of Type III and M' has a cyclic vector, then (M, M') cannot be W^*-strictly local in the sense of vector states. Therefore W^*-strict locality in the sense of vector states is strictly stronger than W^*-strict locality. Moreover, by Florig-Summers Theorem M and M' are W^*-independent and we see thereby that W^*-independence does not imply W^*-strict locality in the sense of vector states in general. By Roos Theorem the C^*-algebra $C^*(M \cup M')$ generated by M and M' is isomorphic to some tensor product $M \otimes_\beta M'$. This product need not always be minimal. This fact is connected with the concept of semidiscretess of von Neumann algebras. A von Neumann algebra is called *semidiscrete* if the identity map of M can be approximated in the topology of simple σ-weak convergence by normal completely positive contractions of finite ranks. It has been proved by Effros and Lance [130] that a factor M on a Hilbert space H is semidiscrete if, and only if, the natural map $\nu\colon\ M \otimes_{alg} M' \to B(H)$ given by $\nu(x \otimes y) = x\,y$ is isometric with respect to the minimal tensor norm on the algebraic product $M \otimes_{alg} M'$. The remarkable result of Connes [83] says that a factor M on a separable Hilbert space is semidiscrete if, and only if, it is hyperfinite, i.e. a weak*-closure of increasing net of finite-dimensional factors. Therefore any nonsemidiscrete factor M produces an example of S-independent algebras that do not generate the minimal tensor product. An analogous example can be found in the category of normal state spaces. Suppose that M is a Type III factor. Then the W^*-tensor product $M \overline{\otimes} M'$ has to be of Type III, while the algebra $W^*(M, M')$ is the whole of $B(H)$, i.e. a Type I factor. In this case M and M' constitute an example of W^*-independent algebras (if the Hilbert space H is separable) that do not generate the W^*-tensor product. This situation is also interesting in the following respect. Namely, it has been proved by Takesaki [305] that the existence of only one normal product state across factors M and M' implies that $W^*(M, M')$ is isomorphic to $M \overline{\otimes} M'$. Hence, all normal partial states on M and M' have

normal extensions none of which can be a product state. On the cther hand, by the Theorem 11.2.17 all partial states have product extensions none of which is allowed to be normal.

Besides the independence conditions treated in previous sections another type of independence, called the *split property*, have been widely studied in quantum field theory as well as pure operator algebras theory [42, 45, 43, 46, 90, 44, 98, 99, 219]. Let A and B be commuting von Neumann algebras in a von Neumann algebra M. We say that a pair (A, B) is *split* if there is a Type I factor F such that $A \subset F \subset B'$. In other words, $A \subset F$ and $B \subset F'$. It can be verified quickly (see e.g [302]) that (A, B) is split if, and only if, $W^*(A, B)$ is spatially isomorphic to the W^*-tensor product $\pi_1(A)\overline{\otimes}\pi_2(B)$, where π_1 and π_2 are some normal faithful representations of A and B, respectively. As a consequence, the split property implies W^*-independence and it is the strongest independence conditions we know. Indeed, it is strictly stronger then W^*-independence as it can be demonstrated again by the example (M, M'), where M is a Type III factor on a separable Hilbert space. As seen before, such algebras are W^*-independent but not isomorphic to their W^*-tensor product.

Among others, the independence is a useful tool for investigating Bell's inequalities in operator algebras (see [274, 302] and the references therein). We shall briefly comment on this link.

11.4.5. Definition. Let A and B be commuting C^*-subalgebras in a C^*-algebra D. The maximal Bell correlation of the pair (A, B) in the state $\varphi \in D^*$ is the number

$$\beta(\varphi, A, B) = \frac{1}{2} \sup \varphi(a_1\,(b_1 + b_2) + a_2\,(b_1 - b_2))\,,$$

where the supremum is taken over all self-adjoint contractions $a_1, a_2 \in A$, $b_1, b_2 \in B$.

It has been proved by Summers and Werner [303] that

$$0 \leq \beta(\varphi, A, B) \leq \sqrt{2}\,, \qquad \text{for all states } \varphi\,.$$

If the correlation $\varphi(a\,b)$ between observables $a \in A$ and $b \in B$ can be described by a single probability space, then

$$0 \leq \beta(\varphi, A, B) \leq 1\,. \tag{11.10}$$

The inequality (11.10) is called the *Bell inequality* and it is a useful quantitative measure for testing the presence of hidden variables. For example, if φ is a product state across local algebras, then the Bell inequality is satisfied. The following result due to Landau [217] shows that the Bell inequalities are always maximally violated for noncommutative local algebras in some state.

11.4.6. Theorem. *Let A and B commuting von Neumann subalgebras in a von Neumann algebra M. Let neither A nor B be abelian. Then there is a normal state φ on M and self-adjoint contractions $a_1, a_2 \in A$; $b_1, b_2 \in B$, such that*

$$\varphi\left(\frac{1}{2}(a_1\,(b_1 + b_2) + a_2\,(b_1 - b_2))\right) = \sqrt{2}\,.$$

Outline of the proof: As A and B are nonabelian they have to contain a copy of $M_2(\mathbb{C})$. It can be shown that passing to suitable hereditary subalgebra of M one may assume that both A and B are copies of $M_2(\mathbb{C})$. Note that in this situation A and B are C^*-independent (see the end of Section 11.3) and generate finite-dimensional subalgebra isomorphic to $M_2(\mathbb{C}) \otimes M_2(\mathbb{C}) = M_4(\mathbb{C})$. Let us take projections p, q in A (identified with $M_2(\mathbb{C})$) given by

$$p = \begin{pmatrix} 1/2 & 1/2 \\ 1/2 & 1/2 \end{pmatrix}, \quad q = \begin{pmatrix} 1/2 & 1/2\,i \\ -1/2\,i & 1/2 \end{pmatrix}.$$

Thus

$$p\,q - q\,p = \begin{pmatrix} -1/2\,i & 0 \\ 0 & 1/2\,i \end{pmatrix}.$$

Then $\|p\,q - q\,p\| = \frac{1}{2}$. In a similar way, let us choose projections r, s in B with $\|r\,s - s\,r\| = \frac{1}{2}$. Set now $a_1 = 2\,p - 1$, $a_2 = 2\,q - 1$, $b_1 = 2\,r - 1$, $b_2 = 2\,s - 1$, and $z = a_1\,(b_1 + b_2) + a_2\,(b_1 - b_2)$. It can be verified by a straightforward calculation that $z^2 = 4 + 16\,[p, q]\,[r, s]$. Note that by C^*-independence of algebras A and B, $\|[p, q]\,[r, s]\| = \frac{1}{4}$. Therefore, we can always find a normal norm-one functional ψ on M such that $\psi([p, q]\cdot[r, s]) = 1/4$. Consequently,

$$\psi(z^2) = 4 + 16\,\frac{1}{4} = 8\,.$$

Thus, $\|z^2\| = \|z\|^2 \geq 8$, which means that $\|z\| \geq 2\sqrt{2}$. Again, since z is an element in a finite-dimensional algebra, we can finally find a normal state φ such that $|\varphi(z)| = \|z\| \geq 2\sqrt{2}$. By changing the signs of a_1 and a_2 if need be, we obtain $\frac{1}{2}\varphi(z) = \sqrt{2}$. □

Let us remark in this connection that it was established by Summers and Werner [302] that algebras of observables associated with complementary wedge regions in quantum field theory are maximally violating Bell inequalities in every normal state.

The importance of independence conditions for investigating structures of subalgebras in operator algebras has been recently recognized. It is not know, e.g., what conditions in addition to C^*-independence imply mutual commutativity of partial algebras. This problem, together with the study of the uniqueness of simultaneous extensions and their relationship to the conditional expectations, will certainly play an important role in the future.

Bibliography

[1] J.F.Aarnes: *Physical states on a C^*-algebra,* Acta Math. **122**, (1969), 161–172.

[2] J.F.Aarnes and R.V.Kadison: *Pure states and approximate identities*, Proc. Amer. Math. Soc. **21**, (1969), 749–752.

[3] F.J.Aarnes: *Quasi-states on C^*-algebras*, Trans. Amer. Math. Soc., **149**, (1970), 601–625.

[4] J.F.Aarnes: *Quasi-states and quasi-measures*, Advances in Mathematics, Vol. **86**, No. 1, (1991), 41–67.

[5] J.F.Aarnes: *Pure quasi-states and extremal quasi-measures*, Math. Ann. **295**, (1993), 575–588.

[6] J.F.Aarnes: *Image transformations, attractors and invariant non-subadditive measures*, Department of Mathematics, University of Trodheim, Preprint mathematics No. 2, (1994).

[7] J.F.Aarnes: *Construction of non-subadditive measures and discretization of Borel measures*, Fundamenta Mathematicae, **147**, (1995), 213–237.

[8] C.A.Akemann: *Interpolation in W^*-algebras*, Duke Math. J. **35**, (1968), 525–523.

[9] C.A.Akemann: *The dual space of an operator algebra*, Trans. Amer. Math. Soc., **126**, (1967), 286-302.

[10] C.A.Akemann: *The general Stone-Weierstrass problem*, J. Funct. Anal. **4**, (1969), 277–294.

[11] C.A.Akemann: *Approximate units and maximal abelian C^*-subalgebras*, Pacific Journal of Mathematics, **Vol 33,** No. 3, (1970), 543-550.

[12] C.A.Akemann and S.M.Newberger: *Physical states on a C^*-algebra*, Proc. Amer. Math. Soc., Vol. **40**, Nu. 2, (1973), 500.

[13] C.A.Akemann, J.Anderson and G.K.Pedersen: *Excising states on C^*-algebras,* Can. J. Math., Vol. XXXVIII, No. 5, (1986), 1239–1260.

[14] C.A.Akemann, J.Anderson and G.K.Pedersen: *Approaching infinity in C^*-algebras,* J. Oper. Theory, **21**, (1989), 255–271.

[15] C.A.Akemann and J.Anderson: *Lyapunov Theorems for Operator Algebras,* Memoirs of the Amer. Math. Soc., **458**, (1991).

[16] V.Alda: *On 0-1 measure for projectors*, Aplik. Math. **25**, (1980), 373–374.

[17] V.Alda *On 0-1 measures for projectors II*, Applik. Math., **26**, (1981), 57–58.

[18] E.M.Alfsen: *Compact Convex Sets and Boundary Integrals*, Springer Verlag, New York, Heidelberg, Berlin (1971).

[19] E.M.Alfsen and F.W.Schultz: *Non-Commutative Spectral Theory for Affine Functions Spaces on Convex Sets,* Memoirs of the Amer. Math. Soc, **172**, (1976).

[20] E.M.Alfsen and F.W.Schultz: *State spaces of Operator Algebras,* Birkfäuser, Boston, Basel, Berlin, (2001).

[21] A.Amann: *Jauch-Piron states in W^*-algebraic quantum mechanics,* J.Math.Phys. **28**, (10), (1989), 2384-2389.

[22] I.Amemiya and H.Araki: *A remark on Piron's paper*, Publ. Res. Inst. Math. Sci. A, **2** (3), (1957), 423–427.

[23] J.Anderson: *Extensions, restrictions, and representations of states on C^*-algebras,* Trans. Amer. Math. Soc., Vol. **249**, No. 2, (1979), 383–400.

[24] J.Anderson: *Pathalogy in the Calcin algebra*, J. Operator Theory **2** (1979), 159–167.

[25] B.A.D'Andrea, P. De Lucia and P.Morales: *The Lebesgue decompositions and the Nikodym convergence theorem on orthomodular posets*, Atti Sem. Mat. Fis. Univ. Modena, **39**, (1991), 137-158.

[26] R.J.Archbold: *Extensions of states of C^*-algebras,* J. London Math. Soc. (2), (1980), 351-354.

[27] R.J. Archbold: *Extensions of factorial states of C^*-algebras*, J. Funct. Anal. **63**, (1985), 86-100.

[28] W.Arveson: *Subalgebras of C^*-algebras,* Acta Math. **123**, (1969), 141–224.

[29] V.Bargmann: *Note on Wigner theorem on symmetry operatios,* J. Math. Phys. **5**, (1964), 862-868.

[30] B.A.Barnes: *Pure states with the restriction property*, Proc. Amer. Math. Soc., Vol **33**, Number **2**, June (1972), 277–294.

[31] W.C.Bell: *A note on pure charges,* Bolletino U.M.I. **7**, (1989) 147–154.

[32] J.S.Bell: *On the problem of hidden variables in Quantum Mechanics*, Review of Modern Physics, XXXVIII, (1966), 447–475.

[33] B.V.Rajarama Bhat: *On a characterization of velocity maps in the space of observables*, Pacific J. of Mathematics, Vol. **152**, No. 1, (1992), 1-14.

[34] G.Birkhoff and J. von Neumann: *The logic of quantum mechanics*, Annals of Mathematics **37**, (1936), 823-824.

[35] B.Blackadar, *K-theory of Operator Algebras*, Mathematical Sciences Research Institute Publications, Springer-Verlag, (1986).

[36] B.Blackadar, *Comparison theory for simple C^*-algebras*, Operator Algebras and Applications, London Math. Soc. Lecture Notes, Vol. **135**, (1988), 21–54.

[37] B.Blackadar, *Projections in C^*-algebras*, Contemporary Mathematics, Vol. **167**, (1994), 131–149.

[38] B.Blackadar and D.Handelman: *Dimension functions and traces on C^*-algebras*, J. Funct. Anal., **45**, (1982), 297–340.

[39] O.Bratteli and D.W.Robinson: *Operator Agebras and Quantum Statistical mechanics*, Springer Verlag, (1979).

[40] L.G.Brown and G.K.Pedersen, *C^* algebras of real rank zero*, J. Funct. Anal., Vol. **99**, no. 1, (1991), 131–149.

[41] J.Bub: *Interpreting the Quantum World,* Cambridge University Press, Cambridge, (1997).

[42] D.Buchholz: *Product states for local algebras*, Commun. Math. Phys **36**, (1974), 287-304.

[43] D.Buchholz, C. D'Antoni and R.Longo: *Modular maps and modular structures I. General properties.* J. Funct. Anal. **88**, (1990) 233-250.

[44] D.Buchholz, C D'Antoni and R.Longo: *Modular maps and modular structures II. Applications to quantum field theory.* J. Funct. Anal. **88**, (1990) 233-250.

[45] D.Buchholz, C.D'Antoni and K.Fredenhagen: *The universal structure of local algebras*, Commun. Math. Phys. **111**, (1987) 123-135.

[46] D.Buchholz and E.H.Wachmann: *Causal independence and the energy-level density of states in local quantum field theory*, Commun. Math. Phys. **106**, (1986) 321-344.

[47] D.Buhagiar: *On weak states on the orthogonally closed subspaces of inner product space,* Int. J. Theor. Phys. (2003), to appear.

[48] D.Buhagiar and E.Chetcuti: *On complete-cocomplete subspaces of an inner product space*, to appear.

[49] J.Bunce: *Characters on singly generated C^*-algebras*, Proc. Amer. Math. Soc., **25**, (1970), 297-303.

[50] L.J.Bunce and J.Hamhalter: *Jauch–Piron states on von Neumann algebras*, Math. Zeitschrift, **215**, 491–502, (1994).

[51] L.J.Bunce and J.Hamhalter: *Traces and subadditive measures on projections in JBW-algebras and von Neumann algebras*, Proc. Amer. Math. Soc. Vol. **123**, No.1, (1995), 157-160.

[52] L.J.Bunce and J.Hamhalter: *Countably additive homomorphisms between von Neumann algebras*, Proc. Amer. Math. Soc., **123**, 3437–3441, (1995).

[53] L.J.Bunce and J.Hamhalter: *Extension of Jauch–Piron states on Jordan algebras*, Math. Proc. Camb. Phil. Soc., **189**, 279–286, (1996).

[54] L.J.Bunce and J.Hamhalter: *Jauch-Piron states and σ-additivity*, Reviews in Mathematical Physics, Vol. **12**, No. 6, (2000), 767–777.

[55] L.J.Bunce and J.D.M.Wright: *Quantum measures and states on Jordan algebras*, Commun. Math. Phys. **98**, (1985), 187–202.

[56] L.J.Bunce and J.D.M.Wright: *Continuity and linear extensions of quantum measures on Jordan operator algebras*, Math. Scand. **64**, (1989), 300–306.

[57] L.J.Bunce, M.Navara, P.Pták and J.D.M.Wright: *Quantum logics with Jauch-Piron states*, Quaterly Journal of Oxford, (2), **36**, (1985), 261-271.

[58] L.J.Bunce and J.D.M.Wright: *Complex measures on projections in von Neumann algebras*, J. London Math. Soc. (2), **46**, (1992), 269-279.

[59] L.J.Bunce and J.D.M.Wright: *The Mackey-Gleason problem*, Bull. Amer. Math. Soc., Vol. **26**, Nu. 2, (1992), 288–293.

[60] L.J.Bunce and J.D.M.Wright: *On Dye's theorem for Jordan operator algebras*, Expo Math. **11**, (1993), 91-95.

[61] L.J.Bunce and J.D.M.Wright: *The Mackey-Gleason problem for vector measures on projections in von Neumann algebras*, J. London Math. Soc. (2), **49**, (1994), 133-149.

[62] L.J.Bunce and J.D.M.Wright: *Velocity maps in von Neumann algebras*, Pacific Journal of Mathematics, Vol. **170**, No. 2, (1995), 421–427.

[63] L.J.Bunce and J.D.M.Wright: *Approximately additive states on Type I C^*-algebras*, Quart. J. Math. Oxford, (**2**), 47, (1996), 269–277.

[64] L.J.Bunce and J.D.M.Wright: *The quasi-linearity problem for C^*-algebras*, Pacific Journal of Mathematics, Vol. **172**, No. 1, (1996), 41–47.

[65] L.J.Bunce and J.D.M.Wright: *A topological characterization of linearity for quasi traces*, Proc. Amer. Math. Soc., **124**, (1996), no. 8, 2377-2381.

[66] L.J.Bunce, K.Saito and J.D.M.Wright: *Completely additive measures on von Neumann algebras.*, Expositiones Mathematicae, 16, (1998), 185-189.

[67] L.J.Bunce, C.-H. Chu and M.Kusuda: *Unique extensions of pure states of von Neumann algebras*, to appear.

[68] L.J.Bunce and C-H Chu: *Unique extensions of pure states of C^*-algebras*, J. Operator Theory, **39** (1998), 319-338.

[69] A.Canetti and G.Marini: *Weights, states and completeness of pre-Hilbert spaces*, to appear.

[70] L.Caruana: *John von Neumann's 'Impossibility Proof' in a historical perspective*, Physics, Nuova Ser. **32**, No. 1, (1995), 109–124.

[71] G.Cassinelli, E. De Vito, A.Levrero and P.Lahti: *Symmetry Groups in Quantum Mechanics and the Theorem of Wigner on the Symmetry transformations*, Rev. Math. Phys., no. 8, (1997), 921-941.

[72] G.Cattaneo, G.Franco and G.Marino: *Ordering of families of subspaces of pre-Hilbert spaces and Dacey pre-Hilbert spaces*, Bolletino U.M.I. (7), 1-B, (1987), 167-183.

[73] G.Cattaneo and G.Marino: *Completeness of inner product spaces with respect to splitting subspaces*, Letters in Mathematical Physics, **11**, (1986), 15-20.

[74] G.Cattaneo and J.Hamhalter: *De Morgan property for effect algebras of von Neumann algebras*, Letters in Mathematical Physics, Vol. **59**, No. 3, (2002), 243-252.

[75] E.Chetcuti: *Completeness criteria for inner product spaces*, MSc. Thesis, University of Malta, (2001).

[76] E.Chetcuti and P.Pták: *On the (non) existence of states on orthogonally closed subspaces in an inner product space*, Technical Report, University of Malta, (2002).

[77] E.Chetcuti and A.Dvurečenskij: *The existence of finitely additive states on orthogonally closed subspaces of incomplete inner product spaces*, to appear.

[78] E.Chetcuti and A.Dvurečenskij: *A finitely additive state criterion for the completeness of inner product spaces*, to appear.

[79] G.Chevalier, A.Dvurečenskij and K.Svozil: *Piron's and Bell's geometric lemmas and Gleason's theorem*, Foundations of Physics, **Vol. 30**, No. 10, (2000), 1737–1755.

[80] E.Christensen: *Measures on projections and physical states*, Commun. Math. Phys., **86**, (1982), 529–538.

[81] E.Christensen: *Comments to my paper: Measures on projections and physical states*, private communication, (1985) (published by S.Maeda).

[82] C.-H. Chu and Kusuda: *On factor states of C^*-algebras and their extensions*, Proc. Amer. Math. Soc., Vol. **124**, Nu. 1, January (1996).

[83] A.Connes: *Classification of injective factors*, Ann of Math. **104**, (1976), 73-116.

[84] A.Connes: *Noncommutative Geometry*, Academic Presss, (1994).

[85] T.A.Cook: *The Nikodým-Hahn-Vitali-Saks Theorem for states in quantum logics*, in ed. A.R.Marlow "Mathematical Fondations of Quantum Theory", Acad. Press., London (1978), 275-285.

[86] R.Cooke, M.Keane and W.Moran: *An elemetary proof of Gleason's theorem*, Math. Proc. Camb. Phil. Soc., **98**, (1985), 117–128.

[87] J.Cuntz: *On the continuity of semi-norms on operator algebras*, Math. Anal. **220**, (1976), 171-183.

[88] J.Cuntz: *K-theory for certain C^*-algebras*, Ann. of Math., **113**, (1981), 181–197.

[89] J.Cuntz: *The structure of addition and multiplication in simple C^* algebras*, Math. Scand. **40**, (1977), 215–233.

[90] C.D'Antoni and R.Longo: *Interpolation by type I factors and the flip automorphism*, J Funct. Anal. **51**, (1983) 361-371.

[91] K.Davidson: *C^*-Algebras by Example*, The Fields Institute for Research in Mathematical Sciences, American Mathematical Society, Vol. 6, (1996).

[92] B. De Facio: *Causal independence in algebraic quantum field theory*, Foundations of Physics, Vol. **5**, No. 2, (1975), 229-237.

[93] A. De Simone: *Decomposition theorems in orthomodular posets: The problem of uniqueness*, Proc. Amer. Math. Soc. **120**, (1998), 2919-2926.

[94] A. De Simone and M.Navara: *Yosida-Hewitt and Lebesgue decomposition of states on orthomodular posets*, J. Math. Anal. Appl. **255**, (2001), 74-104.

[95] A. De Simone and M.Navara: *On the Yosida-Hewitt decomposition and Rüttimann decomposition of states*, Scientiae Mathematicae Japonicae Online, Vol. **6**, (2002), 49-62.

[96] A. De Simone, M.Navara and P.Pták: *The Vitali-Hahn-Saks theorem for the product of quantum logics,* Demonstration Math XXXV, No. 4, (2002), 717–725.

[97] J.Dixmier: *C^*-Algebras,* North Holland, (1977).

[98] S.Doplicher: *Local aspects of superselection rules*, Commun. Math. Phys. **85**, (1982), 73-86.

[99] S.Doplicher and R.Longo: *Standard and split inclusions of von Neumann algebras*, Invent. Math, **7**, (1984), 493-536.

[100] S.Dorofeev and A.N.Sherstnev: *Frame-type functions and their applications,* Izv. vuzov matem., no. 4 (1990), 23-29 (in Russian).

[101] S.Dorofeev: *On the problem of boundedness of a signed measure on projections of a von Neumann algebra,* J. Funct. Anal., **103**, (1992), 209-216.

[102] S.Dorofeev: *On the problem of boundedness of a signed measure on projections of a Type I von Neumann algebra,* Proc. of Higher Educational Institutions, Mathematics, Vol. **3**, (1990), 67-69.

[103] S. Dorofeev: *The solution to a problem posed by V.Palko*, Proc. Winter School on Measure Theory, Liptovský Ján, January 7-12, (1990), 222-223.

[104] T.Drisch: *Generalization of Gleason's theorem,* International Journal of Theoretical Physics, Vol. **18**, (1979), 239–243.

[105] N.Dunford and J.T.Schwarz: *Linear Operators*, Part I, Interscience Publishers, New York (1958).

[106] A.Dvurečenskij: *On the convergence of signed measures*, Math. Slovaca, **28**, (1978), 33–40.

[107] A.Dvurečenskij: *New Look at Gleason's theorem for signed measures*, International Journal of Theoretical Physics, Vol. **26**, No. 3, (1987), 295–305.

[108] A.Dvurečenskij: *Note on a construction of unbounded measures on a separable Hilbert space quantum mechanics,* Ann. Inst. Henri Poincaré, Vol. **48**, no. 4, (1988), 297–310.

[109] A.Dvurečenskij and L.Mišík, Jr. : *Gleason's theorem and completeness of inner product spaces*, International Journal of Theoretical Physics, Vol. **27**, No. 4, (1988), 419-427.

[110] A.Dvurečenskij: *Generalization of Maeda's theorem,* International Journal of Theoretical Physics, Vol. **25**, No. 10, (1986), 1117–1124.

[111] A.Dvurečenskij and S.Pulmannová: *A signed measure completeness criterion,* Letters in Mathematical Physics **17**, (1989), 253–261.

[112] A.Dvurečenskij: *States on families of subspaces of pre-Hilbert spaces,* Letters in Mathematical Physics, **17**, (1989), 19-24.

[113] A.Dvurečenskij: *Regular measures and inner product spaces,* International Journal of Theoretical Physics, Vol. **31**, No. 5, (1992), 889-905.

[114] A.Dvurečenskij: *Quantum logics and completeness criteria of inner product spaces,* International Journal of Theoretical Physics, Vol. **31**, No. 10, (1992), 1899-1907.

[115] A.Dvurečenskij: *Regular charges and completeness of inner product spaces,* Atti. Sem. Mat. Fis. Univ. Modena, XLI, (1993), 269-285.

[116] A.Dvurečenskij: *Gleason's Theorem nad Its Applications*, Kluwer Academic Publishers, Dordrecht, Boston, London, (1993).

[117] A.Dvurečenskij and S.Pulmannová: *Test spaces, Dacey spaces, and completeness of inner product spaces,* Letters in Mathematical Physics **32**, (1994), 299-306.

[118] A.Dvurečenskij: *Frame functions and completeness of inner product spaces,* Ann. Inst. Henri Poincaré, **Vol. 62**, (1995), 429–438.

[119] A.Dvurečenskij: *Test spaces and characterizations of quadratic spaces,* International Journal of Theoretical Physics, Vol. **35**, No. 10, (1996).

[120] A.Dvurečenskij: *A note on the Nikodým boundedness theorem,* Tatra Mountains Math. Publ., **8**, (1996) 231-240.

[121] A.Dvurečenskij: *Remarks on Nikodým boundedness theorem,* Novi Sad J. Math., Vol. **26**, No. 1, (1996), 25-51.

[122] A.Dvurečenskij: *A measure-theoretic characterization of inner product spaces,* Tatra Mountains Math. Publ. **10**, (1997), 273–280.

[123] A.Dvurečenskij: *Measures and $\perp$-decomposable measures on effects of a Hilbert space,* Atti Sem. Mat. Fis. Univ. Modena, XLV, (1997), 259-288.

[124] A.Dvurečenskij: *A new algebraic criterion for completeness of inner product spaces,* Letters in Mathematical Physics, Vol. **58**, (2001), 205-208.

[125] A.Dvurečenskij: *Boundedness of regular charges and the completion of inner product spaces*, to appear.

[126] A.Dvurečenskij and P.Pták: *On states on orthogonally closed subspaces of an inner product space,* Letters in Mathematical Physics, **62**, (1), (2002), 63-70.

[127] H.A.Dye: *On the geometry of projections in certain operator algebras.* Ann Math. **61**, (1955), 73–89.

[128] C.M.Edwards and G.T.Rütimann: *The lattice of weak*-closed inner ideals in a W^*-algebra,* Commun. Math. Phys. **197**, (1998), 131–166.

[129] E.G.Effros: *Order ideals in a C^*-algebra and its dual,* Duke Math. J., **30**, 391–412, (1963).

[130] E.G.Effros and E.C.Lance: *Tensor products of Operator Algebras,* Advances in Mathematics, **25**, (1977), 1-34.

[131] M.Eilers and E.Horst: *The theorem of Gleason for nonseparable Hilbert space.* International Journal of Theoretical Physics, Vol. **13**, (1975), 419–424.

[132] D.R.Farenick and P.B.Morenz: *C^*-extreme points in the generalized state spaces of a C^*-algebra,* Trans. Amer. Math. Soc., Vol. **37**, May (1997), 1725–1748.

[133] D.R.Farenick and H.Zhou: *The structure of C^*-extreme points in spaces of completely positive linear maps on C^*-algebras,* Proc. Amer. Math. Soc., Vol. **126**, Number 5, (1998), 1467–1477.

[134] J.Feldmann and J.M.G.Fell: *Separable representations of rings of operators,* Ann. of Math. (2), **65**, (1957), 241-249.

[135] P.A.Fillmore: *On product of symmetries,* Canad J. Math **18** (1966), 897-900.

[136] M.Florig and S.J.Summers: *On the statistical independence of algebras of observables,* J. Math. Phys., **38**, (3), (1997), 1318-1328.

[137] D.J.Foulis, R.Greechie amd M.K.Bennet: *Sums and products of interval algebras,* International Journal of Theoretical Physics, no. 11, **33**, (1994), 2119-2136.

[138] M.Gell-Mann and J.Hartle: *Quantum mechanics in the light of quantum cosmology,* Complexity, Entropy and the Physics of Information, SFI Studies in the Science of Complexity, Addison-Wesley, Reading (1990), Vol. VIII, 425–458.

[139] M.Gell-Mann and J.Hartle: *Alternative decohering histories in quantum mechanics,* Proc. 25th Int. Conf. on High Energy Physics, August, 2-8, (1990), World Scientific, Singapore, (1990).

[140] A.M.Gleason: *Measures on the closed subspaces of a Hilbert space*, Journal of Mathematics and Mechanics, **Vol. 6**, No. 6, (1957), 885–893.

[141] J.Glimm and R.V.Kadison: *Unitary operators in C^*-algebras*, Pacific J. Math. **10**, (1960), 547-556.

[142] J.Glimm: *A Stone-Weierstrass theorem for C^*-algebras*, Annals of Mathematics, Vol. 72, No. 2, September (1960), 216–244.

[143] S.Goldstein: *Stationarity of operator algebras*, J. Funct. Anal. **118**, 275–308, (1993).

[144] S.Goldstein and A.Paszkiewicz: *Orthogonally additive functions on operator algebras*, International Journal of Theoretical Physics, **34**, (1995), 1409-1416.

[145] S.Goldstein, A.Luczak and I.F.Wilde: *Independence in operator algebras*, Foundations of Physics, **29**, no. 1, (1999), 79-89.

[146] R.J.Greechie: *Orthomodular lattices admitting no states*, J. Combinatorial Theory, Ser. A, **10**, (1971), 119-132.

[147] R.B.Griffiths: *Consistent histories and the interpretation of quantum mechanics*, J. Stat. Phys., **36**, (1984), 219-272.

[148] H.Gross and A.Keller: *On the definition of Hilbert space,* Manuscripta Math. **23**, (1977), 67-90.

[149] S.Gudder: *Plane frame functions and pure states in Hilbert space*, International Journal of Theoretical Physics, Vol. **5**, (1972), 369–375.

[150] S.P.Gudder: *Inner product spaces,* Amer. Math. Monthly, **81**, (1974), 29-36.

[151] S.P.Gudder: *Correction to "Inner product spaces",* Amer. Math. Monthly, **82**, (1975), 251-252.

[152] S.P.Gudder and S.Holland: *Second correction to "Inner product spaces",* Amer. Math. Monthly, **82**, (1975),

[153] S.P.Gudder, Stochastic Methods in Quantum Mechanics, Elsevier, North Holland, (1979).

[154] J.Gunson: *Physical states on quantum logics I*, Ann. Inst. Henri Poincaré, Vol. **XVII**, No. 4, 295–311.

[155] R.Haag and D. Kastler: *An algebric approach to quantum field theory*, Journal of Mathematical Physics, Vol. **5**, Nu. 7, (1964), 848-861.

[156] U.Haagerup: *Grothendieck Inequality for bilinear forms on C^*-algebra,* Advances in Mathematics **56**, (1985) 93–116.

[157] J.Hamhalter and P.Pták: A completeness criterion for inner product spaces, Bull. London Math. Soc. **19**, (1987), 259-263.

[158] J.Hamhalter: *On modular spaces,* Bulletin of the Polish Academy of Sciences Mathematics, **37**, (1989) 647-653.

[159] J.Hamhalter: *Closed subspaces in topological linear spaces*, Ph.D. Thesis, Czech Technical University of Prague (1989).

[160] J.Hamhalter: *A representation of finitely-modular AC-lattices,* Mathematishe Nachrichten, **147**, (1990), 281-284.

[161] J.Hamhalter: *States on W^*-algebras and orthogonal vector measures,* Proc. Amer. Math. Soc., **110**, (1990), 803-806.

[162] J.Hamhalter: *Orthogonal vector measures on projection lattices in Hilbert space,* Comment. Math. Univ. Carolinae **31**, (1990), 655-660.

[163] J. Hamhalter and M. Navara: *Orthosymmetry and modularity in ortholattices*, Demonstratio Mathematica, Vol **24** (1991), 323-329.

[164] J. Hamhalter and P. Pták : *Hilbert-space-valued measures on Boolean algebras (extensions)*, Acta Math. Univ. Comenianae, Vol **LX, 2** (1991), 219-224.

[165] J. Hamhalter and P. Pták : *Hilbert-space-valued states on quantum logics*, Applications of Mathematics, Vol. **37**, No.1 (1992), 51-56.

[166] J. Hamhalter: *Additivity of vector Gleason measures,* Int. J. Theor. Physics, **31**, No. 1, (1992), 1-13.

[167] J. Hamhalter : *On topological linear spaces whose subspace lattices are modular*, Rendiconti del Circolo Matematiko di Palermo, Serie II, Tomo **XXXI** (1992), 213-240.

[168] J.Hamhalter: *Centrally determined states on von Neumann algebras*, Mathematica Bohemica, Vol. **117**, No. 2, (1993), 1889-1892.

[169] J.Hamhalter, *Pure Jauch–Piron states on von Neumann algebras*, Ann. Inst. H.Poincaré, Phys. Théor. **58**, (1993), 173–187.

[170] J. Hamhalter : *States on projection logics of Neumann algebras*, International Journal of Theoretical Physics, Vol. 32, No. 10 (1993), 1889-1892.

[171] J.Hamhalter: *Extension properties of states on operator algebras*, International Journal of Theoretical Physics, **34**, No. 8, (1995), 1431-1437.

[172] J.Hamhalter, M.Navara and P.Pták : *States on orthoalgebras*, International Journal of Theoretical Physics, **34**, No. 8, (1995), 1439–1465.

[173] J.Hamhalter: *Gleason property and extensions of states on projection logics*, Bull London Math. Soc. **26**, (1994), 367-372.

[174] J.Hamhater: *The Nikodým Boundedness Theorem*, Tatra Mountains Math. Publ. **10**, (1997), 179-181.

[175] J.Hamhalter: *Statistical independence of operator algebras*, Ann. Inst. Henri Poincaré , Vol. **67**, no. 4, (1997), 447-462.

[176] J.Hamhalter: *Restricting pure states on JB algebras to maximal associative subalgebras*, Tatra Mountains Math. Publ., **15** (1998), 165–171. Special Issue: Quantum Structures II, Dedicated to Gudrun Kalmbach.

[177] J.Hamhalter: *Universal state space embeddability of Jordan-Banach algebras*, Proc. Amer. Math. Soc., Vol. **127**, Num 1, (1999), 131-137.

[178] J.Hamhalter: *Supporting sequences of pure states on JB algebras*, Studia Mathematica **136**, (1), (1999), 37–47.

[179] J.Hamhalter: *Restricting and extending states and positive maps on operator algebras*, Int. J. Theor. Phys., Vol. **39**, No. 3, (2000), 721–726.

[180] J.Hamhalter: *Pure states on Jordan algebras*, Mathematica Bohemica, No. 1, **126** (2001), 81-91.

[181] J.Hamhalter: *Multiplicativity of extremal positive maps on abelian parts of operator algebras*, J. Operator Theory, **48** (2002), 369-383.

[182] J.Hamhalter: *C^*-independence and W^*-independence of von Neumann algebras*, Mathematische Nachrichten, **239-240**, (2002), 146-156.

[183] J.Hamhalter: *Orthogonal pure states in operator theory*, Proceedings of the Quantum Structures Conference in Vienna (2002), to appear.

[184] J.Hamhalter: *Dispersions of states and traces on operator algebras*, to appear.

[185] J.Hamhalter: *States and structure of von Neumann algebra,* International Journal of Theoretical Physics, to appear.

[186] J.Hamhalter: *Noncommutative phenomena in measure theory on operator algebras,* Rendiconti Dell'Universita Di Trieste, Nuova Serie, to appear.

[187] H.Hanche-Olsen and E.Stormer: *Jordan Operator Algebras*, Pitman Publishing, Boston, London, Melbourne, (1984).

[188] J.Harding: *Regularity in quantum logic*, International Journal of Theoretical Physics, no. 4, **37**, (1998), 377-399.

[189] J.Hartle: *Space time graining in nonrelativistic quantum mechanics.* Phys. Rev. D, **44**, (1991), 3173–3195.

[190] G.Hellman: *Gleason's theorem is not constructively provable*, Journal of Philosophical Logic, **22**, (1993), 193–203.

[191] I.Herstein: *Jordan derivations on prime rings*, Proc. Amer. Math. Soc., **8**, (1957), 1104–1110.

[192] S.S.Holland, Jr.: *Orthomodularity in infinite dimensions, a theorem of M.Soler,* Bull. Amer. Math. Soc., Vol. **32**, Nu. 2, (1995).

[193] A.Horn and A.Tarski: *Measures in Boolean algebras*, Trans. Amer. Math. Soc. **64**, (1948), 467–497.

[194] C.J.Isham: *Quantum logic and histories approach to quantum theory,* J.Math.Phys. **35**, (5), (1994).

[195] C.I.Isham, N.Linden and S.Schreckenberg: *The classification of decoherence functionals: An analog of Gleason's theorem,* J. Math. Phys. **35**, (12), (1994).

[196] C.I.Isham and N.Linden: *Information entropy and the space of decoherence functions in generalized quantum theory*, Physical Review A, Vol. **55**, Nu. 6, June (1997), 4030–4040.

[197] J.Jajte and A.Paszkiewicz: *Vector measures on the closed subspaces of a Hilbert space,* Studia Math. LXII, (1978), 229–251.

[198] R.Jajte: *Strong Limit Theorems in Non-Commutative Probability*, Lecture Notes in Mathematics, **1110**, Springer Verlag, (1985).

[199] M.Jauch and C.Piron, *Can hidden variables be excluded in quantum mechanics ?*, Helvetica Physica Acta, **36**, (1963), 827–837.

[200] M.Jauch and C.Piron: *On the structure of quantum proposition systems,* Helv. Phys. Acta **42**, (1969), 827-837.

[201] M.Jauch: *Foundations of Quantum Mechanics*, Reading, MA, Addison-Wesley (1968).

[202] R.V.Kadison: *On the additivity of the trace in finite factors*, Proc. Nat. Acad. Sci. U.S.A. **41**, (1955), 385–387.

[203] R.V.Kadison and I.M.Singer: *Extensions of pure states*, Amer. J. Math. **81**, (1959), 384–400.

[204] R.V.Kadison: *States and representations,* Trans. Amer. Math. Soc., **103**, 304–319, (1962).

[205] R.V.Kadison: *Remarks on the type of von Neumann algebras of local observables in quantum field theory*, J. Math. Phys. **4**, (1963), 1511-1516.

[206] R.V.Kadison and J.R.Ringorse: *Fundamentals of the Theory of Operator Algebras*, Vol. I, II, III, IV, Academic Press, (1986).

[207] R.V.Kadison: *Transformations of states in operator theory and dynamics*, Topology **3**, (1965), 177–198.

[208] R.V.Kadison: *Diagonalizing matrices*, Amer. J. Math., **106**, (1984), 1451–1468.

[209] S.Kakutani and G.W.Mackey: *Two characterizations of real Hilbert space*, Ann. of Math. (2), **45**, (1944), 50-58.

[210] S.Kakutani and G.W.Mackey: *Ring and lattice characterizations of complex Hilbert space*, Bull. Amer. Math. Soc., Vol. **52**, (1946), 727-733.

[211] A.Kaplan: *Continuous quasi-states on C^*-algebras*, J. Funct. Anal. **86**, (1989), 111-126.

[212] H.A.Keller: *Measures on orthomodular vector space lattices*, Stud. Math., **88**, (1988), 183-195.

[213] H.A.Keller: *Measures on infinite-dimensional orthomodular spaces*, Found. Phys., **20**, (1990), 575-604.

[214] H.A.Keller: *Ein nicht-klassicher Hilbertischen Raum*, Math. Z. **272**, (1980), 42-49.

[215] S.Kochen and E.P.Specker: *The problem of hidden variables in quantum mechanics*, J. Math. Mech. **17**, (1967), 59–87.

[216] K.Kraus: *General quantum field theories and strict locality*, Zeitschrift für Physics, **181**, (1964), 1-12.

[217] L.J.Landau: *On the violation of Bell's inequality in quantum theory*, Phys. Lett. A **120**, (1987), 54-56.

[218] A.L.Licht: *Strict localization*, Journal of Mathematical Physics, Vol. **4**, Nu. 11, (1963), 1443-1447.

[219] R.Longo: *Simple injective factors*, Advances in Mathematics **63**, 152-171 (1987).

[220] J.S.Lomout and P.Mandelsom: *The Wigner unitary-antiunitary theorem*, Ann. Math. **78**, (1963), 548-559.

[221] P. de Lucia and P.Pák:*A comleteness criterion for inner product spaces*, Acta Polytechnica-Math, Czech Technical University, (1993), 71–72.

[222] A.Luczak: *Banach-space-valued measures on von Neumann algebras*, Probability theory on vector spaces IV, Lancut, June 87, Springer's LNM, (1991), 178–190.

[223] G.D.Lugovaya: *On unbounded orthogonal measures on orthoprojectors in Hilbert space*, Russian Mathematics, (Iz. Vuzov), Vol. **36**, 43–46, (1992).

[224] G.W.Mackey: *The Mathematical Foundations of Quantum Mechanics*, New York, Benjamin, (1963).

[225] S.Maeda: *Probability measures on projections in von Neumann algebras*, Reviews in mathematical Physics, **1**, (1990),
235–290.

[226] M.Matvejchuk: *Several methods for solving the Gleason problem*, Measures on Projections and Orthomodular Posets, Kazan Mathematical Society, (1998), 14–37.

[227] M.Matvejchuk: *Description of finite measures in semifinite algebras*, Vol. **15**, (1981), 41–53.

[228] M.Matvejchuk: *Probability measures in W^* J-algebras in Hilbert spaces with conjugation*, Proc. Amer. Math. Soc. Vol. **126**, Nu. 4, (1998), 1115-1164.

[229] M.Matvejchuk: *Semicinstant measures on hyperbolic logics*, Proc. Amer. Math. Soc., Vol. **125**, Nu 1, (1997), 245-250.

[230] M.Matvejchuk: *Gleason's theorem in space with indefinite metric*, Math. Nachr. **184**, (1997), 229–243.

[231] A.Mayet-Ippolito: *Generalized orthomodular posets*, Demonstration Mathematica, Vol. XXIV, No. 1–2, (1991), 263–274.

[232] N.D.Mermin: *Hidden variables and two theorems of John Bell*, Rev. Mod. Phys. **65**, (1993), 803–815.

[233] B.Misra, *When can hidden variables be excluded in quantum mechanics?*, Nuovo Cimento **47A**, (1967), 841–859.

[234] L.Molnar: *Wigner's unitary-antiunitary theorem via Herstein theorem on Jordan homomorphisms*, J. Nat. Geom. **10**, (1996), 137–148.

[235] L.Molnar: *A generalization of Wigner's unitary-antiunitary theorem to Hilbert modulus,* J. Math. Phys. **40**, (1), (1999), 5544-5554.

[236] R.P.Morash: *Angle bisection and orthoautomorphisms in Hilbert lattices*, Canadian Journal of Mathematics, **25**, (1973), 216-272.

[237] G.J.Murphy: *C^*-algebras and Operator theory*, Academic Press, Inc. (1990).

[238] F.J.Murray and J von Neumann: *On rings of operators,* Ann. Math., Vol. **37**, 116-229.

[239] D.Mushtari: *Logics of projectors in Banach spaces*, Izv. Vysch. Uchebn., Zaved. Mat., Vol. **33**, No. 8, (1989), 44–52.

[240] D.H.Mushtari and M.Matvejchuk: *Charges on the logic of skew projections*, Soviet. Math. Dokl., Vol. **32**, (1985), 43-46.

[241] D.Mushtari: *Orthoadditive functions on idempotents*, Measures on Projections and Orthomodular Posets, Kazan Mathematical Society, (1998), 38-83.

[242] L.Nachbin: *A theorem of the Hahn-Banach type for linear transformations*, Trans. Amer. Math. Soc., **68**, (1950), 265-274.

[243] K.Napiorkowski: *On the independence of local algebras*, Rep. Math. Phys. **3**, (1972), 33-35.

[244] M.Navara: *Quantum logics with the Radon Nikodym property,* Order **4**, (1988), 387-395.

[245] M.Navara: *Piron's and Bell's geometrical lemmas*, to appear.

[246] M.Navara: *Easy proofs of some consequences of Gleason's theorem,* to appear.

[247] M.Navara and P.Pták: *For $n \geq 5$ there are no nontrivial Z_2-measures on $L(\mathbb{R}^n)$*, Int. J. Theor. Phys., to appear.

[248] J. von Neumann: *Mathematical Foundations of Quantum Mechanics*, Princeton, New Jersey: Princeton Univ. Presss, (1955).

[249] H.Nishimura: *A Boolean-valued approach to Gleason's theorem*, Institute of Mathematics, University of Tskuba, Ibaraki, Japan.

[250] T.Ogasawara: *A theorem on operator algebras*, J. Sci. Hiroshima Univ. **18**, (1955), 307-309.

[251] R.Omnes: *Logical reformulation of quantum mechanics I, Foundations,* J. Stat. Phys. **36**, (1984) 219–272.

[252] R.Omnes: *Logical reformulation of quantum mechanics II, the Einstein-Podolsky-Rosen experiment,* J. Stat. Phys. **53**, (1990), 354–447.

[253] R.Omnes: *Logical reformulation of quantum mechanics III, classical limit and irreversibility,* J. Stat. Phys. **53**, (1988), 933-935.

[254] R.Omnes: *From Hilbert space to common sense: A synthesis of recent progress in the interpretation of quantum mechanics,* Ann. Phys. (NY), **201**, (1988), 933-935.

[255] R.Omnes: *Consistent interpretations of quantum mechanics* Rev. Mod. Phys. **64**, (1992) 339–382.

[256] K.R.Parthasarathy: *An Introduction to Quantum Stochastic Calculus*, Birkh"auser Verlag, Basel (1992).

[257] A.Paszkiewicz: *On homomorphisms of projective lattices in compex Hilbert space,* Colloquium Mathematicum, Vol. XLIII, (1980), Fasc. 2, 271–280.

[258] A.Paszkiewicz: *Measures on projections in W^*-factors*, J. Funct. Anal. **62**, (1985), 87–117.

[259] G.K.Pedersen: *C^*-Algebras and their Automorphisms Group*, New York, London, Academic Press, (1979).

[260] A.Peres: *Two simple proofs of Kocher-Specker theorem*, J. Phys. Math. Rev. Gen. A **24**, (1991), 175-178.

[261] A.Peres: *Quantum Theory, Concepts and Methods,*, Kluwer Academic Publishers, Dordrecht, (1995).

[262] C.Piron: *Axiomatique quantique,* Helv. Phys. Acta **37**, (1964), 439-468.

[263] C.Piron: *Foundations of Quantum Mechanics,* Addison-Wesley Publishing Company, Reading, Massachussettts, (1968).

[264] I.Pitowski: *Infinite and finite Gleaso'n theorems and the logic of indeterminacy,* J. Math. Phys. **39**, January 1998, 218-228.

[265] I.Pitowski: *Constructive quantum mechanics, Gleason's theorem and the indeterminacy relations*, Tatra Mt. Math. Publ. **15**, (1998), 155-164.

[266] R.J.Plymen, *Dispersion free normal states*, Il Nuovo Cimento, Vol. LIV A, N. 4, (1968), 862–870.

[267] P.Pták: *FAT–CAT (in the state state space of quantum logics)*, Proceedings of Winter School on Measure Theory, Liptovský Ján, Czechoslovakia, (1988).

[268] P.Pták and S.Pulmannová: *Orthomodular Structures as Quantum Logics*, Academic Publishers, Dordrecht, (Boston), London, (1991).

[269] P.Pták and H.Weber: *Two remarks on the subspace poset of an inner product space,* Contributions to General Algebra, 10, Proceedings of the Klagenfurt Conference, May 1997, Verlag Johannes Heyn, Klagenfurt, (1998).

[270] P.Pták: *On extensions of states on logics*, Bull. Polish Acad. Sci. Math. **33**, (1985) 493-497.

[271] P.Pták and H.Weber: *Lattice properties of subspace families in an inner peoduct space*, Proc. Amer. Math. Soc., Vol. **129**, Num 7, (2001), 2111-2117.

[272] M.Redei: *Logical independence in quantum logic*, Foundation of Physics, **25**, (1995), 411-422.

[273] M.Redei: *Logically independent von Neumann lattices*, International Journal of Theoretical Physics, **34**, (1995), 1711-1718.

[274] M.Redei: *Quantum Logic in Algebraic Approach*, Kluwer Academic Publishers, Fundamental Theories of Physics, Dordrecht, Boston, London, (1998).

[275] F.Richman: *A constructive proof of Gleason's theorem*, J. Funct. Anal., **162**, (1999), 287–312.

[276] J.R.Ringrose: *Linear functionals on operator algebras and their abelian subalgebras*, J. London Mathematical Soc., **7**, (1973), 553-560.

[277] H.Roos: *Independence of local algebras in quantum field theory*, Commun. Math. Phys. **37**, (1974), 273–286.

[278] W.Rudin: *Real and Complex Analysis*, McGraw-Hill, New York (1987).

[279] O.Rudolph and J.D.M.Wright: *The multi-form generalized Gleason theorem*, Commun. Math. Phys. **198**, (1998), 705-709.

[280] O.Rudolph and J.D.M.Wright: *Homogeneous decoherence functionals in standard and history quantum mechanics,* Commun. Math. Phys. **204**, (1999) 249–267.

[281] O.Rudolph: *The representation theory of decoherence functionals in history quantum theories,* International Journal of Theoretical Physics, Vol. **39**, No. 3, (2000), 871–884.

[282] O.Rudolph and J.D.M.Wright: *On tracial representations of quantum decoherence functionals*, J. Math. Phys. **38**, (1997), 5643–5652.

[283] O.Rudolph and J.D.M.Wright: *On unentangled Gleason theorems for quantum information theory,* Letters in Mathematical Physics, **52**, (200), 239–245.

[284] O.Rudolph: *Temporal quantum theory,* arXiv: math-ph 9 807034, Jul (1998).

[285] G.T.Rüttimann: *Decompositions of cones of measures*, Atti Sem. Mat. Fiz. Univ. Modena **38**, (1990), 267-279.

[286] G.T.Rüttimann: *Weakly purely finitely additive states*, Canad. J. Math. **46**, (1994), 872–885.

[287] G.Rütimann: *Jauch-Piron states*, J. Math. Phys. **18** (2), (1997) 189-193.

[288] K.Saito: *Non-commutative extensions of Lusin's theorem.*, Tohoku Math. J. **19**, (1967), 332–340.

[289] K.Saito: *Diagonalizing normal operators in properly infinite von Neumann algebras and the continuity of probability measures*, (1985).

[290] S.Sakai: *C^*-algebras and W^*-algebras*, Ergebnisse der Mathematik **60**, Springer-Verlag, Berlin, Heidelberg, New York, (1971).

[291] S.Sakai: *Operator Algebras in Dynamical Systems,* Encyklopdia of Mathematics and its application **41**, Cambridge University Press, (1991).

[292] L.S.Sharma and D.F. Almeida: *A direct proof of Wigner's theorem on maps which preserve transition probabilities between pure states of quantum systems*, Ann. Phys. **197**, (1990), 300-309.

[293] A.N.Sherstnev: *On the notion of charge in the noncommutative scheme of measure theory*, Verajatostnye Metody i Kybernetika, Nos. 10-11, (1974), 68-72.

[294] A.N.Sherstnev and E.A.Turilova: *Classes of subspaces affiliated with a von Neumann algebra*, Russian Journal of Mathematical Physics, Vol. **6**, No. 4, (1999), 420-434.

[295] B.Simon: *Quantum dynamics: from automorphism to hamiltonian*, Studies in Mathematical Physics. Essays in honour of Valentine Bargmann, eds. E.H.Lieb, B.Simon, A.S.Whightman, Princeton University Press, Princeton, New Jersay, (1976), 327-349.

[296] A.M.Sinclair: *Jordan homomorphisms and derivations on semi-simple Banach algebras*, Proc. Amer. Math. Soc. **24**, (1970), 209–214.

[297] M.P.Soler: *Characterization of Hilbert spaces by orthomodular spaces*, Communications in algebra, **23**, (1995), 219-243.

[298] W.Stinespring: *Positive functions on C^* algebras*, Proc. Amer. Math. Soc. **6**, (1955), 211–216.

[299] E.Stormer: *A characterization of pure states of C^*-algebra*, Proc. Amer. Math. Soc., **1**, (1968), 1000-1102.

[300] E.Stormer: *Irreducible Jordan algebras of self-adjoint operators*, Trans. Amer. Math. Soc., **130**, (1968), 153-166.

[301] S.Stratilla and L.Zsido: *Lectures on von Neumann Algebras*, Tunbridge Wells: Abacus Press, (1979).

[302] S.J.Summers: *On the independence of local algebras in quantum field theory*, Rev. Math. Physics, Vol **2**, No. 2, (1990), 201-247.

[303] S.J.Summers and R.Werner: *Bell's inequalities and quantum field theory I, general setting*, J.Math. Phys. **28**, (1987), 2440-2447.

[304] K.Svozil and J.Tkadlec: *Measures and the Kochen-Specker theorem*, J. Math. Phys. **37**, (1996), 5380–5401.

[305] M.Takesaki: *On the direct product of W^*-factors*, Tohoku Math. J. **10**, (1958), 116-119.

[306] M.Takesaki: *On the nonseparability of singular representation of operator algebra*, Kodai Mat. Sem. Rep. **12**, (1960), 102-108.

[307] M.Takesaki: *Theory of Operator Algebras I*, Berlin, Heidelberg, New York, Springer (1979).

[308] M.Takesaki: *Theory of Operator Algebras II, III*, Berlin, Heidelberg, New York, Springer (2002).

[309] J.Tisher: *Gleason's Theorem for Type I von Neumann algebras*, Pacific J. Math. **100**, (1982), 273–288.

[310] M.Tomita: *Spectral Theory of Operator Algebras I*, Math. J. Okayama Univ. **10**, (1960), 19-60.

[311] D.Topping: *Jordan Algebras of Self-Adjoint Operators*, Memoirs of the Amer. Math. Soc., **53**, (1965).

[312] T.Turumaru: *On the direct product of operator algebras, IV*, Tohoku Math. J. **8**, (1956), 281-285.

[313] U.Uhlhorn: *Representation of symmetry transformations in quantum mechanics*, Arkiv Fyzik, **23**, (1962), 307–340.

[314] V.S.Varadarajan: *Geometry of Quantum Theory*, Vol. I, Van Nostrand, Princeton (1968).

[315] V.S.Varadarajan: *Probability in physics and a theorem of simultaneous observability*, Comm. Pure Appl. Math. **15**, (1962), 189–217.

[316] N.R.Wallach: An unentangled Gleason's theorem, Contemporary Mathematics, Vol. **305**, (2002), 291-298.

[317] N.Weaver: *Mathematical Quantization*, Studies in Advanced Mathematics, Chapman & Hall/CRC, (2001).

[318] S.Weinberger: *The Quantum Theory of Fields*, Cambridge, USA, (1995)

[319] E.P.Wigner: *Group Theory and Its Applications to the Quantum Theory of Atomic Spectra*, Acedemic Press Inc., New York (1959)

[320] E.P.Wigner: *The problem of measurement*, Amer. J. Phys., **31**, 6-15, (1963).

[321] W.J.Wilbur: *Quantum logic and locally convex spaces*, Transactions of the American Mathematical Society, **207**, (1975), 343-360.

[322] W.J.Wilbur: *On characterizing the standard quantum logics*, Transactions of the American Mathematical Society, **233**, (1977), 265-282.

[323] J.D.M. Wright: *An extension theorem and dual proof of a theorem of Gleason*, J. London Math. Soc. **43**, (1968), 699-702.

[324] J.D.M.Wright: *The structure of decoherence functionals for von Neumann quantum histories*, J. Math. Phys. **36**, (10), (1995), 5409–5431.

[325] J.D.M.Wright: *Decoherence functionals for von Neumann quatum histories: Boundedness and countable additivity*, Commun. Math. Phys. **191**, (1998), 493–500.

[326] J.D.M.Wright: *Linear representations of bilinear forms on operator algebras,* Exposition Math. **16**, no. 1, (1968), 75-83.

[327] J.D.M.Wright: *An invariant Haagerup-Pisier-Grothendieck inequality,* Expositiones Mathematicae, **17**, (1999), 245–254.

[328] J.D.M.Wright: *Physical symmetries of quantum histories,* Commun. Math. Phys. **201**, (1999), 507–517.

[329] F.J.Yeadon: *Measures on projections in W^*-algebras of Type II_1*, Bull. London Math. Soc. **15**, (1983), 139–145.

[330] F.J.Yeadon: *Finitely additive measures on projections in finite W^*-algebras,* Bull. London Math. Soc., **16**, (1984), 145–150.

[331] K.Ylinen: *Vector measures on the projections of a W^*-algebra,* Ann. Univ. Turkuensis, Ser. A, **186**, (1984), 129–135.

[332] K.Ylinen: *Vector Gleason measures and their Fourier transforms*, Lecture Notes in Mathematics, **1132**, Operator Algebras and Their Connections with Topology and Ergodic theory, 589-594.

[333] K.Y.Ylinen: *Representations of bimeasures,* Studia Mathematica **104** (3), (1993), 269-278.

[334] K.Y.Ylinen: *Positive operator bimeasures and a noncommutative generalization,* Studia Mathematica **118**, (1996), 157-168.

Index

Fundamental Theories of Physics

94. V. Dietrich, K. Habetha and G. Jank (eds.): *Clifford Algebras and Their Application in Mathematical Physics.* Aachen 1996. 1998 ISBN 0-7923-5037-5
95. J.P. Blaizot, X. Campi and M. Ploszajczak (eds.): *Nuclear Matter in Different Phases and Transitions.* 1999 ISBN 0-7923-5660-8
96. V.P. Frolov and I.D. Novikov: *Black Hole Physics.* Basic Concepts and New Developments. 1998 ISBN 0-7923-5145-2; Pb 0-7923-5146
97. G. Hunter, S. Jeffers and J-P. Vigier (eds.): *Causality and Locality in Modern Physics.* 1998 ISBN 0-7923-5227-0
98. G.J. Erickson, J.T. Rychert and C.R. Smith (eds.): *Maximum Entropy and Bayesian Methods.* 1998 ISBN 0-7923-5047-2
99. D. Hestenes: *New Foundations for Classical Mechanics (Second Edition).* 1999 ISBN 0-7923-5302-1; Pb ISBN 0-7923-5514-8
100. B.R. Iyer and B. Bhawal (eds.): *Black Holes, Gravitational Radiation and the Universe.* Essays in Honor of C. V. Vishveshwara. 1999 ISBN 0-7923-5308-0
101. P.L. Antonelli and T.J. Zastawniak: *Fundamentals of Finslerian Diffusion with Applications.* 1998 ISBN 0-7923-5511-3
102. H. Atmanspacher, A. Amann and U. Müller-Herold: *On Quanta, Mind and Matter Hans Primas in Context.* 1999 ISBN 0-7923-5696-9
103. M.A. Trump and W.C. Schieve: *Classical Relativistic Many-Body Dynamics.* 1999 ISBN 0-7923-5737-X
104. A.I. Maimistov and A.M. Basharov: *Nonlinear Optical Waves.* 1999 ISBN 0-7923-5752-3
105. W. von der Linden, V. Dose, R. Fischer and R. Preuss (eds.): *Maximum Entropy and Bayesian Methods Garching, Germany 1998.* 1999 ISBN 0-7923-5766-3
106. M.W. Evans: *The Enigmatic Photon Volume 5: O(3) Electrodynamics.* 1999 ISBN 0-7923-5792-2
107. G.N. Afanasiev: *Topological Effects in Quantum Mecvhanics.* 1999 ISBN 0-7923-5800-7
108. V. Devanathan: *Angular Momentum Techniques in Quantum Mechanics.* 1999 ISBN 0-7923-5866-X
109. P.L. Antonelli (ed.): *Finslerian Geometries A Meeting of Minds.* 1999 ISBN 0-7923-6115-6
110. M.B. Mensky: *Quantum Measurements and Decoherence Models and Phenomenology.* 2000 ISBN 0-7923-6227-6
111. B. Coecke, D. Moore and A. Wilce (eds.): *Current Research in Operation Quantum Logic.* Algebras, Categories, Languages. 2000 ISBN 0-7923-6258-6
112. G. Jumarie: *Maximum Entropy, Information Without Probability and Complex Fractals.* Classical and Quantum Approach. 2000 ISBN 0-7923-6330-2
113. B. Fain: *Irreversibilities in Quantum Mechanics.* 2000 ISBN 0-7923-6581-X
114. T. Borne, G. Lochak and H. Stumpf: *Nonperturbative Quantum Field Theory and the Structure of Matter.* 2001 ISBN 0-7923-6803-7
115. J. Keller: *Theory of the Electron.* A Theory of Matter from START. 2001 ISBN 0-7923-6819-3
116. M. Rivas: *Kinematical Theory of Spinning Particles.* Classical and Quantum Mechanical Formalism of Elementary Particles. 2001 ISBN 0-7923-6824-X
117. A.A. Ungar: *Beyond the Einstein Addition Law and its Gyroscopic Thomas Precession.* The Theory of Gyrogroups and Gyrovector Spaces. 2001 ISBN 0-7923-6909-2
118. R. Miron, D. Hrimiuc, H. Shimada and S.V. Sabau: *The Geometry of Hamilton and Lagrange Spaces.* 2001 ISBN 0-7923-6926-2

Fundamental Theories of Physics

119. M. Pavšič: *The Landscape of Theoretical Physics: A Global View*. From Point Particles to the Brane World and Beyond in Search of a Unifying Principle. 2001 ISBN 0-7923-7006-6
120. R.M. Santilli: *Foundations of Hadronic Chemistry*. With Applications to New Clean Energies and Fuels. 2001 ISBN 1-4020-0087-1
121. S. Fujita and S. Godoy: *Theory of High Temperature Superconductivity*. 2001 ISBN 1-4020-0149-5
122. R. Luzzi, A.R. Vasconcellos and J. Galvão Ramos: *Predictive Statitical Mechanics*. A Nonequilibrium Ensemble Formalism. 2002 ISBN 1-4020-0482-6
123. V.V. Kulish: *Hierarchical Methods*. Hierarchy and Hierarchical Asymptotic Methods in Electrodynamics, Volume 1. 2002 ISBN 1-4020-0757-4; Set: 1-4020-0758-2
124. B.C. Eu: *Generalized Thermodynamics*. Thermodynamics of Irreversible Processes and Generalized Hydrodynamics. 2002 ISBN 1-4020-0788-4
125. A. Mourachkine: *High-Temperature Superconductivity in Cuprates*. The Nonlinear Mechanism and Tunneling Measurements. 2002 ISBN 1-4020-0810-4
126. R.L. Amoroso, G. Hunter, M. Kafatos and J.-P. Vigier (eds.): *Gravitation and Cosmology: From the Hubble Radius to the Planck Scale*. Proceedings of a Symposium in Honour of the 80th Birthday of Jean-Pierre Vigier. 2002 ISBN 1-4020-0885-6
127. W.M. de Muynck: *Foundations of Quantum Mechanics, an Empiricist Approach*. 2002 ISBN 1-4020-0932-1
128. V.V. Kulish: *Hierarchical Methods*. Undulative Electrodynamical Systems, Volume 2. 2002 ISBN 1-4020-0968-2; Set: 1-4020-0758-2
129. M. Mugur-Schächter and A. van der Merwe (eds.): *Quantum Mechanics, Mathematics, Cognition and Action*. Proposals for a Formalized Epistemology. 2002 ISBN 1-4020-1120-2
130. P. Bandyopadhyay: *Geometry, Topology and Quantum Field Theory*. 2003 ISBN 1-4020-1414-7
131. V. Garzó and A. Santos: *Kinetic Theory of Gases in Shear Flows*. Nonlinear Transport. 2003 ISBN 1-4020-1436-8
132. R. Miron: *The Geometry of Higher-Order Hamilton Spaces*. Applications to Hamiltonian Mechanics. 2003 ISBN 1-4020-1574-7

KLUWER ACADEMIC PUBLISHERS – DORDRECHT / BOSTON / LONDON